ENGINEERING DRAWING

ENGINEERING DRAWING

RATHIN BANDYOPADHYAY
Ex-Lecturer, A. P. C. Roy Polytechnic, Jadavpur, Kolkata

NEW CENTRAL BOOK AGENCY (P) LTD
8/1 Chintamoni Das Lane, Kolkata 700 009
INDIA

ENGINEERING DRAWING

First Published: July 2007

Publication, Distribution, and Promotion Right reserved by
New Central Book Agency (P) Ltd

Publisher: New Central Book Agency (P) Ltd
8/1 Chintamoni Das Lane, Kolkata 700 009

Typesetter: Anin
BC 97, Sector I, Salt Lake, Kolkata 700 064

Printer: New Central Book Agency (P) Ltd
Web-Offset Division, Dhulagarh, Sankrail, Howrah

Cover Designer: Soumen Paul

Cover Printer: Biswanath Studio
3 Garanhatta Street, Kolkata 700 006

Proof-checking: Dipan Roy and the Author

Project Team: Prabhat Jas and Subir Pal

ISBN: 81-7381-544-5

Price: Rs. 235.00 [Rupees Two Hundred and Thirty-five only]

Dedicated

to

My Parents

Contents

Preface

This book is meant for the students of Engineering Colleges and Polytechnics of India on the subject '**Engineering Drawing**'. It will also be helpful for the students who will appear in the A.M.I.E. (Section A) examination, conducted by The Institute of Engineers (India), on the above subject.

Latest standards of Engineering Drawing have been followed as per the Bureau of Indian Standards (BIS). As such, First-Angle Projection Method has been used. Only a few problems have been solved by the Third-Angle Projection Method for the benefit of the students.

The author has tried his best to explain the subject-matter in a lucid way, so that the students can follow the topics easily. To solve each problem, step by step construction-procedure has been followed. Thus, the students can proceed smoothly from stage to stage during the time of drawing.

Emphasis has been laid on the basic principle of all the topics—in order to enable the students to get a clear conception of the parts, as well as the whole.

Solutions of some problems from A.M.I.E. (Section A) Examination, and from West Bengal State Council Examination (Engineering and Technical) have been provided on the subject.

Practice-problems have been given at the end of each chapter.

Practical applications of different curves, development and intersection of surfaces have been mentioned in the respective chapters.

A chapter on 'Building Drawing' (for Civil Engineering students) and another on 'Perspective Projection' (for students of Architecture) have been included in the text. The last chapter is on latest 'Computer Graphics', which is very useful nowadays.

I wish to thank Sri T. N. Chakravorty, ex-Lecturer in Civil Engineering at North Calcutta Polytechnic, who has carefully revised the chapter on 'Building Drawing'.

Thanks are due to Sri S. P. Bhattacharya, ex-Lecturer in Mechanical Engineering of the above Polytechnic, who has helped in many ways in the production of this book.

Thanks are also due to Mrs. Rita Sarkar, Junior Lecturer in Architectural Engineering of the said Polytechnic, for contributing the chapter on 'Perspective Projection'.

I wish to thank Mr. Ranaprotap Chatterjee and Mr. Ashis Bhattacharya, Librarian and Assistant Librarian, respectively, at Sir R. N. Mukherjee Engineering Information Service Centre of The Institute of Engineers (India), Kolkata, for their valuable suggestions and furnishing important data on 'Computer Graphics'.

Thanks are also due to Sri Suryabrata Mukherjee for the solution of some A.M.I.E. Questions on 'Computer Graphics' and furnishing information on various 'Commands' in a computer.

Lastly, I would like to thank Mr. Amitabha Sen, Director, and his efficient production team of New Central Book Agency (P) Ltd. for publishing the book with both elegance and meticulous care.

Any corrections and suggestions for the improvement of the book will be thankfully received.

Rathin Bandyopadhyay

3 June 2007
46/7 Babu Bagan Lane
Dhakuria
Kolkata 700 031

Chapter 1

Drawing-Instruments and their Uses

1.1 Introduction

Before starting a drawing, the students should know its meaning. In general, there are two types of drawings—(1) Standard and (2) Non-standard. The first one is known as *Technical* and the second is *Artistic*. Both the above types of drawings have got universal appeal. Their common factor is beauty, which is perceived by our eyes.

Artistic drawings have sensitivity and emotion, so that these represent some thoughts and ideas of persons or other things in nature (birds, flowers, trees, rivers, seas, sunrise, mountains, etc.). But technical drawing has no such sensitivity and emotion. In all countries of the world including India, almost the same procedure is being followed during the preparation of a technical drawing. Hence, technical drawing is the language of all types of Engineers, and it is an important subject in Engineering and Technology.

1.2 Technical Drawings

These are prepared by different instruments, such as drawing board, scale, set-squares, compass, etc. Naturally, the quality of a drawing depends upon the quality and accuracy of the instruments used. All engineering students must choose good quality instruments. These instruments will be the main tools throughout their student and professional lives.

1.3 Drawing Board

Bureau of Indian Standards (BIS) has recommended the size of standard Board as $B_0(1000 \times 1500)$, $B_1(700 \times 1000)$, $B_2(500 \times 700)$ and so on, and the sizes are expressed in mm. The student will have to purchase a board of size 810×580 approximately, as available in the market. It is essential to check up the straightness of the ebony edge of the board (inserted at the left side edge) with the help of the T-square and set-squares.

1.4 Set-Squares (a) 45°, and (b) 30°-60°

One set-square is used to draw lines at angles of 30° and 60° to the horizontal. The other is used to draw lines inclined at 45° to the horizontal. The two set-squares can be combined together in different ways to draw lines inclined at angles of 15°, 75°, 105° to the horizontal. Perpendiculars can be drawn with the help of one set square and the T-square.

The student should purchase thick quality set-squares. Bevelled-edge should be avoided, which will slip into the gap between the bottom-surface of T-square and the top-surface of the Drawing sheet.

1.5 T-Square

The T-square is used to draw one or more horizontal lines. Its accuracy must be tested with the help of set-squares.

There are two types of T-squares—one is made of wood and the other in plastics. In case of plastics T-square, the working edge should be thick, otherwise set-squares may slip-in into the T-square at the time of drawing lines.

Instead of a T-square, a drafter (a drafting device) may also be used for quick operation in drawing.

1.6 Large and Small-Size Compasses

The compass is used to draw circles and arcs of a circle. There are pencil and ink-compasses, but the pencil-compass is widely used. The pencil-compass consists of two legs, one is·equipped with a steel needle, and the other has a socket to hold lead for marking on the paper or drawing sheet. A compass is usually made of steel.

1.7 Divider

It is used to divide straight or curved lines, or to transfer dimensions from one drawing to another, from a measuring scale to a drawing or from drawing to scale. It consists of two legs with two steel needles. A divider is generally made of steel.

1.8 Diagonal Scale

It is used to measure linear dimensions up to 140 mm, and fractional part of an inch or mm with the help of diagonal divisions.

Instead of using a protractor, a diagonal scale may also be used to draw or measure angles with an accuracy up to 1 degree (1°).

1.9 French Curve

The French Curves are made in various shapes, and are specified in number such as Curve no. 1, 2, 3, 4, 5, etc. For the first year students, curve nos. 5 and 6 will serve the purpose.

1.10 Drawing Sheet

According to *BIS, the size of the drawing sheet is standardised as A_0(841 × 1189 mm), A_1(549 × 841 mm), A_2(420 × 594 mm) and so on. The drawing sheets are available in two types—trimmed and untrimmed.

*Bureau of Indian Standards

The students should select A_1 size drawing sheet, which will be milky white, non-glossy and non-pinkish. Sometimes the students are misguided with the poster sheet as engineering drawing sheet, as are available in the local market. Drawing sheets should be purchased from proper dealer. *Always avoid glossy sheet.*

1.11 Drawing Pencil

Drawing-pencils are specified as ... ... 3B, 2B, 1B, HB, 1H, 2H, 3H, etc., in the order from softer to harder. Students should purchase one HB and one 1H pencils (preferably, Nataraj or any other brand). Micro-tip pencil will also do, if its lead is of "1H" quality.

Length of exposed part of pencil will be 40 mm from fine tip to its maximum diameter, as shown in the Fig. 1-1. In order to sharpen the pencil, a blade or sand-paper of 0-0 (zero-zero) quality may be used. Too much pressure should not be applied on the pencil, in order to avoid breakage of the lead (i.e., the tip of pencil).

Fig. 1-1

1.12 Eraser

Milky-white eraser should be used, because this will remove the unwanted pencil-marks clearly from drawing sheet, not the paper-fibre. An eraser should be rubbed lightly on the drawing-sheet to remove the unwanted pencil-marks only, without damaging the sheet.

1.13 Erasing Shield

A thin sheet of plastics (non-metallic) perforated in different shapes may be used to erase some portion of a drawing without erasing the relevant portions.

1.14 Drawing Clips

In order to fix the drawing sheet to the board, drawing clips (4 nos. in a set) are used. U-shaped steel clips should be used with different limb-lengths. Plastic clips should be avoided, as they will break due to constant use. For fixing the sheets to board, pins should be avoided, because it will not only spoil the board, but also make holes in the drawing sheet, which will create error in the drawing.

1.15 Cloth-Duster

It is a piece of clean cloth, which is used from time to time for cleaning the deposited dirts from the drawing sheet, T-square, set-squares, diagonal scale, french curve and our hands during the period of drawing.

1.16 Thick Sheet

A thick sheet of paper of board-size is to be placed between the drawing sheet and the board. This is done in order to avoid the impressions of the wooden fibres of the board on the drawing sheet. Newspaper should not be used, as the ink of the paper will spoil the drawing sheet. Always cover the drawing with foolscap paper to avoid dirt due to friction between the T- and set-squares.

1.17 Care and Maintenance of Drawing Instruments

The 'Compass' and the 'Divider' are usually made of steel. These should be cleaned and smeared with machine-oil periodically, in order to avoid their corrosion and jamming at the joints. Care should be taken so that they may not fall from a height to the floor, otherwise they will be damaged. After the use of a compass or a divider, it should be kept within the instrument box.

Nowadays, T-square, set-squares, protractor, Diagonal scale, 15 cm or 30 cm-long, French curves are made of plastics, which is a brittle substance. Care should be taken, so that they may not fall to the floor from a height, otherwise cracks will be developed in them.

After drawing, the 'Drawing Board' should be covered with a cloth or newspaper on the top in order to avoid deposition of dusts. Also the drawing board should be kept on the table (after drawing) in order to avoid its bending.

Objective-Type Questions

1. Choose the correct answer among the three answers suggested for each question :

(i) Divider is used for

(a) drawing circles and curves (b) transferring dimensions

(c) measuring length. **(Summer, 2005)**

Ans. (b)

(ii) The drawing board should be placed on the table always with the working-edge at

(a) Right side (b) Left side (c) Any side. **(Summer, 2006)**

Ans. (b), i.e. left side of the person who draws.

Chapter 2

Types of Lines

Every subject has its own language and notations. Similarly, **one should know the definite meanings and purposes of various types of lines, which are used in engineering drawing practice.**

The different types of lines, their thickness and specific uses are given below in a tabular form. For more details, the students should refer to the publication SP46 : 1988 which has been published by Bureau of Indian Standards (BIS).

Table 2-1

Sl. No.	Type	Sample	Thickness	Use	Remark
1	Thick continuous		0.6 mm designated at 't'	Visible edges and surface of object	
2	Thin continuous		t/3 = 0.2 mm	Extention, dimension lines and also for hatching in sections	For hatching lines being spaced depending upon the area of cross-section
3	Short dash		t/2 = 0.3 mm	Hidden outlines and intersection	Legth of dash varies from 2 to 8 mm
4	Chain thin		t/3 = 0.2 mm	To indicate the axes of cylindrical, conical or spherical objects.	Length of short dash 1.5 mm space between dash 1 mm known as centre line
5	Thick thin thick		0.6 & 0.2 mm	Cutting plane line	When an object is cut by an imaginary cutting plane
6	Wavy continuous		t/2 = 0.3 mm	To show a short break or irregular boundaries	
7	Long Zig-Zag		t/2 = 0.3 mm	Long breaks	To show a part of an object

Important Notes

1. The symbols D, d, ϕ and R, r are generally used to represent **diameters** and **radii** respectively.
2. **Outlines or principal lines** are drawn as **'continuous thick' lines**, as shown in serial No. 1, in the Table 2-1.

3. **An imaginary cutting plane** is represented by a **'Thick-Thin-Thick' line**, as shown in serial No. 5 (Table 2-1).
4. The proportion of length to width of an arrow-head is 3 : 1, as shown in Fig. 2-1.

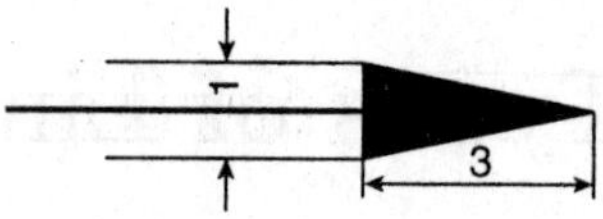

Fig. 2-1

5. Dimension lines should touch the outlines of a drawing.
6. Dimension lines should not cross each other.
7. **The two methods of dimensioning** a drawing are **unidirectional** and **aligned systems** and are shown in Fig. 2-2 and Fig. 2-3 respectively.

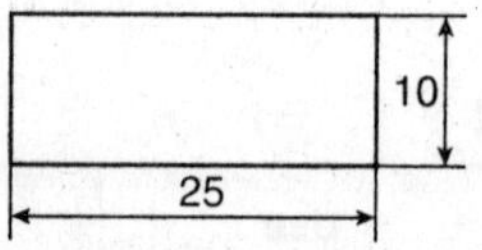

Fig. 2-2 Unidirectional System

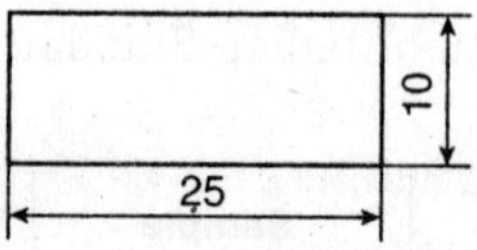

Fig. 2-3 Aligned System

Objective-Type Questions

1. Choose the correct answer :

   (i) *B.I.S. recommends the following unit for all dimensions in any one drawing

   (a) m (b) cm (c) mm. **(Summer, 2005)**

   Ans. (c)

   (ii) A line, consisting of short dashes throughout its length, represents the following

   (a) invisible edge of a solid (b) axis of a solid (c) trace of a cutting plane. **(Summer, 2005)**

   Ans. (a)

*Bureau of Indian Standards

Chapter 3

Lettering

3.1 Introduction

'Letters' may be said to be the ornaments of a drawing. So, the drawing should be carefully ornamented by proper lettering in order to write titles, dimensions, notes and other particulars. Neat and attractive lettering is very important. The English poet John Keats said 'A thing of beauty is a joy for ever.' Letters are divided into two classes : (a) *Single-stroke* and (b) *Double-stroke*.

The students should learn to draw the single-stroke letters first, and then the double-stroke. In the Technical Institutions, the usual practice is to draw double-stroke letters at the top of a sessional drawing sheet, for the purpose of ornamentation. It is better to draw single-stroke letters first, leaving the space at the top of the sheet for double-stroke letters.

3.2 Parameters of Lettering

As far as practicable, lettering should be drawn so that letters will be (a) *neat,* (b) *legible* (i.e., easily readable), (c) *uniform* or equal in sizes, (d) *rapid* (i.e., quickly written), (e) *pleasing* (good-looking), (f) *ease* (upper and lower case) and (g) *free hand* (without using instrument).

3.3 Single-Stroke Letters

'Single-stroke lettering' means drawing of english alphabet and numerals (1, 2, 3, 4, 5, 6, 7, 8, 9, 0) in single line. These letters are divided into two classes : (i) Vertical or upright and (ii) Inclined or slant, in both capitals and italics. Sloping letters are inclined to the right and the inclination varies from 67.5° to 75° with the horizontal. But the usual practice is to draw 60° inclined letters with the help of a 30°–60° set-square in order to avoid complication. **The "size" of a letter means its height**. For different items, the letter sizes are given in the Table 3-1.

Table 3-1

Item	Height in mm	Width
1. Title	10 to 15	Generally kept equal
2. Sub-title	3 to 6	to height for item
3. Dimension	3 to 5	nos 1,2,3

Over and above, the student may select any size, according to his artistic sense. It is better to utilise set-square in case of large letters.

Examples

1. Single-Stroke Vertical Capital

In drawing these letters, the steps are as follows (see Fig. 3-1) :

(a) Draw two horizontal lines *A* and *B* on given height with the help of T-square.

(b) Divide the height into two halves and draw the line *C*.

(c) Draw vertical lines with given width for letters, leaving the breathing space very little space between the two. Now, draw the letters. The gap between the two letters should be so spaced that they do not appear too close or too far. As for example, say 'ACHARYYA', if we leave equal breathing space between two letters, then last 'A' will be too far from last 'Y'. This idea will be clear, if we look into the lettering of signboards of the shops, etc.

Fig. 3-1

(d) In Fig. 3-2, letters and figures are shown. The first two letters are shown on graph, but it is not essential for single-stroke letters.

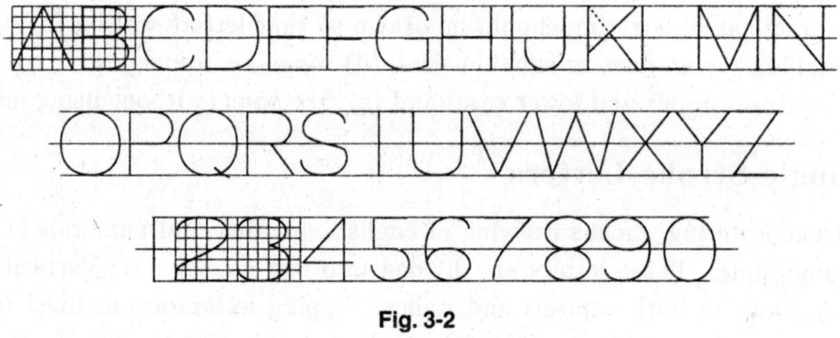

Fig. 3-2

(e) The student should practise free-hand lettering as shown in Fig. 3-3 and gradually make it a habit.

ABCDEFGHIJKLMNOPQRSTUVWXYZ

1234567890

A QUICK BROWN FOX JUMPS

RIGHT OVER THE LAZY DOG

Fig. 3-3

2. Single-Stroke Vertical Italics

Refer to Fig. 3-4.

(a) Draw two horizontal lines 'C', 'D' of equal space above and below the lines 'A' and 'B'. Now, start drawing the letters.

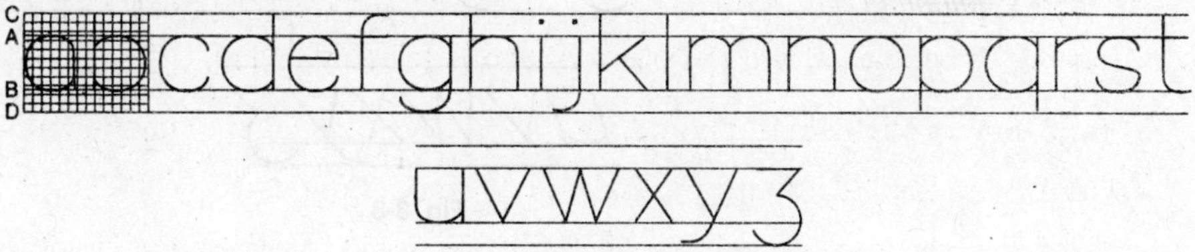

Fig. 3-4

(b) Practise free-hand as shown in Fig. 3-5.

abcdefghijklmnopqrstuvwxyz

Drawing is the language of Engineers

Fig. 3-5

3. Single-Stroke Inclined Capital

(a) The same procedure is followed as in vertical capital, but make it 60° inclined. See Fig. 3-6.

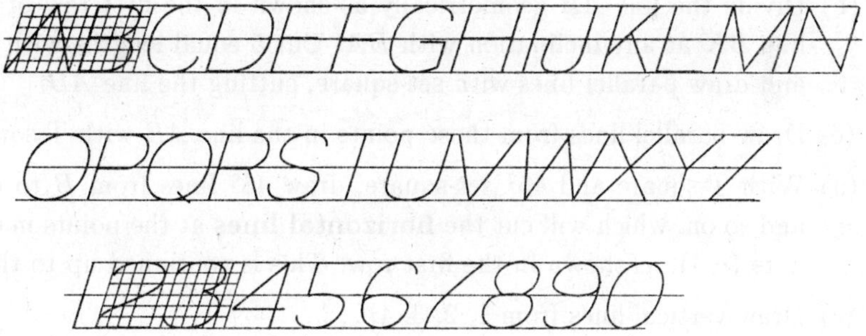

Fig. 3-6

(b) Practise free-hand, as in Fig. 3-7.

ABCDEFGHIJKLMNOPQRSTUVWXYZ

1234567890

EACH SOUL POTENTIAL DIVINE

Fig. 3-7

4. Single-Stroke Inclined Italics

(a) The same procedure is followed as in inclined capital. Refer to Fig. 3-8.

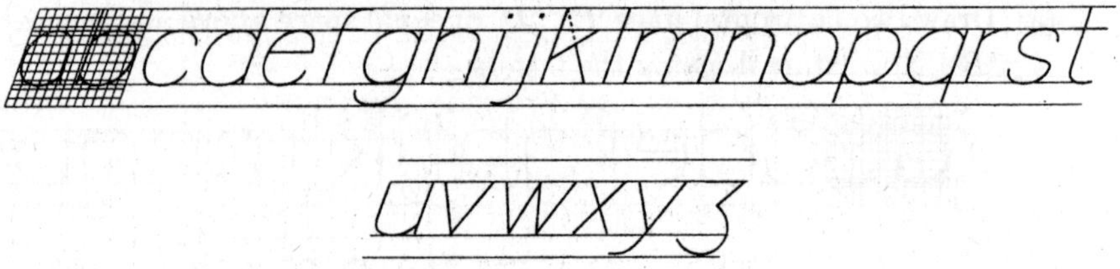

Fig. 3-8

(b) Practise free-hand as in Fig. 3-9.

abcdefghijklmnopqrstuvwxyz
Cowards can never be moral

Fig. 3-9

3.4 Double-Stroke or Gothic or Block Lettering

It is used for the heading of a topic or chapter. In drawing these letters, the steps are as follows :

(a) Draw two straight lines AC and BD (See Fig. 3-10) at a given height.

(b) Divide the line AB geometrically as shown in the first row of Fig. 3-10. Draw a line BK at any inclination with BA. Cut 5 equal lengths with compass. Join AK and draw parallel lines with set-square, cutting the line AB.

(c) Draw parallel lines from those points in the line AB with T-square.

(d) With T-square and 45° set-square, draw 45° lines from B to C, C to D, D to E and so on, which will cut the **horizontal lines** at the points marked by bigger size dots $(\cdots)$, as shown in the first row. This is continued up to the end.

(e) Draw vertical lines from 1, 2, 3, 4, ...

(f) Now the graph is ready to draw the letters.

(g) Construction lines should be light and letters will be drawn prominent.

(h) Generally, height-width ratios are 5 : 4 or 7 : 4. If the ratio is 7 : 4, students will have to divide the vertical line in 7 equal parts and other procedures are the same.

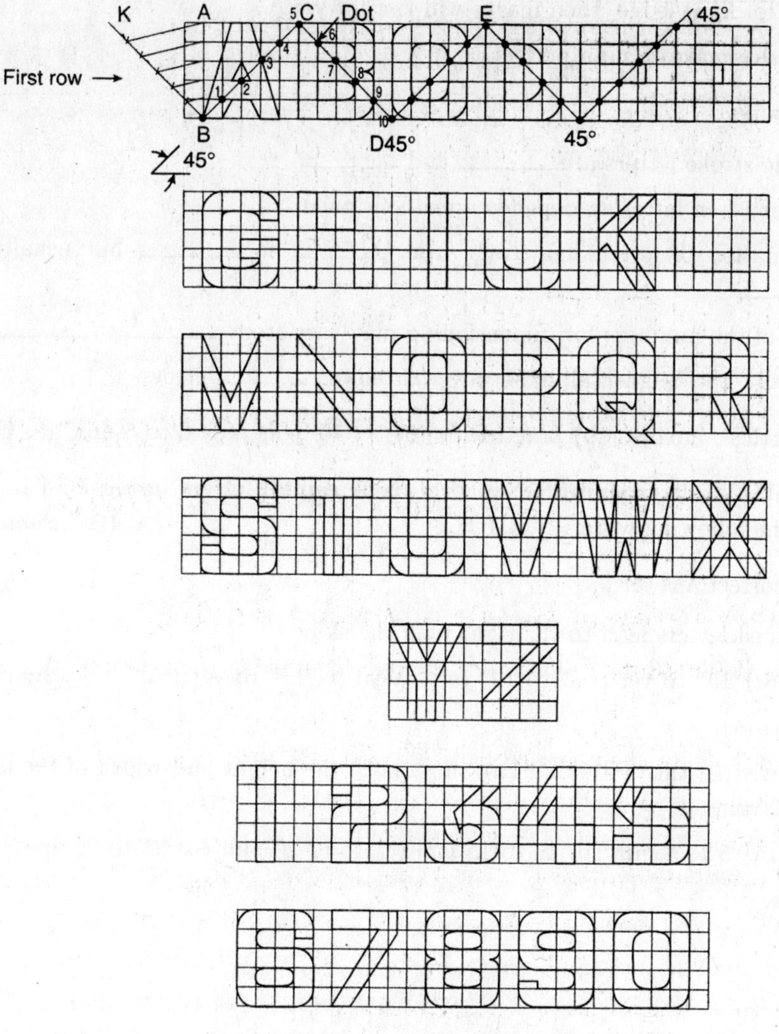

Fig. 3-10

Problems

1. Draw single-stroke capital, height 12 mm (vertical and 60° inclined)

   (a) POLYTECHNIC, (b) STATE COUNCIL, (c) CONSTITUTION, (d) PUBLIC SERVICE COMMISSION, (e) WORLD CUP.

2. Draw BLOCK CAPITAL—height 15 mm in 5 : 4 and 7 : 4

   (a) ENGINEERING, (b) WORK SHOP, (c) DRAWING, (d) DIPLOMA.

3. Draw italics, vertical and inclined, 6 mm height, the following sentences :

   (a) A good drawing is spoiled by bad lettering.

(b) Run for knowledge, then marks will run for you.

(c) Practice makes a man perfect.

4. Fill in the blanks :

(a) Single-stroke letters are ______ and ______.

(b) Expertise in lettering depends upon continued ______.

(c) Inclination of letters may vary from ______ to ______, but usually done in ______.

(d) Size of the numbers for dimensioning may vary from ______ to ______.

(e) Usually, the English alphabet are written in ______ letters.

Ans. (a) Vertical, inclined (b) practice, (c) 67.5°, 75°, 60°, (d) 3 to 5 mm, (e) Capital.

5. Write in double-stroke, round-type, vertical block capital letters, using 7 : 4 ratio and 27 mm height the following word : STORM. **(AMIE—Summer, 2001)**

6. Choose the correct answer :

(i) Inclined letters lean to the right with the slope

(a) 15° to vertical (b) 12° to vertical (c) 17° to vertical. **(Summer, 2006)**

Ans. (a)

(ii) For writing the 'Title of a Drawing', both the height and width of the letters may vary from

(a) 5 to 8 mm (b) 10 to 15 mm (c) 16 to 20 mm (d) 21 to 25 mm.

Ans. (b)

Chapter 4

Scales

4.0 Purpose of a Scale

'Scale' is the relation of measurement or size between the object and its corresponding size in the drawing. According to purpose, types of scales may be—

(a) *Same* or *full-size*, (b) *Reduced-size* and (c) *Enlarged-size.*

According to the Bureau of Indian Standards (BIS), the sizes of drawing sheets are A_0 (841 × 1189 mm), A_1 (549 × 841 mm), A_2 (420 × 594 mm) etc.

We can easily draw the views of a 12 cm long object on a drawing sheet (e.g., pen, pencil, book, etc.), but cannot draw full size views of a large building for it cannot be accommodated within the size of a drawing sheet. Similarly, we will have to enlarge a small part (like the gear of a wrist watch, small letters in a book etc.). This enlargement or reduction should be expressed in a definite ratio.

4.1 Parameters of Scale

The parameters of a scale are :

(i) Length of the scale, (ii) Representative fraction (RF)

and (iii) Maximum length to be measured.

The relation between the above three is

Length of the scale = Representative fraction (RF) × max. length to be measured

4.1.(i) Length of the Scale

We cannot choose any length of scale, and it should be limited within 30 cm for convenience of drawing on a sheet of paper.

4.1.(ii) Representative Fraction (R.F.) or Scale-factor

In R.F., both the numerator and denominator should be expressed in the same units of length (mm or cm or m, etc.). Since R.F. is a ratio, it has no unit.

When R.F. is less than 1, the actual object is drawn to a reduced scale. When R.F. is greater than 1, the actual object is drawn to an enlarged scale.

$$\text{R.F.} = \frac{\text{Length of the drawing}}{\text{Actual length of the object}}.$$

4.1.(iii) Maximum Length to be Measured

We know that in a foolscap paper (about 330 × 200 mm), maps of Calcutta, West Bengal, India, Asia, etc. may be drawn, but max. length to be measured is not the same for each of the above maps. Max. length to be measured in the above maps may be mentioned as 10 km, 300 km, 10 000 km, 20 000 km. Similarly, in engineering, max. length to be measured is different for different drawings. Divisions from zero to max. should be mentioned in the drawing.

4.2 Types of Scale

There are different types of scales, such as plain scale, diagonal scale, vernier scale, scale of chords, etc. Two types of scales are explained here only to cover the new syllabus, i.e., plain scale and diagonal scale. The standardisation of scales is as follows :

1 : 1, i.e., scale full size, 1 : 2 scale half full size and so on. Reduced scales, such as 1 : 2, 1 : 2.5, 1 : 5, 1 : 10, 1 : 20, 1 : 100, 1 : 200, 1 : 500, etc.

Enlarged scales, such as 2 : 1, 5 : 1, 10 : 1, etc.

Scales for Civil and Architectural engineering are as under :

1 : 1... ... 1 : 200, 1 : 1000, 1 : 2000 and so on.

4.2.1 Plain Scale

Let us take one example and try to follow the instructions. Then, everything will be clear about scale. In plain scale, we can measure up to two dimensions or units, such as metres and decimetres, decimetres and centimetres etc. One thing is important to solve the problem on a scale, which is artistic patience.

Problem 1. Construct a plain scale of R.F. 1 : 5 to show decimetres and centimetres and long enough to measure up to 1 metre. Mark a length of 8.6 dm on it.

Solution : Students should show the calculations and R.F. on the drawing sheet in an examination.

Construction : Draw step by step as follows :

(a) Length of the scale = R.F. × max. length to be measured

$$= \frac{1}{5} \times 100 \text{ cm} = 20 \text{ cm}.$$

(b) Take a 20 cm long line CD. Divide the like line geometrically into ten equal parts as shown in Fig. 4-1. Keep in mind that, all the construction lines will be very light. Only the scale will be prominent (i.e., deeply drawn in Fig. 4-1). Each small division represents 1 decimetre.

(c) Mark '0' (zero) on the second point from the left and then write 1, 2, 3 and so on the right from '0'.

(d) Now, divide the first division into ten equal parts geometrically. Then each small division will represents one centimetre.

(e) Mark the division-points from '0' to the left as 1, 2, 3 and so on (if the space between two divisions is small, then we mark as 0, 2, 4 and so on).

(f) Write **CENTIMETRE** and **DECIMETRE** at the bottom of the scale. This is done free-hand by drawing two parallel faint lines with proper height.

(g) Now, draw three lines above CD, 2 mm apart.

(h) Draw vertical lines from specified points.

(i) Cover with black colour all alternate central rectangular divisions with pencil for easy identification of measurements.

(j) Mark the distance 8.6 decimeter between the points A and B.

(k) Do not erase construction lines, which are within the answer.

(l) Write RF = 1/5 on the bottom-left or right-hand side of the scale.

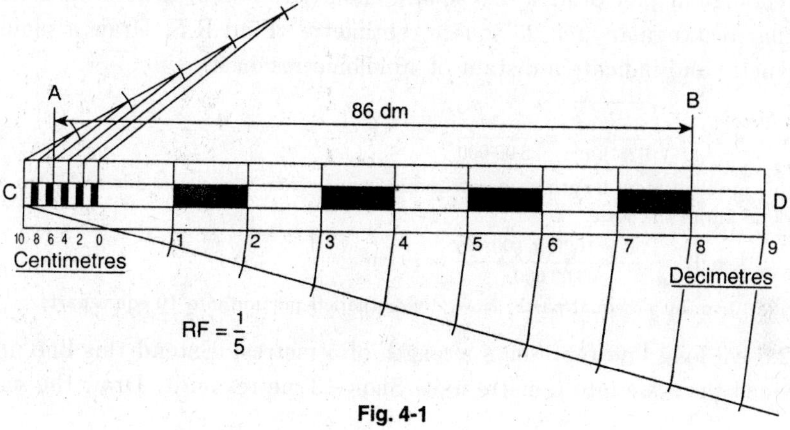

Fig. 4-1

Problem 2. Construct a plain scale in a given R.F. = 1/60. Show metres and decimetres, if maximum length to be measured is 6 metres. Indicate, a distance of 4.7 on the scale.

Solution : Length of the scale = R.F. × max. length to be measured

$$= \frac{1}{60} \times 600 \text{ cm} = 10 \text{ cm}.$$

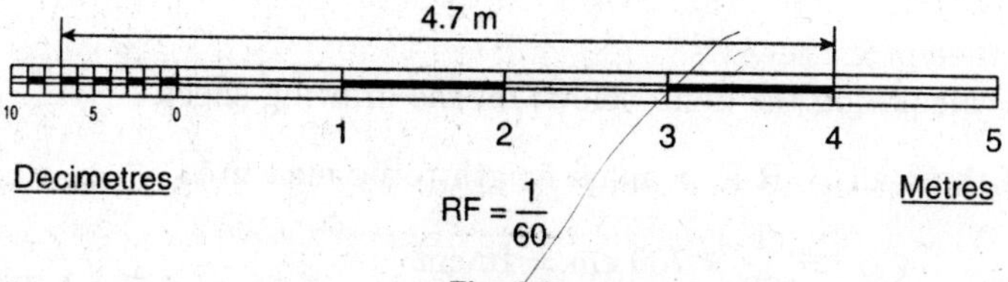

Fig. 4-2

To draw the scale, follow the same instructions as in Problem 1.

Problems on Plain Scale

(1) Construct a plain scale of R.F. = $\frac{1}{50}$ to read metres and decimeters and long enough to measure up to 8 m. Mark a distance of 5.3 m on the scale.

Hint. Length = $\frac{1}{50} \times 800$ cm = 16 cm.

Divide 16 cm into 8 equal parts geometrically.

(2) In a map, a distance of 7.2 km is represented by 12 cm. Draw a plain scale to scale to show kilometres and hectometres and long enough to measure up to 9 km. Find R.F. Mark a distance of 6.4 km on the scale.

Hint. R.F. = $\frac{12 \text{ cm}}{7.2 \text{ km}} = \frac{12 \text{ cm}}{7.2 \times 1000 \times 100 \text{ cm}} = \frac{1}{60\,000}$.

Length = $\frac{1}{60\,000} \times 9 \times 1000 \times 100$ cm = 15 cm.

Divide 15 cm into 9 equal parts geometrically, each part will represent 1 km.

(3) A rectangular plot of area 625 square kilometres is represented on a certain map by a similar rectangular area 25 square centimetre. Find R.F. Draw a plain scale to show kilometre and indicate a distant of 46 kilometres on it.

Hint : R.F. = $\sqrt{\frac{25 \text{ cm}^2}{625 \text{ km}^2}} = \frac{1}{5\,00\,000}$

In this problem max. length to be measured is not given. Let us take it as 50 km as we will have to mark 46 km in the scale.

Length = $\frac{50 \times 1000 \times 100 \text{ cm}}{5\,00\,000}$ = 10 cm.

Divide 10 cm into 5 equal parts. Now, divide the left portion into 10 equal parts.

(4) A 2.5 cm long line represents a length of 5 metres. Extend this line up to 20 metres. Divided the scale into 5 metre unit. Show 13 metres on it. Draw the scale.

4.2.2 Diagonal Scale

We shall solve a problem first and then explain the theory. (Please see Art. 4.2.3) In a diagonal scale, we can measure maximum three dimensions or units, such as metres, decimetres and centimetres.

Problem 1. Construct a diagonal scale 1 : 70 showing metres, decimetres and centimetres and to measure up to 7 metres. Show a distance of 4.66 metres on it.

Solution :

(a) Show the calculation (as done below) on the drawing sheet.

(b) Length of the scale = R.F. × max. length to be measured

$$= \frac{1}{70} \times 700 \text{ cm} = 10 \text{ cm}$$

This calculation is to be shown at the top or bottom of the scale on the Drawing sheet in an examination.

(c) Take a 10 cm long line AB and divide it into 7 equal parts geometrically, as shown in Fig. 4-3. Then, one small division represents a metre.

(d) Divide the left-hand-side-division into ten equal parts geometrically, as shown in Fig. 4-1. One small division represents one decimetres, because 1 m = 10 decimetres (dm).

Up to this stage, all procedures are same as in plain scale.

(e) Complete the plain scale, AB line.

(f) Now, we are proceeding to divide the smallest division into ten equal parts in order to get centimetres. But it is physically impossible to divide the same by ordinary method.

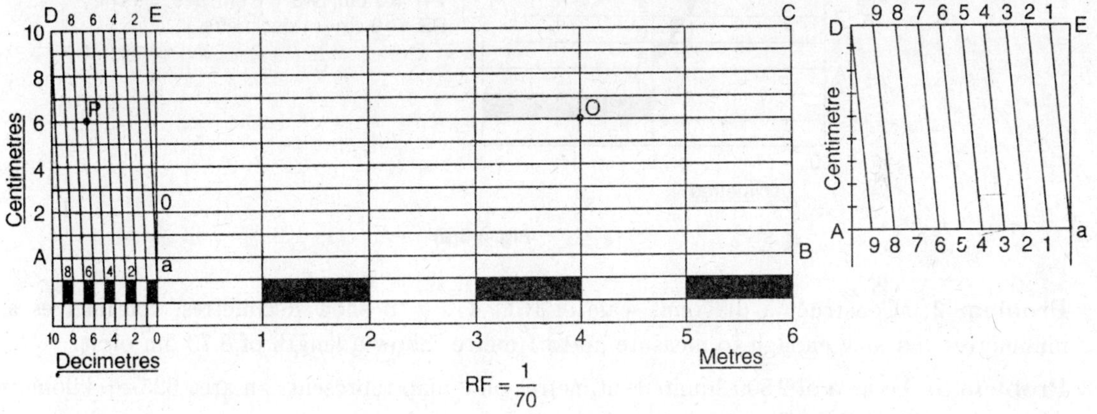

Fig. 4-3

Diagonal Method

We are to take advantage of the diagonal method. Let us imagine the last small division. Putting the same in an enlarged form, draw a rectangle above 10-9 division, i.e., A-a-E-D in Fig. 4-3(a). Now, divide AD into ten equal parts geometrically. Draw diagonal Da. Draw perpendiculars from the above points (1 to 9) intersecting the diagonal Da. The distance b–b' represents $\frac{1}{10}$th of Aa, which will indicate 1 cm, because $\frac{1}{10} \times$ (1 decimetre) $= \frac{1}{10} \times 10$ cm = 1 cm.

(g) Now, complete the rectangle $ABCD$ (Fig. 4-3) above AB line. AB = 7 metres. The height AD varies from 25 to 30 mm. There is no rule for this height, main factor is convenience. Take AD as 30 mm. Divide the line AD into ten equal parts geometrically.

(h) Draw horizontal parallel lines by the T-square from those points (1 to 10). Be careful while drawing all parallel lines. To avoid deflection, always keep the pencil in the same inclination from start to finish. (Fig. 4-3)

(i) Mark those vertical points on the line AD as 1,2,3,. . . ,10 which are known as centimetres.

(j) Draw vertical lines on the metre-measuring side from points 1,2,3,4,5,6 up to the horizontal line DC (Fig. 4-3).

(k) Write **METRES, DECIMETRES** and **CENTIMETRES** in proper positions (Fig. 4-3).

(l) The distance *PQ* represents 4 metres, 6 decimetres and 6 centimetres, i.e., 4.66 metres (Fig. 4-3).

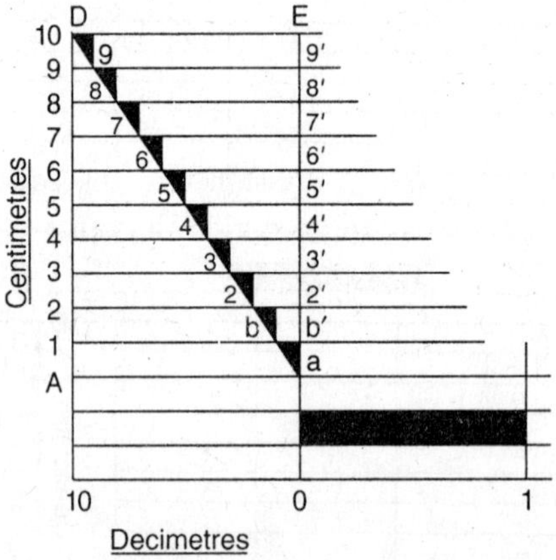

Fig. 4-3(a)

Problem 2. Construct a diagonal scale of R.F. 1 : 5 to show decimetres, centimetres and millimetres and long enough to measure up to 1 metre. Show a length of 8.73 dm on it.

Problem 3. An area of 2 500 square centimetres on a map represents an area 625 sq. kilometres on the field. Find R.F. Draw a diagonal scale to show kilometres, hectometres and decametres and to measure up to 5 kilometres. Indicate a distance of 3.46 km on the scale.

Solution : In case of an area, we are to take the square-root of both numerator and denominator to find the linear distance.

(a) $\text{R.F.} = \sqrt{\dfrac{2500\text{ cm}^2}{625\text{ km}^2}} = \sqrt{\dfrac{2500\text{ cm}^2}{625 \times (1000 \times 1000)(100 \times 100)\text{ cm}^2}} = \dfrac{1}{50\,000}$

[625 sq. km is to be converted into sq. cm.

1 km = 1000 m, 1 m = 100 cm

$\therefore$ 625 sq. km = 625 × 1000 × 1000 sq. m = (625 × 1000 × 1000) × (100 × 100 sq. cm]

(b) Length of the scale = R.F. × max. length to be measured

$= \dfrac{1}{50\,000} \times 5000 \times 100\text{ cm} = 10\text{ cm}.$

(c) Divide 10 cm long *AB* line into five equal parts geometrically as shown in Fig. 4-4. Each small division represents 1 km.

(d) Divide left division into ten equal parts. Each division represents 1 hectometre.

(e) Follow the same procedure as in problem No. 1.

(f) The distance 3.46 km is shown in Fig. 4-4 with circle marks at the two points (Fig. 4-4).

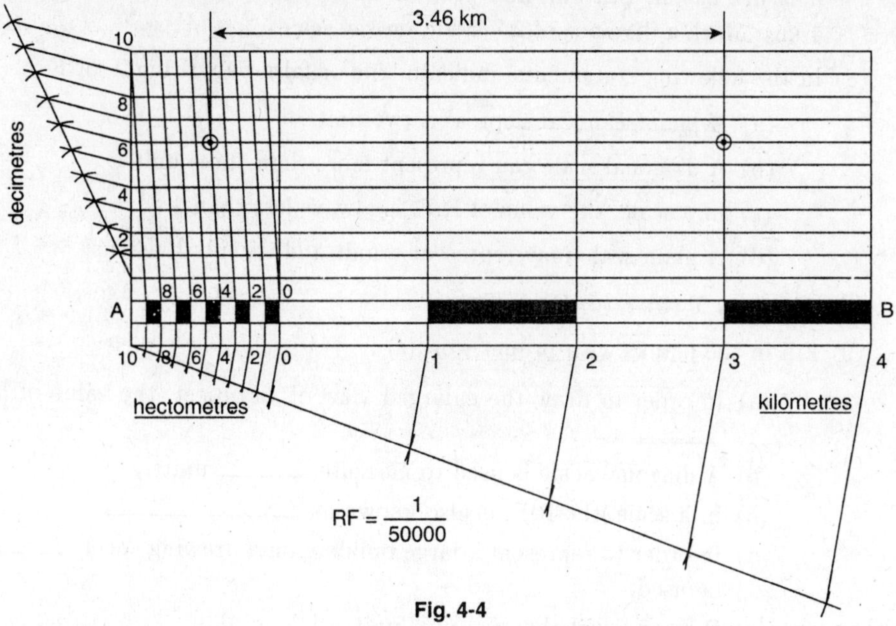

Fig. 4-4

Problems on Diagonal and Plain Scales

(1) Construct a diagonal scale of R.F. = $\frac{1}{40}$ to read metre, decimetre and centimetre, and long enough to measure up to 6 metres. Mark a distance of 4.58 m on it.

(2) The distance between Howrah and Kharagpur is 90 km, which is represented by 3 cm in a Railway Map. Find R.F. and draw a diagonal scale to read a minimum distance of 1 km, and long enough to measure up to 500 km. Mark a distance of 335 km on the scale.

(3) A rectangular plot of land, of area 4500 square metres, is represented on a map by a similar rectangle of area 5 square centimetres. Find R.F. and draw a diagonal scale to read up to single metre, and long enough to measure up to 400 m. Mark a distance of 325 m on the scale.

(4) An area of 4000 square kilometres is represented by a rectangular map of dimensions 200 cm × 80 cm. Construct a plain scale to read hectometre and kilometre, and long enough to measure up to 7 km. Mark a distance of 4.7 km on the scale.

(5) The distance between Howrah and Bardhaman is 107 km, which is represented by 16 cm in a map. Find the R.F. and draw a diagonal scale to show a minimum distance of 1 km, and long enough to read up to 100 km. Mark a distance of 82 km on the scale.

(6) The shortest distance between Sealdah and Baghajatin Railway stations is 10 km, which is represented in a map by 16 cm. Find the R.F. and draw a diagonal scale to read km, hm and dm. Mark a distance of 9.47 km in it.

(7) A distance of 1 km is represented by 4 cm in a map. Construct a diagonal scale to measure in km, $\frac{1}{8}$th km, and $\frac{1}{40}$th of a kilometre, and long enough to measure up to 4 km. Mark a distance of 2.775 km on the scale.

(8) In the following statements, indicate true or false by writing T or F :

(a) A plain scale can represent two units.

(b) A diagonal scale can represent more than three units.

(c) In a scale, the value of R.F. is always less than 1.

(d) A plain scale may represent a unit and its sub-divisions.

Ans. (a) T (b) F (c) F (d) T.

(9) Fill in the blanks with proper word(s) :

(a) In order to draw the enlarged view of an object, the value of R.F. is ______ ______ 1.

(b) A diagonal scale is used to measure ______ units.

(c) In a scale, the R.F. is also known as ______ ______.

(d) In order to represent a large building on a drawing sheet, ______ scale should be used.

(e) R.F. of a scale has ______ unit.

Ans. (a) more or greater, than (b) three, (c) scale factor, (d) reduced, or diminished, (e) no.

State Council Model Questions (West Bengal)

1. Construct a plain scale to show metre when 1 centimetre represents 4 metres and long enough to measure up to 50 metres. Find R.F. and mark on the scale a distance of 36 metres.

Hint. R.F. $= \frac{1 \text{ cm}}{4 \text{ m}} = \frac{1}{400}$, Length $= \frac{1}{400} \times 50 \times 100 \text{ cm} = 12.5 \text{ cm}$.

Take 12.5 cm line and divide it into 5 equal parts. One small division represents 10 metres.

2. A room of 1000 m^3 volume is represented by a block of 125 cm^3 volume. Find the R.F. and construct a plain scale to measure up to 30 metres. Measure a distance of 18 metres on the scale.

Hint. R.F. $= \sqrt[3]{\frac{125 \text{ cm}^3}{1000 \text{ m}^3}} = \frac{1}{200}$, Length $= \frac{1}{200} \times 30 \text{ m} = 15 \text{ cm}$.

Divide the 15 cm line into 3 equal parts and one small division represents 10 m.

3. A rectangular plot of land 25 sq km is represented on a certain map by a similar rectangle of area 1 sq cm. Find R.F. Draw a plain scale to show in kilometres. Indicate a distance of 55 km on it.

Hint. R.F. $= \sqrt{\frac{1 \text{ cm}^2}{25 \text{ km}^2}} = \sqrt{\frac{1}{25 \times (1000 \times 1000)(100 \times 100)}} = \frac{1}{5\,00\,000}$.

Let us imagine the maximum length to be measured as 60 km because we will have to show 55 km in the scale.

$$\text{Length of the scale} = \frac{60 \times 1000 \times 100}{5\,00\,000} = 12 \text{ cm.}$$

Divide the 12 cm length line into 6 equal parts where one small division represents 10 km.

4. An area of 4 000 sq km is represented by a regular map 200 cm × 80 cm. Construct a simple scale to read hectometre and kilometre up to a length of 7 kilometres. Show a distance of 4 km and 7 hectometres on the scale.

Hint. $\text{R.F.} = \sqrt{\dfrac{200 \times 80 \text{ sq cm}}{4000 \text{ sq km}}} = \sqrt{\dfrac{200 \times 80}{4000 \times 1000 \times 1000 \times 100 \times 100}} = \dfrac{1}{50\,000}.$

$$\text{Length of the scale} = \frac{1}{50\,000} \times 7 \text{ km} = \frac{7 \times 1000 \times 100 \text{ cm}}{50\,000} = 14 \text{ cm.}$$

Divide a line of 14 cm in length into 7 equal parts geometrically and one small division represents 1 kilometres

5. Construct a diagonal scale of R.F. = $\frac{1}{2500}$ to show in metres and long enough to measure up to 300 metres. Show a distance of 287 metres on your scale.

Hint. Length of the scale $= \dfrac{1}{2500} \times 300 \times 100 \text{ cm} = 12 \text{ cm.}$

Divide a line 12 cm long into 3 equal parts. One small division represents 100 metres. Divide the left hand division into 10 equal parts and one small division represents 10 metres. Divide the small division diagonally and that will represent 1 metre.

6. On a building plan, a line 22 cm long represents a distance of 440 metres. Construct a diagonal scale for the plan to calculate its R.F. Indicate clearly on the scale the lengths 8 metres, 43 metres and 150 metres.

Hint. $\text{R.F.} = \dfrac{22 \text{ cm}}{440 \text{ m}} = \dfrac{22}{440 \times 100 \text{ cm}} = \dfrac{1}{2\,000}.$

Max. length to be measured is not given, but as we have to measure max. 150 metres, we are taking it to be 160 metres.

$$\text{Then, length of the scale} = \frac{1}{2\,000} \times 160 \text{ m} = \frac{16\,000 \text{ cm}}{2\,000} = 8 \text{ cm.}$$

Divide a line 8 cm long into 4 equal parts. One small division represents 40 metres. Divide the left division into 4 equal parts, then one small division will be 10 metres. Diagonally divide the line into ten equal parts, that will represent as 1 m.

7. Construct a diagonal scale to show single metre when 15 cm on the drawing represents 250 metres.

Hint. Length of the scale = 15 cm.

Divide a line 15 cm long into 5 equal parts which will represent 50 metres. Divide the left division into 5 equal parts and one small division will represent 10 metres. Diagonally divide one small division in ten equal parts which will represent 1 metre.

8. On a certain map a distance of 5 km is represented by a length of 15 cm. Construct a diagonal scale to read up to distance of 0.01 km and mark on it a distance of 3.64 km. Find also the R.F. of the scale.

Hint. R.F. $= \dfrac{15 \text{ cm}}{5 \text{ km}} = \dfrac{15}{5 \times 1000 \times 100} = \dfrac{3}{100\,000}$.

Let us take max. length to be measured as 4 km as we will have to measure 3.64 km.

Length of the scale $= \dfrac{3}{100\,000} \times 4 \text{ km} = \dfrac{3 \times 4 \times 1000 \times 100}{1,00,000} = 12$ cm.

Divide a line 12 cm long into 4 equal parts, one division represents 1 km. Divide the left division into ten equal parts and one small division will represent as 100 metres, i.e., 0.1 km. Divide one small division diagonally into 10 equal parts, each of which will represent 10 metres, i.e., 0.01 km.

9. What is meant by representative fraction (RF)?

Ans. R.F. $= \dfrac{\text{Length of the Drawing (mm) or cm}}{\text{Actual length of the object (mm) or cm}}$

10. Differentiate between plain and diagonal scale.

Ans. In a plain scale, two consecutive units can be measured. In a diagonal scale, three consecutive units can be measured.

11. Name the different types of scales in engineering practice, on the basis of R.F.

Ans. Full-sized scale, Reduced-size scale, Enlarged-size scale.

12. Fill in the blanks with appropriate words :

(a) The ratio of length of the drawing of the object to the actual length of the object is called ______.

(b) For drawing of small object, ______ scale is always used.

(c) When measurements are required in three units, the scale used is called ______.

(d) Building drawings are drawn, using ______ scale.

(e) The relative values of enlarged, full size and reducing scales are ______, ______ and ______ respectively.

(f) If 1 cm on drawing represents 1 metre length of an object, then R.F. of the scale is ______.

(g) Civil Engineers and Architects generally use ______ scale.

(h) When drawings are made larger than their real size, the scale is known as ______ scale.

(i) The R.F. of full size scale is ______.

(j) If length of the scale is 15 cm and max. length 5 dm, the R.F. of the scale is ______.

Ans. (a) Representative Fraction, (b) enlarged, (c) diagonal, (d) reducing, (e) greater than unity, equal to unity, less than unity, (f) $\frac{1}{100}$, (g) reducing scale, (h) Enlarged (i) 1 : 1, (j) $\frac{3}{10}$.

13. On a building plan a line 10 centimetres long represents a distance of 5 m. Construct a diagonal scale for the plan to read up to 6 cm, showing metres, decimetres and centimetres. Indicate on your scale a length of 3.24 m. **(AMIE—Summer, 2001)**

14. Choose the correct answer from the following alternatives :

(a) A room of volume 1000 m^3 is represented by a block of volume 125 cm^3 in a map. The R.F. of the scale used for the map is

(a) 1/100 (b) 1/150 (c) 1/200. **(AMIE—Summer, 2000)**

Ans. (c)

N.B. We are to take the cube-root of a cubical dimension (m^3, cm^3, mm^3 etc.) in order to get a linear dimension (m, cm, mm etc.)

15. Fill in the blank :

An area of 0.45 hectare is represented by an area of 5 sq. cm in a map. The representative fraction is ______. **(AMIE—Winter, 2000)**

Ans. $\frac{1}{3000}$, N.B. 1 hectare = 10,000 sq. m.

16. Construct a diagonal scale of RF = $\left(\frac{1}{2.5} \times 10^6\right)$ to read up to a single kilometre and long enough to measure 400 km. Mark a length of 278 km on the scale. **(AMIE—Winter, 2001)**

17. The actual length of 300 m is represented by a line of 10 cm on a drawing. Draw a vernier scale to read up to 600 m. Mark on the scale a length of 343 m and mention the R.F. of the scale. **(AMIE—Summer, 2002)**

18. Fill in the blank :

When measurements are desired in three units, ______ scale is used.

Ans. Diagonal. **(AMIE—Winter, 2001)**

19. Write true or false :

Enlarging scales are usually found in civil engineering practice.

Ans. False, because 'scales' are used in all branches of Engineering, and reducing scale is used in Civil Engineering. **(AMIE—Winter, 2002)**

20. Answer briefly to the point :

How many consecutive units a diagonal scale can measure accurately?

Ans. Three units, such as m, dm, cm or dm, cm and mm etc. **(AMIE—Winter, 2002)**

21. Define Representative Fraction. Write a brief note on scale of chords. **(Summer, 2006)**

22. The map of a big area has a scale in which actual distance between two stations A and B of 150 km is represented by 50 cm. A small portion of this map is to be enlarged to 3 times. To the new map, construct a diagonal scale to read maximum distance of 8 km. Indicate on the scale, distances 1.86 km, 3.72 km and 7.33 km. **(Summer, 2006)**

23. On a building plan, a line 22 cm long represents a distance of 440 metres. Construct a diagonal scale for the plan to read up to 300 metres, showing single metre by diagonal division. Calculate its R.F. Indicate clearly on the scale the lengths of 157, 143 and 8 metres. **(Winter, 2006)**

Chapter 5

Geometrical Constructions

5.1 Purpose of the Chapter

This chapter deals with the problems on different geometrical constructions. Often such constructions are needed in engineering drawing. The students have already acquired primary knowledge on some of the geometrical constructions from their school-geometry. So, this chapter may be said as the recapitulation of bisecting a line and an angle, drawing perpendicular, dividing a line into any number of equal parts, etc. (which has already been shown in Chapter 4).

5.1.1 Construction of Regular Polygons

The word 'Poly' means 'many'. When all the sides of a polygon (many-sided figure) are equal, it is known as a 'regular polygon'*.

5.2 To Draw a Circle within a Regular Polygon of any Number of Sides (Length of the Sides given)

The steps are as follows :

(a) Draw an isosceles triangle $ABCA$ (Fig. 5-1), a square $ABCDA$ (Fig. 5-2), a regular pentagon $ABCDEA$ (Fig. 5-3) and a regular hexagon $ABCDEFA$ (Fig. 5-4).

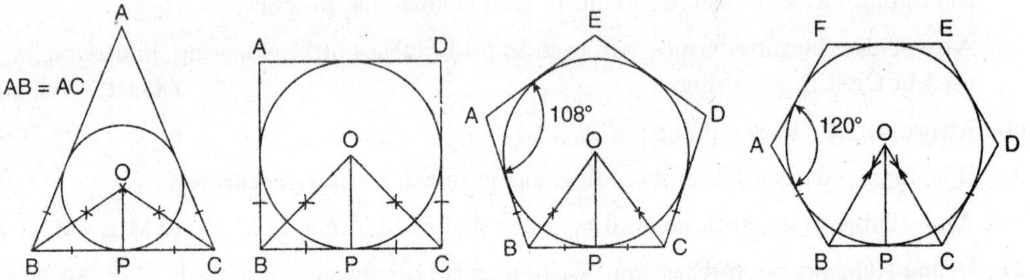

Fig. 5-1 **Fig. 5-2** **Fig. 5-3** **Fig. 5-4**

(b) Bisects any two angles, say $\angle ABC$ and $\angle ACB$ (Fig. 5-1) or $\angle ABC$ and $\angle DCB$ (Fig. 5-2) or $\angle ABC$ and $\angle DCB$ (Fig. 5-3) or $\angle BCD$ and $\angle ABC$ (Fig. 5-4).

---

*According to the number of sides of a polygon (regular or irregular), it is named as under. The number of sides are given within the brackets : (a) Pentagon (5), (b) Hexagon (6), (c) Heptagon (7), (d) Octagon (8), (e) Nonagon (9) and (f) Decagon (10). The triangle, trapezium, square, rectangle are also polygons. A Polygon has three or more number of sides, and it is a closed figure.

(c) All the bisectors meet at O, which will be the centre of the required circle.

(d) Draw a perpendicular OP on the side BC. With centre O and radius OP, draw a circle.

(e) This circle will touch all the sides of the regular polygon. (See figures 5-1 to 5-4)

5.3 To Draw a Regular Polygon

5.3.1 Equilateral Triangle (Length of one side given)

There are two methods (a) Set-square (b) Geometrical.

(a) Set-square Method :

The drawing steps are as follows :

(i) With T-square draw a line PQ (Fig. 5-5).

(ii) Cut the given length AB.

(iii) From the points A and B, draw 60° angles (with 30°–60° set-square), intersecting at C.

(iv) Then $ABCA$ is the required triangle.

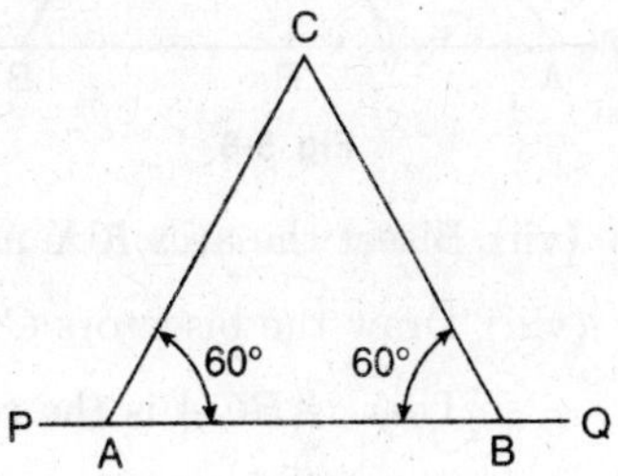

Fig. 5-5

In an equilateral triangle, all the sides are equal, and the three angles are 60° to one another. So, $\angle A = \angle B = \angle C = 60°$. Also $AB = BC = CA$.

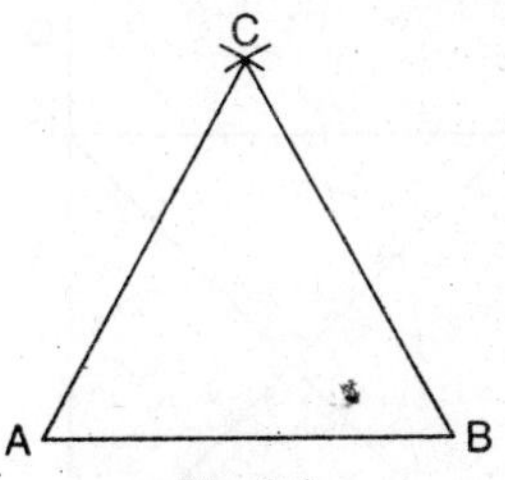

Fig. 5-6

(b) Geometrical Method :

The drawing steps are as follows :

(i) Draw the given line AB (Fig. 5-6).

(ii) With A and B as centres and AB as radius, draw arcs intersecting at C.

(iii) Join AC and BC. Then $ABCA$ is the required triangle.

5.3.2 Equilateral Triangle (altitude h given)

(a) Set-square Method :

The drawing steps are as follows :

(i) With T-square, draw a horizontal line PQ (Fig. 5-7).

(ii) Take any point R on PQ.

(iii) With T-square and set-square, draw a vertical line, and from it cut RC equal to the altitude 'h' of the triangle.

(iv) With 30°–60° set-square and T-square, draw lines from C on both sides of CR such that inclination to vertical is 30° on both sides, and

(v) Lines from C cut PQ at A and B. Then, $ABCA$ is the required triangle.

Here, $AB = BC = CA$. Also, $\angle CAB = \angle CBA = 60°$.

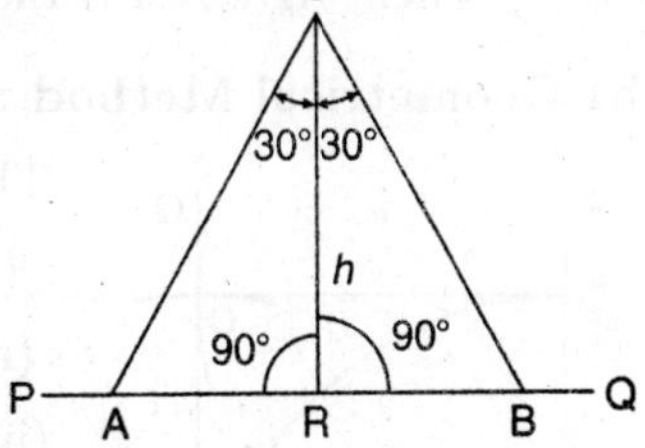

Fig. 5-7

(b) Geometrical Method :

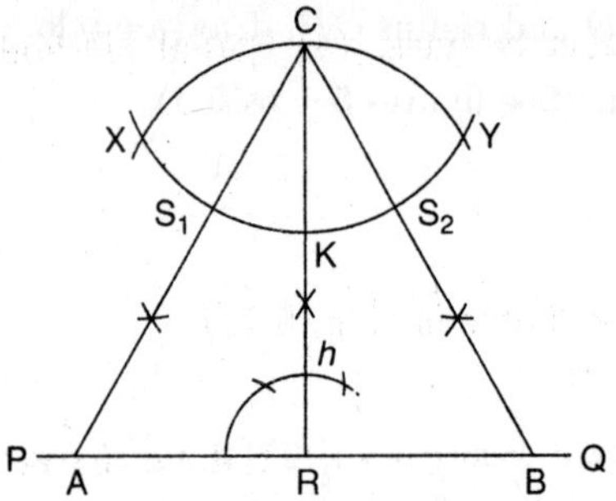

Fig. 5-8

The drawing steps are as follows :

(i) With T-square, draw any line PQ (Fig. 5-8).

(ii) Take any point R on PQ.

(iii) Draw perpendicular from R.

(iv) Cut RC equal to the altitude h.

(v) With centre at C and any radius, draw an arc, cutting CR at K.

(vi) With K as a centre and CK as radius, draw arcs, cutting the previous arc at X and Y.

(vii) Bisect the arcs KX and KY at S_1 and S_2 respectively.

(viii) Draw the bisectors CS_1 and CS_2 and produce them to cut PQ at A and B.

Then, $ABCA$ is the required equilateral triangle.

5.4 Square

(a) Set-square Method :

One side AB is given see Fig. 5-9.

The drawing steps are as follows :

(i) With T-square, draw line AB, equal to given length (See Fig. 5-9).

(ii) With a set-square and T-square, draw vertical lines AP and BQ from A and B respectively.

(iii) With a 45° set-square and T-square, draw 45° lines from A and B, cutting AP and BQ at D and C respectively.

(iv) Join C and D.

Then, $ABCDA$ is the required square.

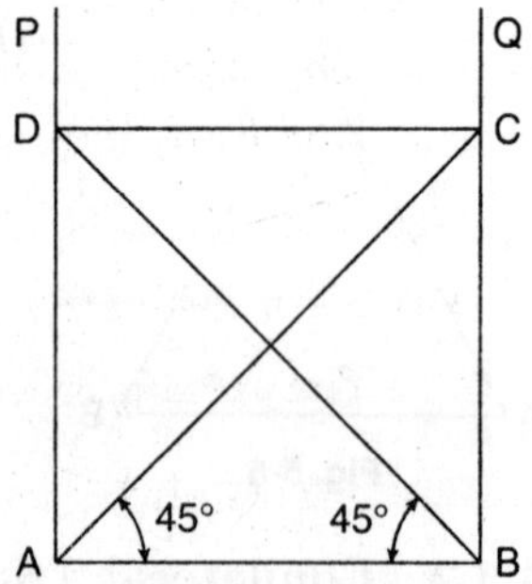

Fig. 5-9

(b) Geometrical Method :

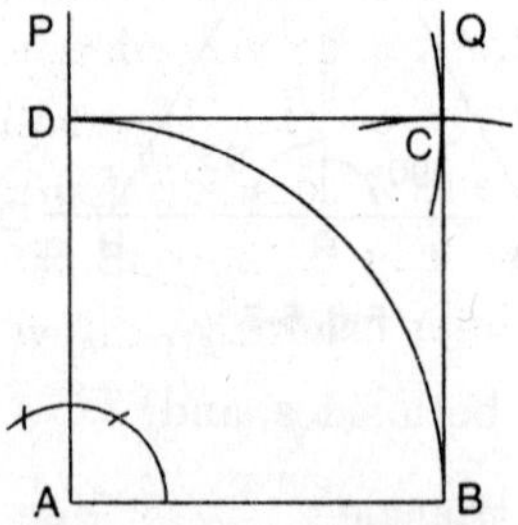

Fig. 5-10

The drawing steps are as follows :

(i) With T-square, draw a straight line AB (Fig. 5-10).

(ii) At A, draw perpendicular AP geometrically.

(iii) With A as centre and radius AB, draw an arc cutting AP at D.

(iv) With B and D as centres and AB as radius, draw two arcs, meeting at C.

(v) Join DC and BC.

Then, $ABCDA$ is the required square.

5.5 Regular Polygon

There are two methods to draw a regular polygon of any number of sides. (a) Radial Method (b) Centre Method.

5.5.1 Radial Method (one side given)

The drawing steps are as follows :

(i) Draw a line AB equal to the length of side of the regular polygon (Fig. 5-11).

(ii) Draw a semi-circle BPH, with A as centre and radius AB.

(iii) Divide the semi-circle into the same number of sides as the polygon by 'Trial and Error' method, as explained in Art. 5.5.1.(a). In this case, the number of sides is 6 for a hexagon. Mark the division points as 1, 2, 3, 4, 5, 6.

(iv) Join $A2$. Always join the centre and the second division-point for any regular polygon. Then join $A3$, $A4$, $A5$ and produce them. The lines $A2$, $A3$, $A4$, etc. are called radial lines. It is a general rule that, $A2$ will be one of the sides of the required regular polygon.

(v) With radius AB and B as centre, draw an arc. This arc cuts the line $A5$ produced at C. Similarly, with centre C and radius AB, draw an arc cutting $A4$ produced at D. The same procedure is followed to get the point E. Mark 2 as F, as they are same points.

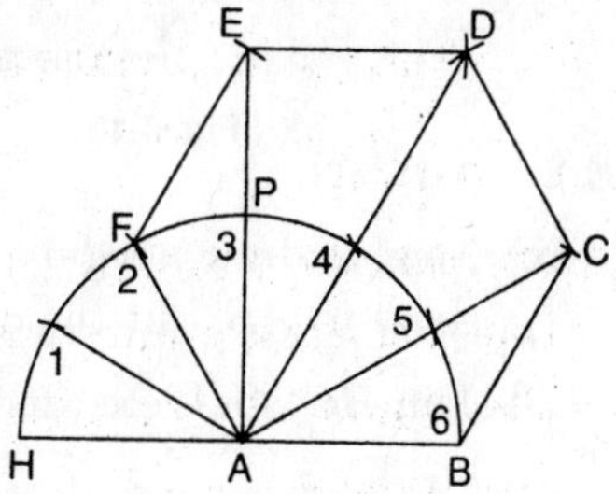

Fig. 5-11

(vi) Join BC, CD, DE and EF.

(vii) Then $ABCDEFA$ is the required regular polygon, which is a hexagon in this case.

The same procedure will be followed to draw a regular polygon of any number of sides.

5.5.1(a) Trial and Error Method

This method is simple, but requires practice. Let AB is an arc of a circle (Fig. 5-12). It is to be divided into a certain number of equal parts (say, five). Taking any radius (according to your choice) and starting from B, cut it into 5 parts (say, 1, 2, 3, 4, 5). Since we get 5 divisions by trial, some errors exist, say $5A$. It is called positive error. We shall have to increase the radius accordingly. Adjusting the error, we shall get another set of points. We shall continue to reduce the error gradually until 5 and A coincide. Error may be negative also, i.e., last division-point may be beyond the point A. Then we shall have to reduce the radius gradually, till we come back to the point A.

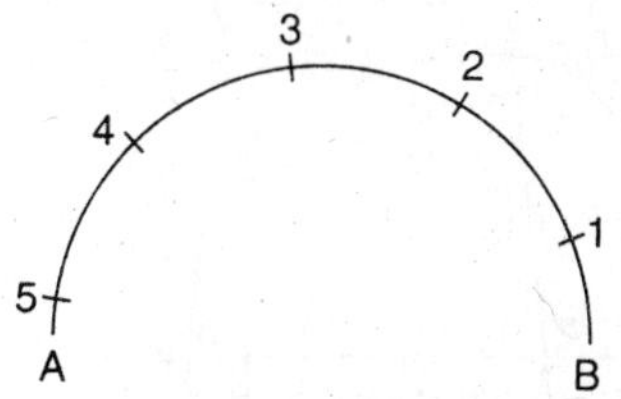

Fig. 5-12

5.5.2 Centre Method (one side given)

The drawing steps are as under :

(i) Draw the given side AB (Fig. 5-13).

(ii) Draw perpendicular bisector of AB geometrically.

(iii) Draw perpendicular at B (geometrically).

(iv) Cut $BC = AB$.

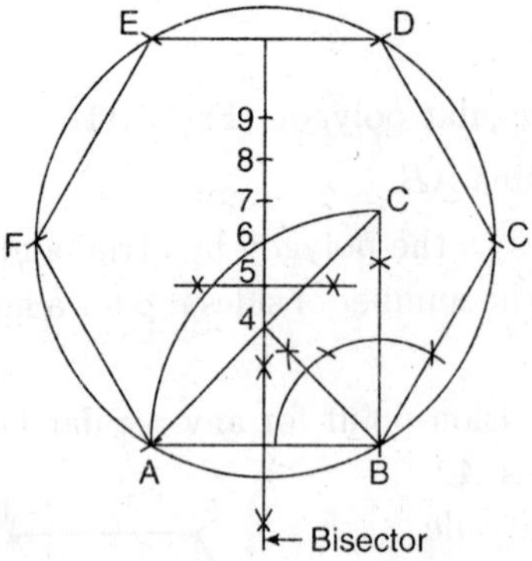

Fig. 5-13

(v) Join AC, which will cut the bisector at 4.

(vi) With B as centre and BA as radius, draw an arc AC, which will cut the bisector at point 6.

(vii) Bisect the line-segment 4-6 at the point 5.

(viii) From the point 6, the vertical line is produced at the points 7, 8, 9, etc., so that the line-segments 6-7, 7-8, 8-9 are all equal to 4-5.

(ix) All the points will be centre of inscribed circle (polygon within a circle) of the regular polygons. Centres 4, 5, 6, 7, etc. will be the centres of square, regular pentagon, hexagon, heptagon, etc.

(x) Now, to draw a regular hexagon, with centre 6 and radius $A6$, draw a circle. With radius equal to AB, cut the circumference into six equal divisions, starting from B.

(xi) Join BC, CD, etc., to draw the required regular hexagon ($ABCDEFA$).

5.5.3 To draw a regular polygon within a circle of given radius

The drawing steps are as follows :

Refer Fig. 5-14.

(i) With O as centre, draw the circle (Fig. 5-14) with given radius. Draw the diameter AK.

(ii) Divide AK (geometrically) into the same number of divisions as the sides of the polygon. Here it is 5 (say), i.e. pentagon.

(iii) Number the points as 1, 2, 3, 4 and 5. Fig. 5-14.

(iv) With centres at A and K and radius equal to AK, draw two arcs, meeting at the point P. Join P-2 and produce it to the circumference at B.

(v) Join A and B. AB is the length of one side of the polygon (Here it is a pentagon, i.e. no. of sides = 5).

(vi) Starting from B and with radius AB, draw arcs to cut the circle at C, D and E.

(vii) Join BC, CD, DE and EA.

Then $ABCDEA$ is the required pentagon.

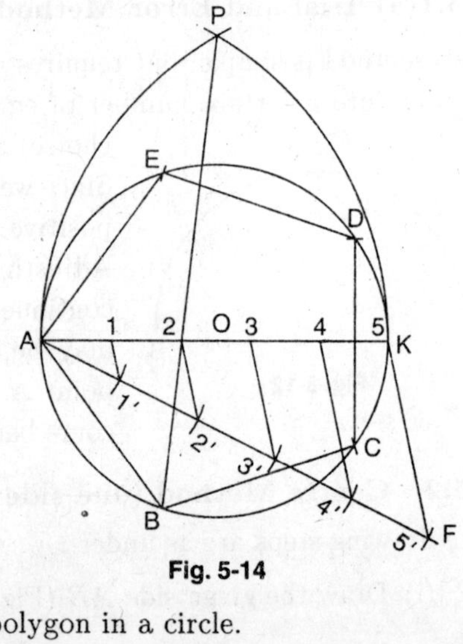

Fig. 5-14

The same procedure will be followed to inscribe any regular polygon in a circle.

5.6 To Draw a Number of Circles in a Regular Polygon

The conditions are as follows :

(i) **The number of circles is equal to the number of sides of a polygon.**

(ii) **Each circle should touch two other circles and one of the side of the polygon.**

The drawing steps are as follows :

(i) Draw the centre of polygon, as shown before (*O* is the centre).

(ii) From the centre, draw lines to join the end points or corners of a polygon.

(iii) We will get same number of circles as the number of sides of the polygons, as shown in the figures 5-15, 5-16, 5-17 and 5-18.

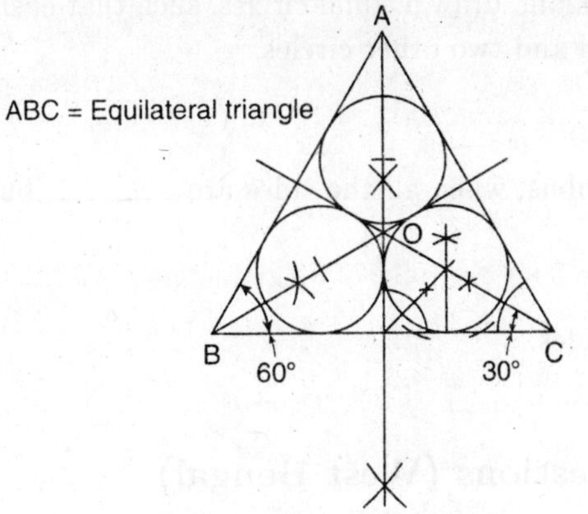

Fig. 5-15 Triangle

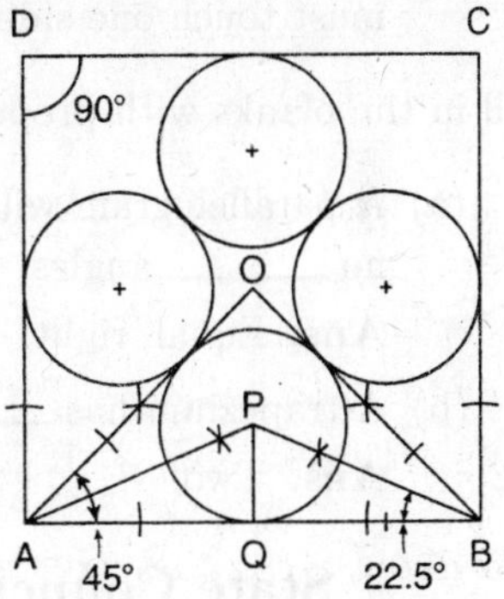

Fig. 5-16 Square

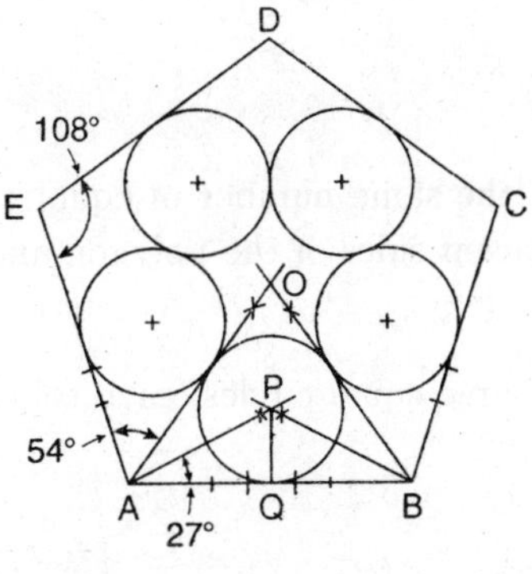

Fig. 5-17 Pentagon

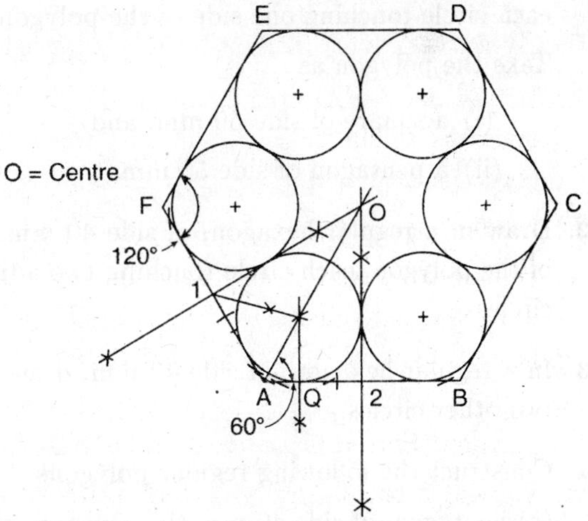

Fig. 5-18 Hexagon

(iv) Draw three circles within a triangle, as shown in the Fig. 5-15. Similarly, draw four circles within the square (Fig. 5-16), five circles within the pentagon (Fig. 5-17) and six circles within the hexagon (Fig. 5-18). Similarly, we can draw 7 circles within a regular heptagon and 8 circles within an octagon.

Problems

1. (a) Divide a line (90 mm long) into 5 equal parts.

(b) Draw a 50° angle with a diagonal scale and trisect it.

(c) Construct regular polygons of 5, 6, 7 and 8 sides, (length of the side 20 mm) by radial or centre method.

(d) Draw regular polygons of 5, 6, 7 and 8 sides within a circle of diameter 60 mm.

(e) Within a regular pentagon of side 50 mm, draw 5 equal circles, such that each circle must touch one side of the pentagon and two other circles.

2. Fill in the blanks with proper words :

(a) A parallelogram will become a rhombus, when all the sides are ______, but with no ______ angles.

Ans. Equal, right.

(b) A trapezium has ______ parallel sides.

Ans. Two.

State Council Model Questions (West Bengal)

1. Draw in a regular polygon, the same number of equal circles as the sides of the polygon, each circle touching one side of the polygon and two of the other circles.

Take the polygon as :

(i) a square of side 50 mm, and

(ii) a pentagon of side 50 mm.

2. Draw in a regular hexagon, of side 40 mm, the same number of equal circles as the sides of the polygon, each circle touching two adjacent sides of the polygon and two of the other circles.

3. In a regular hexagon, of side 40 mm, draw three equal circles, each touching one side and two other circles.

4. Construct the following regular polygons :

(a) heptagon, of side 40 mm (b) hexagon, of side 40 mm (c) pentagon, of side 40 mm.

5. Inscribe the following regular polygons in a circle of diameter 60 mm :

(a) Pentagon (b) Hexagon (c) Heptagon.

Chapter 6

Engineering Uses of Curves

Introduction

Curves are generated in various ways. For example, if we cut a right circular cone in different positions, we get different sectional curves. These may be circle, ellipse, parabola, hyperbola or triangle. Since these curves are obtained by cutting a right cone, these are known as 'conic sections or conics'. These conic sections are used in various engineering components. For example, elliptical petrol tanks, parabolic arches and reflectors, hyperbolic cooling towers, etc.

Also, there are other curves, such as cycloid, involutes, spiral, etc. These curves have different applications in engineering. For example, cycloidal or involute gear-teeth, spiral casing of a centrifugal pump, etc.

The aim of this chapter is to explain the procedures of drawing curves, used in engineering practice. Only the construction procedures are explained here, without mathematical proofs.

6.1 Ellipse

Elliptical mirrors are fitted in some dressing tables. Also, some petrol tanks are elliptical in cross-section. As such, we should know how an ellipse is generated and how it may be drawn. In a right circular cone, if the cutting plane is inclined to the base of the cone at an angle less than the base-angle of the cone, then the actual or true shape of the section is an ellipse. Since the cutting plane does not cut the base of the cone, the ellipse is a closed curve.

The ellipse may also be generated by a point. This point moves in such a way that, the sum of its distances from two fixed points (known as foci) is always constant, and equal to the length of the major axis.

Referring to Fig. 6-1,

AB = Major axis, CD = Minor axis. The two foci are F_1 and F_2.

$F_1C = AB/2$, $F_2C = AB/2$,

$\therefore\ F_1C + F_2C = AB$.

In the same manner, it may be shown that,

$PF_1 + PF_2 = AB$ (See Fig. 6-1).

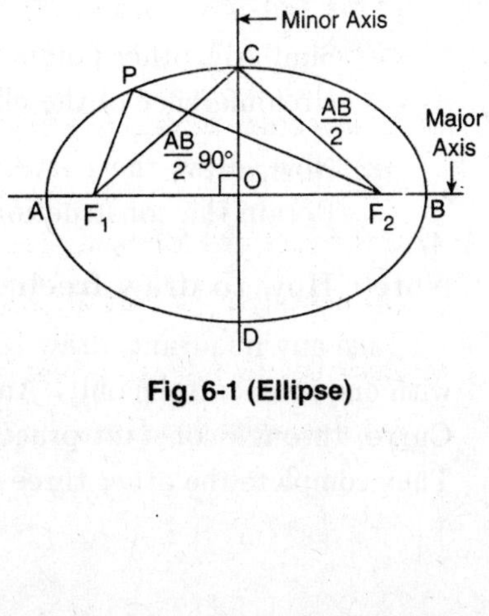

Fig. 6-1 (Ellipse)

There are different methods of construction of an ellipse. Here, we shall concentrate on three methods only (a) oblong method, (b) concentric circle method and (c) arcs of circles method.

6.1(a)(i) Oblong Method (Given major and minor axes or a rectangle)

In order to draw an ellipse, the following procedures are usually maintained :

(i) Draw major axis AB (Fig. 6-2). Draw the minor axis CD, which bisects the major axis AB at right angles at the point O.

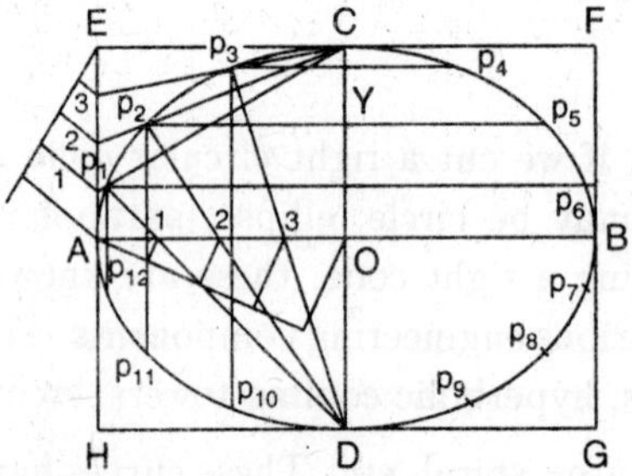

Fig. 6-2

(ii) Complete the rectangle with both the axis as adjacent sides, marked as $EFGH$.

(iii) Divide AO and AE into any equal numbers geometrically, as shown in the Fig. 6-2. It may be marked as 2,3,4 or 5 according to suitability. If we divide AO into 4 parts, the AE will also be divided into 4. If we take higher number of divisions, the ellipse will be more accurate in shape. Here, we have taken 4 divisions only.

(iv) Number the division points as 1,2,3 starting from A as origin.

(v) Join C with 1, 2 and 3 of AE.

(vi) Join D with 1, 2 and 3 of AO and extend them to meet the lines $C1$, $C2$ and $C3$ at the points P_1, P_2 and P_3 respectively. These points are situated on the ellipse.

(vii) To get the other points in other three quadrants, we may follow the same procedure, alternatively.

(viii) Draw vertical and horizontal lines with T-square and set-square from the points P_1, P_2 and P_3 which will cross the major and minor axis.

(ix) Let us take an example, say, from P_2, a horizontal line P_2Y is drawn. Extend P_2Y up to P_5, so that $P_2Y = YP_5$.

Now, draw a vertical line P_2X and extend it to P_{11}, such that $P_2X = XP_{11}$.

Similarly, other points $P_4, P_6, \ldots, P_{12}$ can be found out. These points are situated on the circumference of the ellipse.

(x) Now, draw the curve, passing through these points, by a French Curve or freehand. Retain the construction lines faint, for which credit will be given by the examiner.

Note : How to draw freehand

Take any quadrant, draw faint curves as correct as possible. Then try to complete the curve with one pencil stroke only. Any student can develop this skill of drawing curve (without French Curve) through constant practice. Take any three points and draw curve in the above manner. Then complete the other three quadrants. Try to make it as a hobby of drawing freehand curves.

6.1(a)(ii) Oblong Method (axis inclined)

Let the axis be 60° inclined to each other, as shown in Fig. 6-3. This can also be drawn, if three points are given, i.e., A, B and C, which are not in a straight line. Complete the parallelogram $ABCD$ and follow the same procedure, as above, to draw the ellipse.

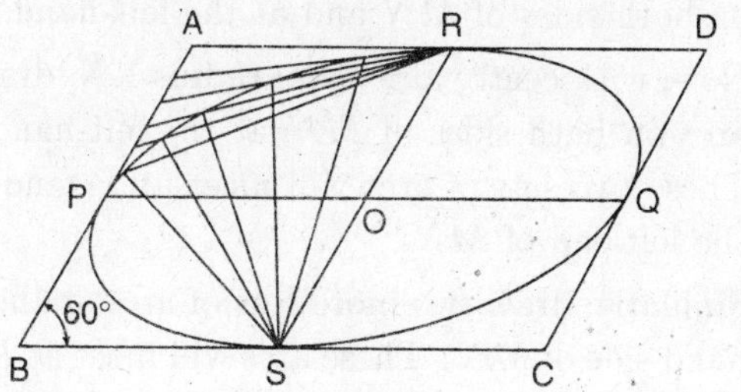

Fig. 6-3

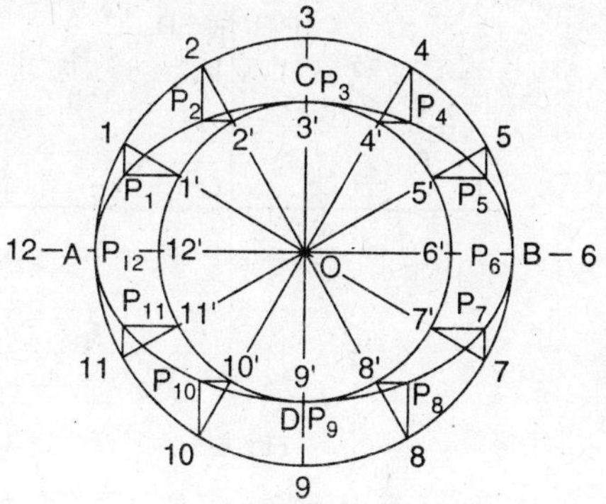

Fig. 6-4

6.1(b) Concentric Circles Method

The method is explained, step by step as follows :

(i) Draw major and minor axes AB and CE (at right angles to each other), meeting at a point O (Fig. 6-4).

(ii) Draw two circles, with O as centre and radii being the semi-major and semi-minor axes.

(iii) Divide the concentric circles into 12 equal parts, with T-square and 30°/60° set-square (i.e. by radial divisions).

(iv) Mark the radial divisions of the major and minor axis of the circles as $1, 2, 3, \ldots, 12$ and $1', 2', 3 \ldots 12'$ respectively.

(v) Draw vertical lines from $1, 2, 3, \ldots, 12$ points and horizontal lines from $1', 2', 3 \ldots, 12'$ parts, which will meet at $P_1, P_2, P_3, \ldots, P_{12}$. All these points are situated on the required ellipse, and

(vi) With a French curve or freehand, draw the ellipse through these points.

Note : If a student forgets wherefrom to draw horizontal and vertical lines, he or she may draw horizontal lines from any system of points. If vertical lines are drawn from $1', 2', 3 \ldots, 12'$, then the line 3-9 will be major axis and AB will be the minor axis. That matters little, because we will get the same ellipse.

6.1(c) Arcs of Circles Method

Refer Fig. 6-5. In this case, the major and minor axes of the ellipse are given. The steps of drawing procedures are as follows :

(i) Draw the major axis MN and the minor axis RS according to given measurements. The axes MN and RS bisect each other at right angle and at the centre O.

(ii) With centre at R and radius MO (i.e., semi-major axis), draw two arcs, which cut MN at F_1 and F_2. These are the foci of the ellipse.

(iii) On the axis MN, mark a few points, such as 1, 2, 3, etc. which may or may not be equal to each other.

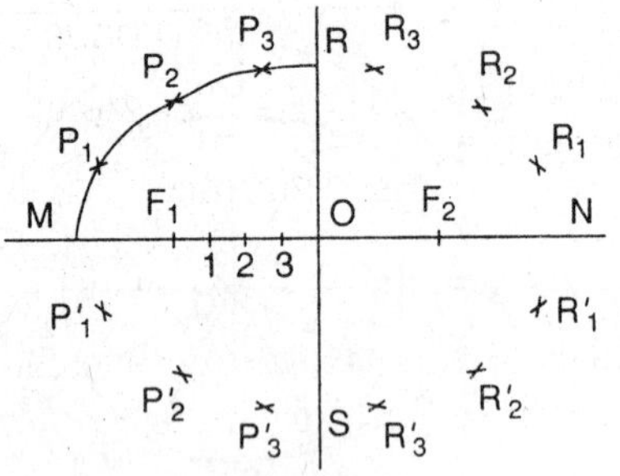

Fig. 6-5

(iv) With centre at F_1, and radius M-1, draw two arcs on both sides of MN and at the left-hand side.

(v) Now, with centre at F_2 and radius 1-N, draw two arcs on both sides of MN at the left-hand side. These two sets of arcs will meet at P_1 and P_1' at the left end of MN.

(vi) Similarly, draw two more sets of arcs at the right hand side of MN. These arcs will meet at R_1 and R_1' at right end of MN.

(vii) These four points (R_1, P_1, P_1', R_1') are situated on the ellipse. This is because $P_1F_1 + P_1F_2 = P_1'F_1 + P_1'F_2 = MN =$ length of major axis. Also $M - 1$ and $1 - N = MN$.

(viii) In a similar manner we get other points, such as $P_2, P_2', P_3, P_3', R_2, R_2', R_3, R_3'$, etc., which are also situated on the ellipse, and

(ix) Through these points $(M, N, R, S, P_1, P_1', P_2, P_2'$, etc.) draw a smooth curve either by a French Curve or freehand. This closed curve is the required ellipse.

The ellipse is to be drawn prominent, and other constructional lines are to be kept as faint.

6.2 Eccentricity, Axis and Vertex of a Conic

It has already been mentioned at the beginning of this chapter that the different sections obtained by cutting a right circular cone are known as 'Conics'.

A conic may also be defined as follows :

The conic is the locus of a point, which moves in a plane. The points move in such a manner that the ratio of its distances from a fixed point (called focus) and fixed straight line (called directrix) is always constant. This ratio (i.e., distance of the point from focus/distance of the point from directrix) is known as eccentricity, denoted by 'e'. The value 'e' is less than 1 for an ellipse, equal to 1 for a parabola, and greater than 1 for a hyperbola.

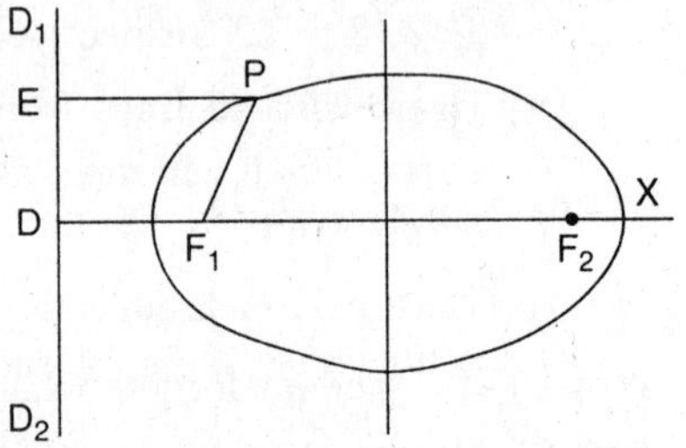

Fig. 6-6

The 'axis' is the line, which passes through the focus and perpendicular to the directrix. The "vertex" is the point at which the conic cuts the axis.

Referring to Fig. 6-6,

Eccentricity $= (VF_1/VD) = (PF_1/PE)$, where V is the vertex, $F1$ and $F2$ are foci, D_1, and D_2 are the directrix and DX is the axis. Here the conic is an ellipse. Hence the eccentricity (VF_1/VD) is less than 1.

6.3 Parabola

A parabola is a plane open curve (Fig. 6-7). The curve is generated by the locus of a point. This point is always equidistant from a fixed point (focus) and a straight line perpendicular to the axis (the directrix).

Some practical examples of parabola are arches, bridges, sound and light reflectors, path of a cannon ball, or projectile, path of falling water from a tap, etc.

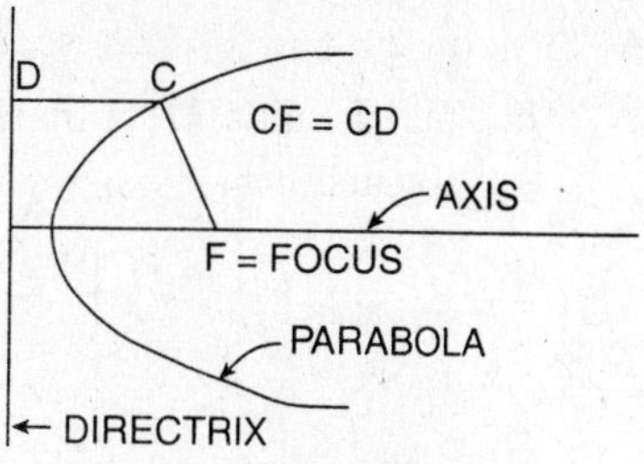

Fig. 6-7 (Parabola)

6.3A. Construction of Parabola

Although there are many methods for drawing a parabola, we are explaining only the rectangle method here, i.e., when the dimension of a rectangle are given.

(A) Rectangle Method (given the length and breadth)

Refer to Fig. 6-8.

(i) Draw the given rectangle $ABCD$ according to dimensions and axis EF, which is perpendicular to AD.

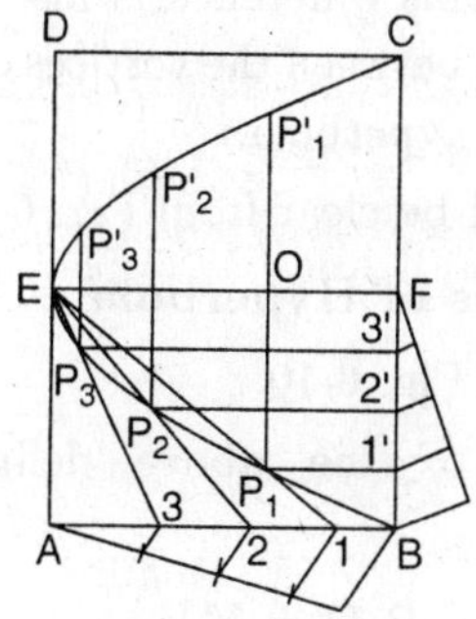

Fig. 6-8 (Parabola)

(ii) Divide AB and FB into same equal numbers. We have taken 4 divisions here.

(iii) Number these points with B as origin, into 1,2,3 towards A, and $1'$, $2'$, $3'$ towards F.

(iv) Join E-1, E-2, E-3.

(v) Draw lines parallel to AB from $1', 2'$ and $3'$. These lines will cut $E1, E2, E3$ at P_1, P_2 and P_3 respectively.

(vi) Points on the parabola are P_3, P_2, P_1 and B.

(vii) draw a smooth curve with the help of a French Curve or freehand, as already explained in case of ellipse.

(viii) Now transmit the points P_1, P_2 and P_3 to P_1', P_2' and P_3' making $P_1O = P_1'O$ and so on. and

(ix) Do not erase the construction lines, including the geometrical divisions of lines AB and BF. The parabola will be drawn prominent and other lines faint.

(B) Parallelogram method (when axis inclined, or given three points, which are not in a straight line)

Refer to Fig. 6-9. The steps are :

(i) Draw the parallelogram $ABCD$ or take A, B, C as given points.

(ii) Other procedures are same as in the rectangle method, explained in (A).

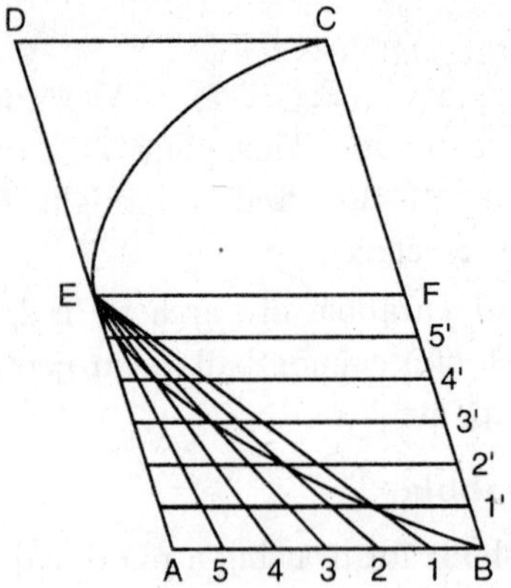

Fig. 6-9

6.4 Hyperbola

The hyperbola is a plane open curve, which is generated by a point. The point moves in such a fashion that, the differences of distances from foci F_1, and F_2 to that point is constant. This difference is also equal to the distance between the vertices of the two branches of hyperbola.

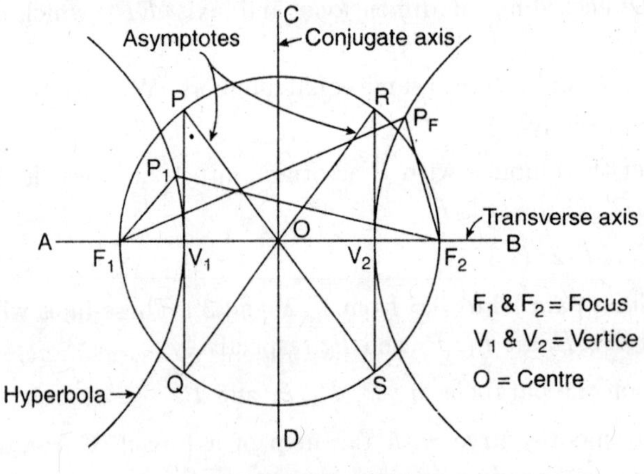

Fig. 6•10 (Hyperbola)

This will be clear from Fig. 6-10

Parameters of Hyperbola

Refer to Fig. 6-10.

According to the above definition of hyperbola,

$$P_1F_2 \sim P_1F_1 = V_1V_2$$

and $P_2F_1 \sim P_2F_2 = V_1V_2$.

Transverse axis $= AB$, conjugate axis $= CD$.

PS and RQ are asymptotes. V_1 and V_2 are vertices. $O =$ centre of hyperbola.

6.4.1 To Draw a Hyperbola

(Given : The positions of foci and distance between the vertices)

Refer to Fig. 6-11.

The construction procedures of a hyperbola are as follows :

(i) Draw horizontal axis AB (transverse axis) and vertical axis CD (conjugate axis) of the hyperbola.

(ii) According to given measurements, mark foci F_1 and F_2 and vertices V_1 and V_2.

(iii) Choose several points on the horizontal axis, say 1,2,3, etc., between points F_2 and B.

(iv) With radius V_1-1 and centres at F_1 and F_2, draw arcs on both sides of AB and CD.

(v) Again draw another set of arcs, radius V_2-1 and centres at F_1, F_2 These arcs will cut the previous arcs at P_1, P_1', and R_1, R_1' points.

(vi) Proceeding as above, we will get the points P_2, P_2'; P_3, P_3'; R_2, R_2'; R_3, R_3' and so on. All these points are situated on the two branches of the hyperbola.

(vii) Through the above mentioned points, draw two curves with the help of French Curve or freehand (Fig. 6-11). These curves are the two branches of the hyperbola.

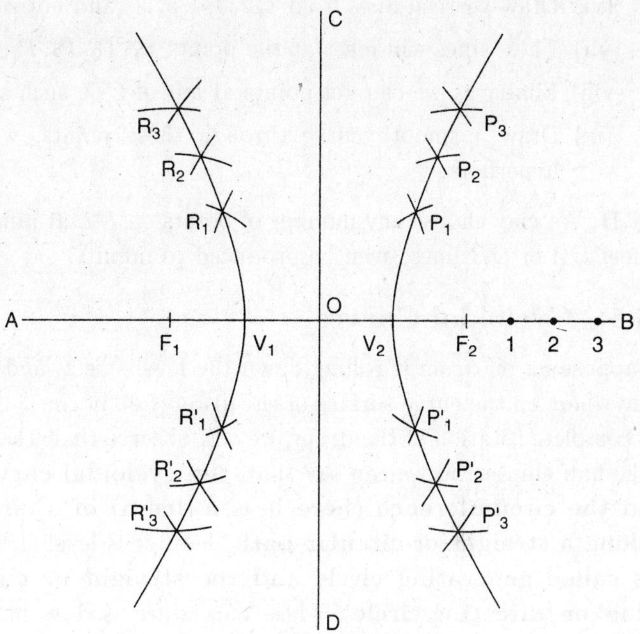

Fig. 6-11 (Hyperbola)

6.4.2 Rectangular Hyperbola

In Fig. 6-10, we have noticed that the asymptotes PS and RQ are not at right angles to each other. When the asymptotes intersect at right angles, the curve is known as a 'rectangular hyperbola' (or equiangular hyperbola). This is the graphical representation of Boyle's Law, PV = constant for gases, where P = Pressure and V = Volume. It is also applicable for design of water channels.

Constructions methods are as follows :

Refer to Fig. 6-12.

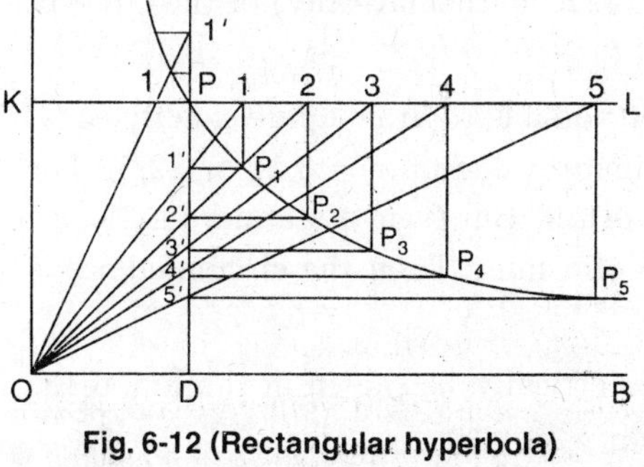

Fig. 6-12 (Rectangular hyperbola)

(i) Draw the axes AO and OB, at right angle to each other, which are known as asymptotes.

(ii) Take any given point P, whose co-ordinates (X, Y) are known.

(iii) Through the point P, draw two lines KL and CD, which are parallel to OB and AO respectively.

(iv) On the line KL, take a few points such as $1, 2, 3, 4, 5, \ldots$ at a distance, which may or may not be equal to each other.

(v) Join $1, 2, 3, 4, 5, \ldots$ with O. These lines will cut the line CD at $1', 2', 3', 4', 5', \ldots$ respectively.

(vi) Draw vertical lines from $1, 2, 3, 4, 5, \ldots$ and horizontal lines from $1', 2', 3', 4', 5', \ldots$.

(vii) These lines will meet at the points P_1, P_2, P_3, P_4, P_5 etc.

(viii) Similarly, we can get points at left of CD, such as P_{-1}, P_{-2} and so on, and

(ix) Draw a smooth curve through these points, which will be the required rectangular hyperbola.

N.B. We may choose any number of points on KL at infinite distance. But, the curve will never meet OA or OB lines, even be produced to infinity.

6.5 Cycloidal Curves

Suppose an oil drum is rolling down the level street, and without slip. A white mark is painted anywhere on the curve surface of the drum (i.e., in rim). If we notice carefully the white mark for a complete rotation of the drum, we will observe that, the mark is tracing approximately a curve like half ellipse. So we can say that, **the cycloidal curves are generated by a fixed point on the circumference (here it is a drum) of a circle. This circle rolls without slip along a straight or circular path** (here, it is level street). **In this case, the rolling drum is called generating circle and the straight or curved path is known as 'directing line' or 'directing circle'.** These curves are used in the teeth-profiles of gear wheels for proper engagement during power transmission.

6.5.1 Cycloid

In the previous article, we have noticed how a cycloid is generated. Now, let us see how a cycloid can be drawn geometrically.

In order to draw a cycloid, the following procedures are generally used :

Here, the radius R of the generating circle is given.

Refer to Fig. 6-13.

(i) Draw the given circle of radius R, of which the centre is C.

(ii) Take one point P on the circumference of the circle. From P, draw a tangent PA with 'T-square'. PA is equal to the circumference ($2\pi R$ mathematically) of the circle of radius R.

[**N.B.** The students should remember that, it is not possible to draw a line exactly equal to the circumference, because we take the approximate value of $\pi = 3.14$ or $22/7$. For more accuracy, we may assume the value of $\pi = 3.14159$. But from a standard scale, we cannot measure even up to two places of decimals of a mm. Then the circumference is drawn approximately.]

(iii) Divide PA (circumference) into 12 equal parts geometrically, as shown in the Fig. 6-13. Now, number the division as $1, 2, 3, 4, \ldots, 12$,

(iv) Divide the circle into twelve divisions with T-square and $30°/60°$ set-square. Number these points anticlockwise as $1', 2', 3', \ldots, 12'$. This is because the circle is rolling in a clockwise direction. These divisions may be any number. But the circumference (PA) and the circle should be divided into same number of parts.

(v) From centre C, draw a line CB parallel to PA, with 'T-square'. As the circle rolls clockwise over the line PA, the centre will be shifted gradually from C to B in one complete rotation. The height of centre remains the same, i.e., $CP = BA$.

(vi) Draw vertical lines from $1, 2, 3, \ldots 12$, which will meet the line CB at C_1, C_2, C_3, C_{12} (i.e., B). These are the different positions of centre.

(vii) Draw horizontal lines from the points $6', 7', 8', 10', 11'$, parallel to PA to meet the line AB produced.

(viii) Imagine the circle is rolling gradually over the line PA. Then, centre C will take the new position at C_1, and point P will rise up to the horizontal line $1'$-$11'$, i.e., at a distance R from C_1, (i.e., new position P_1). Again, when C will move to C_2 the point P will touch the horizontal line $2'$-$10'$, i.e., at a distance R from C_2 (new position P_2). Similarly we will get the remaining points ($P_3, \ldots, P_{12}$) of the cycloid at distance of radius R from $C_3, \ldots, C_{12}$, i.e., on horizontal lines from $3', \ldots, 12'$, i.e., C_3 with $3'$, C_4 with $4'$ and so on. We notice that, 1 meets with 1, 2 with 2 and so on, provided number of division should be 12. Then it will be easy to remember. P_{12} coincides with the point A, which is actually the starting point P.

(ix) Draw a smooth curve through $P_1, P_2, \ldots, P_{12}$ (by French Curve or freehand), which should be prominent. This is the required cycloid.

This curve may be compared with the flute-profile of a twist drill or profile of a spur gear tooth.

(x) All construction lines should be retained faint.

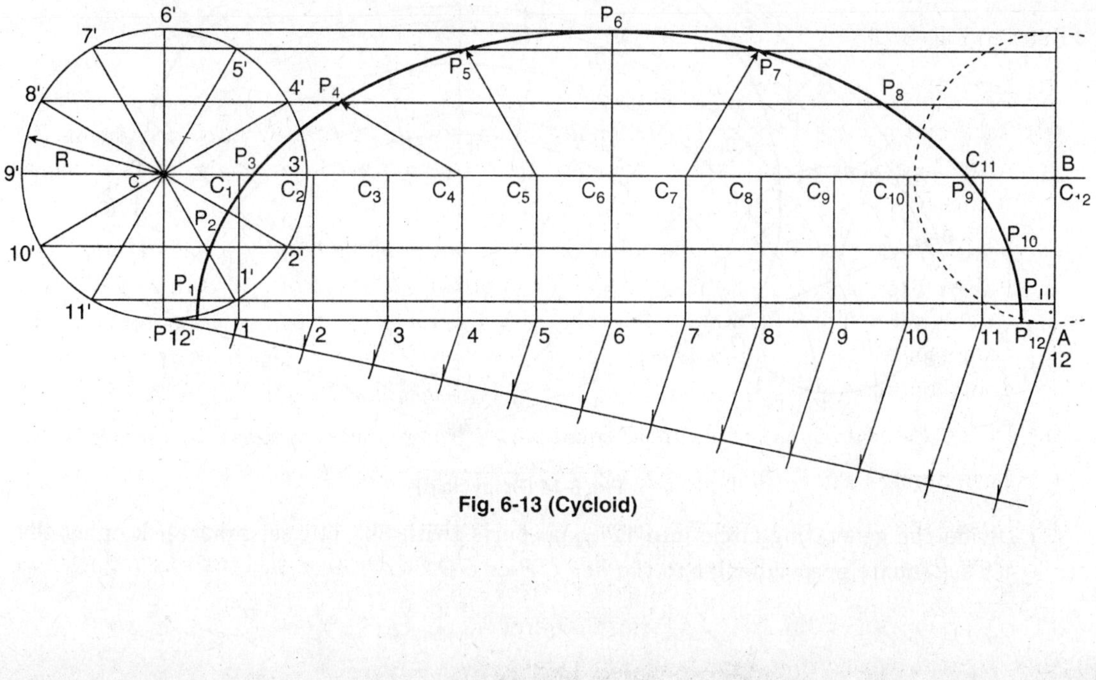

Fig. 6-13 (Cycloid)

6.5.2 Epicycloid

A circle (called rolling or generating circle) rolls over another bigger circle (called the directing circle) on the convex surface without slip. The locus of a point on the circumference of the generating circle is known as the epicycloid for one complete turn of the rolling circle.

Construction of Epicycloid

Here, the radius (r) of the generating circle and the radius (R) of the directing circle are given :

The drawing procedures are same as in the case of a cycloid. The only difference lies in the division of the convex surface. Let us see how this can be done.

We know that, arc of a circle is proportional to its central angle. Referring to Fig. 6-14, the arc PQ is the circumference of the rolling circle, i.e., $2\pi r = PQ$.

The circumference of directing circle $= 2\pi R$

$$(\alpha/2\pi r) = (360°/2\pi R) \text{ or, } \alpha = (360° \times r/R).$$

Now the steps of construction procedures are as follows :

(i) Draw a radial line $OQ = R$ and mark an angle α, so that $\angle POQ = \alpha$ (use protractor or diagonal scale).

(ii) With centre O and radius R, draw an arc which is PQ.

(iii) Join OP and produce it. Make $CP = r$.

(iv) With centre O and radius r, draw a circle, which is the generating or rolling circle.

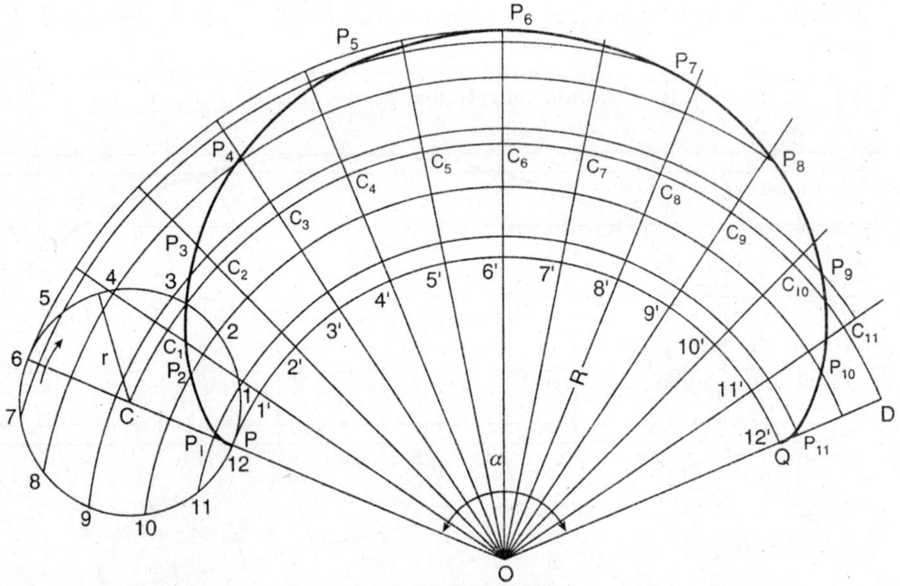

Fig. 6-14 (Epicycloid)

(v) Divide the generating circle into 12 equal parts (with 30°/60° set-square), keeping the 45° set square perpendicular to the line CP or CO.

(vi) Number the division points anticlockwise at $1', 2', 3', \ldots, 12'$. This is because the generating circle rolls in a clockwise direction.

(vii) Draw an arc CD with radius $(R + r)$ and O as centre. D is the point obtained by producing OQ.

(viii) Divide the arc PQ into 12 equal parts. In order to do this, first divide the arc geometrically into four equal parts. Then divide one small division into three equal parts by trial and error method.

(ix) Number the divisions as $1', 2', 3', \ldots, 12'$.

(x) Draw number of arcs with O as a centre and radius O-11-1, O-10-2, O-9-3, O-8-4, O-7-5, O-6 and OC (which will meet the extended OQ at D.

(xi) Join O-1′, O-2′, O-3′,... and O-12′ and extend up to O-6 arc.

(xii) Mark $C_1, C_2, \ldots$ and C_{12}, which are the different positions of the centre of the rolling circle on CD line.

(xiii) Now get the points $P_1, P_2, \ldots$ and P_{12} on respective arcs as in the case of cycloid, i.e., 1 in 1 arc line, 2 in 2 and so on, and

(xiv) Draw a smooth curve through the points $P, P_1, \ldots$ and P_{12} (by French Curve or freehand) which is the required Epicycloid.

6.5.3 Hypocycloid

The generating circle rolls on the concave surface of a bigger circle without slip. The locus of a point on the circumference of the generating circle is known as hypocycloid. This is for one complete turn of the generating circle.

To draw a hypocycloid, all the procedures are same as in Epicycloid. The only difference is that, the generating circle is rolling on the concave surface of the bigger circle P.A. in an anticlockwise direction.

The procedures are as follows :

(i) Get the value of α or the central angle.

(ii) Divide the arc AP and rolling circle into 12 equal parts. Number the points in a clockwise direction.

(iii) Get the points P_1, P_2, ..., P_{12}, and

(iv) Draw the curve through the points P_1, P_2, ... and P_{12}, which is the required hypocycloid.

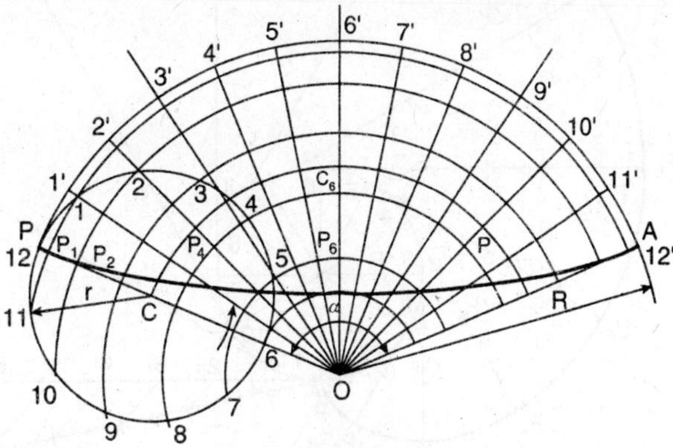

Fig. 6-15 (Hypocycloid)

6.5.4 Involute

The involute is a curve traced out by the end of a piece of thread, unwound or wound from or to a cylinder, circle or a regular polygon .The thread must be kept tight all the time.

In order to draw the involute of a circle, the procedures are as follows :

The diameter of the circle is given.

(i) Draw the given circle of radius R with O as centre (Fig. 6-16).

(ii) Divide the given circle into 12 equal parts with 30°/60° set-square.

(iii) Number the division-points as $1, 2, 3, \ldots$ and 12 in a clockwise direction.

(iv) Draw a tangent AB from the point 12 of the circle. The distance AB is equal to $2\pi R$ (approximate in measurement) of the circumference of the circle.

(v) Divide the line AB into twelve equal parts geometrically and number the points as $1', 2', \ldots$ and $12'$.

(vi) Draw tangent at the points of the circle, i.e., at $1, 2, 3, \ldots, 11$.

(vii) With a compass, take a measurement of one small division of AB (i.e., A-$1'$).

(viii) From the tangent at 1, cut the distance A-$1'$ to get the point P_1. Similarly, from the tangent at 2, cut the distance A-$2'$ (i.e., 2 small divisions) to get the point P_2. In a similar manner, other $P_3, P_4, \ldots$ and P_{11}, are obtained, and

(ix) Through the points $P_1, P_2, P_3, \ldots$ and P_{12} draw a smooth curve (with French Curve or freehand), which is known as involute.

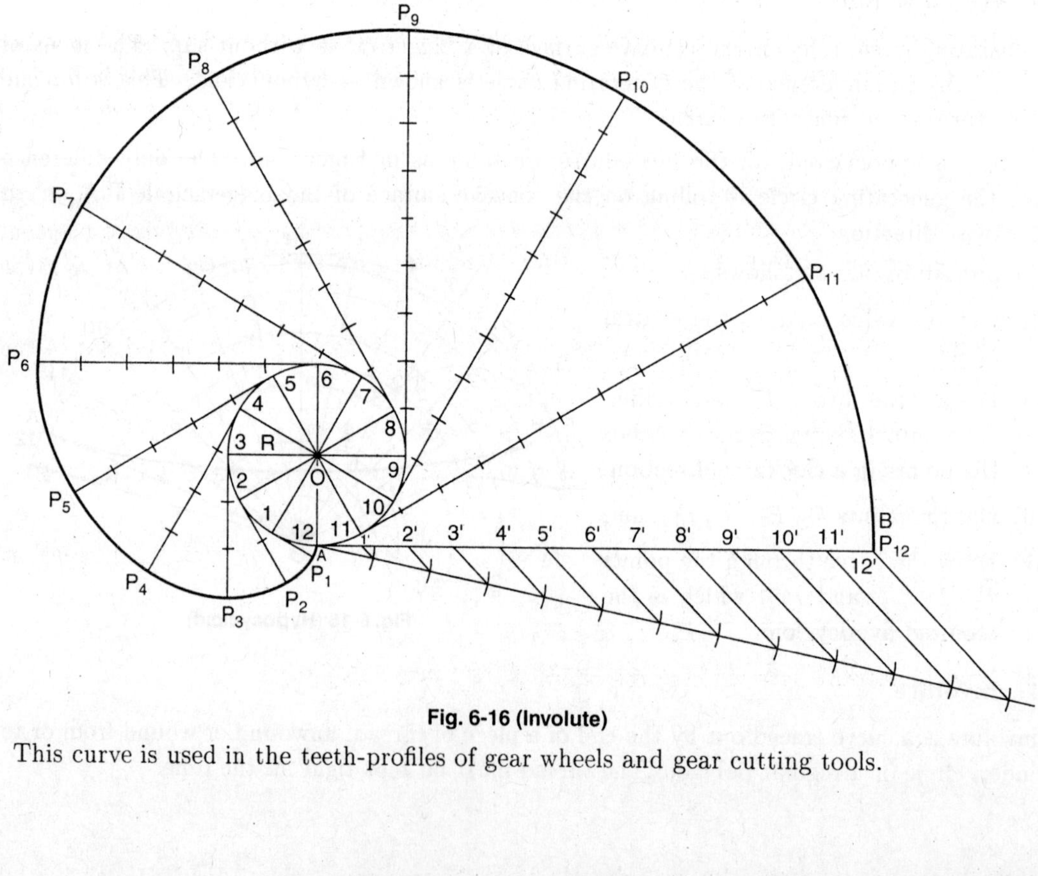

Fig. 6-16 (Involute)

This curve is used in the teeth-profiles of gear wheels and gear cutting tools.

6.5.5 Spiral

A spiral is a plane curve. It is the locus of a point in a straight line. One end of the straight line is fixed. The line rotates in a plane about its fixed end. *A* point moves continuously in one direction along the line. The curve traced by the moving point is called a spiral.

This curve is used in the teeth-profiles of helical gears, profiles of cams, three jaws self-centering chuck, etc.

Parameters of a Spiral

(i) **Pole :** The fixed point of the rotating straight line.

(ii) **Radius Vector :** Direction between any point on the curve and the pole.

(iii) **Vertical Angle :** Angle between original and final position of the line.

(iv) **Convolution :** Revolution of a straight line. One complete revolution is one convolution. It may rotate any number of convolutions.

6.5.6 Archimedean Spiral

This curve is named after the Greek scientist Archimedes. This spiral is a curve traced by a point and a straight line. Both are moving with uniform velocity.

In order to draw the Archimedean Spiral, the procedures are as follows :

The length of the straight line is given. Starting point may be the fixed end or any other points. But for convenience, we choose the starting point as the fixed end of the straight line. If in the problem, it is mentioned that convolution is more than one, the length of the line should be increased according to given condition. Refer to Fig. 6-17.

(i) Draw a circle with centre O with given length as radius in case of one convolution. Let us consider $1\frac{1}{2}$ convolutions. The radius will be $1\frac{1}{2}\times$ given length. Angular movement of the line will be $360° \times 1\frac{1}{2} = 540°$. Now, draw the circle with radius $1\frac{1}{2} \times l$ ($l =$ given length) $= R$.

(ii) Divide the circle into 12 equal parts with $30°/60°$ set-square. Then movement of the straight line will be $12 \times 1\frac{1}{2} = 18$ small divisions of the circle.

(iii) Number the points as $1, 2, 3, 4, \ldots$ and 18 as shown (clockwise).

(iv) Divide the line OP into 18 equal parts geometrically and number them as $1', 2', 3', 4, \ldots$ and $18'$.

(v) Draw arcs with O as centre and $01', 02', 03', \ldots$ and $018'$ as radius to meet at $01, 02, 03, \ldots, 018$ lines.

(vi) Meeting points are $P_1, P_2, P_3, \ldots$ and P_{18}.

(vii) Draw a smooth curve through the points $P_1, P_2, P_3, \ldots$ and P_{18}.

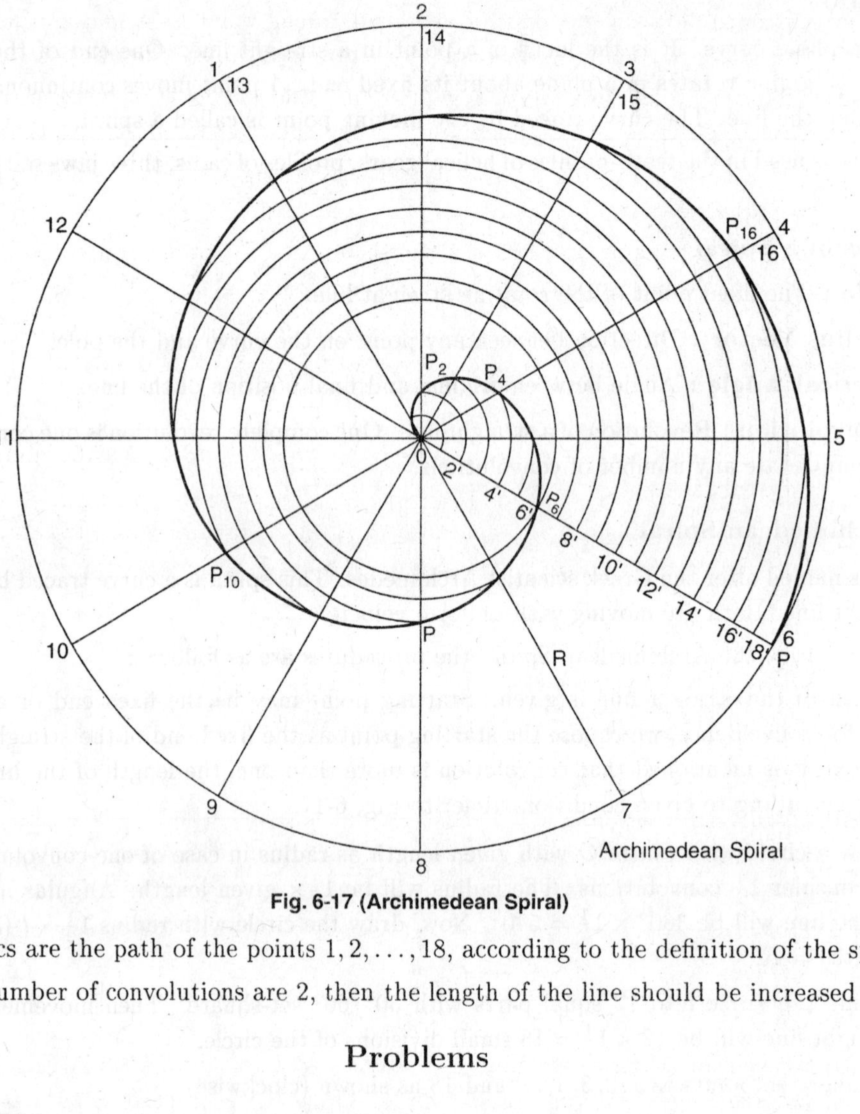

Fig. 6-17 (Archimedean Spiral)

Note : Arcs are the path of the points 1, 2, . . . , 18, according to the definition of the spiral.

If the number of convolutions are 2, then the length of the line should be increased to $2 \times l$.

Problems

1. Ellipse :

(a) The major and minor axes of an ellipse are 100 mm and 70 mm respectively. Draw the ellipse by

(i) Concentric Circles Method. Locate its foci.

(ii) Arcs of Circles Method and

(iii) Oblong Method.

(b) The major and minor axes of an ellipse are 80 mm and 60 mm respectively. Draw its left-half portion by oblong method, and right-half portion by concentric circle method.

(c) Distance between the focus of an ellipse to end point is 15 mm and length of the minor axis is 70 mm. Draw the major axis and complete the ellipse by any method.

(d) The major axis of an ellipse is 100 mm long and the distance between two foci is 70 mm. Find the length of the minor axis and complete the ellipse by oblong method.

(e) Draw an ellipse within a parallelogram having sides 100 mm and 80 mm long and angle between them is 60°.

(f) ABC is a triangle of which the length of AB, BC and CA are 100 mm, 60 mm and 75 mm respectively. Draw an ellipse passing through A, B and C.

Hint : With triangle ABC, draw a parallelogram making any two sides parallel. For simplicity, bisect the biggest side (100 mm) at a point O. Join CO and produce it to D, so that $CO = OD$.

2. Parabola :

(a) The length of the base is 80 mm and the length of the axis is 60 mm. Draw the parabola by the rectangle method.

(b) AB and AC are two adjacent sides of a parallelogram, which makes an angle of 60°. Draw a parabola within this parallelogram when the axis is parallel to the shorter side.

(c) In a test match, Sunil Gavaskar hits a ball to over-boundary reaching at a height of 20 m and fall at a distance 50 m.

Trace the path of the ball, and name the curve.

3. Hyperbola :

(a) Draw two branches of a hyperbola, distance between foci and vertices are 65 mm and 40 mm respectively. Draw the asymptotes and measure the angle between them.

(b) Draw a rectangular hyperbola on a given co-ordinate (25,30), dimensions are in mm.

4. Cycloids :

(a) A cycle of 50 mm diameter rolls along a straight line path without slipping for one complete revolution of the circle. Draw the curve traced by a point on the circumference. Name the curve.

(b) A circle of 40 mm diameter rolls on a horizontal path for half revolution and then on a vertical path for next half. Draw the curve followed by a point at lowermost point of the circumference of the circle.

(c) A circle of 40 mm diameter rolls on a path 30° with the horizontal by a point on the lowermost point of the circumference.

(d) A circle of 40 mm diameter rolls along a circle of diameter 180 mm. Draw the paths of a point on the lowermost point of the circumference of the circle. The rolling circle moves (i) outside and (ii) inside of the circle. Name the curves.

5. Involute :

(a) Draw involute of a triangle, a square, a pentagon, and a hexagon of 20 mm side in each case.

(b) Draw the involute of a circle of 50 mm diameter.

(c) An inelastic string of length 130 mm has its one end attached to the one corner of a hexagon of side of 20 mm. Draw the curve traced by the other end of the string when it is tightly wound around the hexagon.

6. Spiral :

(a) Draw an Archimedean Spiral of $1\frac{1}{2}$ convolution, the diameter of the circle is 30 mm.

7. Fill in the blanks with proper words :

(a) If a right circular cone is cut by planes at different angles with its base, the curves of intersections are called ________.

(b) If the cutting plane is perpendicular to the axis of the right circular cone, the sectional curve is known as a ________.

(c) If the cutting plane is parallel to the axis of the right circular cone, and at a distance from the vertex, the sectional curve is called a ________ ________.

(d) If the cutting plane is parallel to the slant-surface of the cone, the sectional curve is known as a ________.

(e) The sum of the distances from two foci of any point on the ________ is always equal to the length of the ________.

(f) Distance from one of the foci to the one end of the ________ of an ellipse is equal to half the length of the ________.

(g) The locus of a point on the circumference of a circle, rolling along a straight line, without slip, is known as ________.

(h) If the plane cuts the right circular cone at an angle less than the base-angle of the cone, the sectional curve is known as ________.

(i) If the plane cuts the right circular cone at an angle greater than the base-angle of the cone, the sectional curve is known as ________.

(j) If the cutting plane is vertical (perpendicular to the base of the cone) and passes through the vertex (or apex) of the right circular cone, the section may be ________ ________.

Ans. (a) Conic sections, (b) circle, (c) rectangular hyperbola, (d) Parabola, (e) Ellipse, major axis, (f) minor axis, major axis (g) cycloid, (h) ellipse, (i) Hyperbola (j) isosceles or equilateral, triangle.

State Council Model Questions (West Bengal)

1. The semi-major and semi-minor axes of an ellipse are 50 mm and 35 mm respectively. Draw the ellipse separately by each of the following methods :

   (a) Concentric Circles, (b) Rectangular or Oblong, (c) Arcs of circles.

2. Two fixed points P and Q are located on a sheet of paper and are 82.5 mm apart. Another point R moves on the plane in such a way that the sum of the focal distances is always 133 mm. Trace the locus of the point. State its eccentricity and name out the curve. Draw a tangent and normal at any point on the curve.

3. Construct an ellipse when a pair of conjugate diameters AB and CD equal to 110 mm and 50 mm, respectively. The angle between the conjugate diameters is 70°.

4. A straight line and a fixed point are 50 mm apart. Trace the locus of a point P such that the ratio of its distances from the fixed point to the fixed line is 2/3.

5. A fixed point and a fixed line are 50 mm apart. Trace the locus of a point so that the ratio of its distances from the fixed point to the fixed line is unity. Draw a tangent and normal at any point P on the curve.

6. A stone thrown from the top of a building 4 metres high crosses the top of a tree 9 metres high and falls on the ground. The distance between the building and the tree is 3 m. Take a suitable scale and trace the path of the projectile.

7. Construct a hyperbola, when the distance of the focus from the directrix is 65 mm and eccentricity is 3/2.

8. Fill in the blanks with appropriate words :

   (a) The curve generated by a point on the circumference of a circle, rolling along another circle outside it, is called ______.

   (b) Rectangular hyperbola is also called the ______ hyperbola.

   (c) The line joining any point on the spiral with the pole is called ______.

   (d) If the eccentricity is less than unity, the curve is called ______.

   (e) If the eccentricity is greater than unity, the curve is called ______.

   (f) If sum of the focal distances from any point on an ellipse is 110 mm, the length of the major axis will be ______ mm.

   (g) When the diameter of the rolling circle is half the diameter of the directing circle, the hypocycloid is a ______.

   (h) Logarithmic spiral is also called the ______ spiral.

Ans. (a) Epicycloid (b) Equiangular (c) Radius Vector (d) Ellipse (e) Hyperbola (f) 11 (g) Straight line (h) Equiangular.

9. In a slider-crank mechanism, the connecting rod AB is 75 cm. long and the crank OA, which rotates in the clockwise direction is 15 cm long as shown in the figure below. The end B moves along the straight line passing through O. Trace the locus of the point P, 30 cm away from A on AB, for one revolution of the crank OA. Also name the curve.

(AMIE—Summer, 2000)

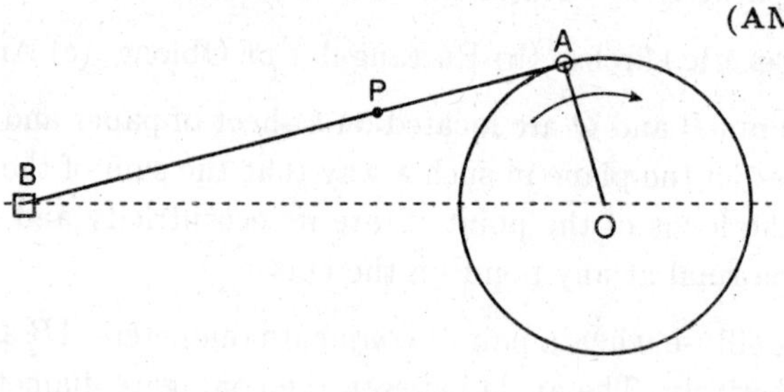

10. Choose the correct answer :

(i) The P-V diagram of a gas is represented by the curve

(a) helix (b) parabola (c) rectangle hyperbola. **(AMIE—Summer, 2000)**

Ans. (c)

(ii) A bicycle wheel, fitted with an inflated tyre, moves on a plane road. The path traced by a point on the rim of the wheel, is

(a) a cycloid (b) a trochoid (c) an epicycloid. **(AMIE—Summer, 2000)**

Ans. (a)

(iii) A solid cone is rotating about its vertical axis at a constant speed. An ant starts from a point on the periphery of the base circle and climbs up the vertex for one revolution of the cone. The path, as traced by the ant, is

(a) a spiral (b) a helix (c) an involute. **(AMIE—Summer, 2000)**

Ans. (b)

(iv) The curve, traced by a fixed point on the circumference of a wheel, which rolls on the outside of another wheel, is called

(a) a cycloid (b) an epicycloid (c) a hypocycloid. **(AMIE—Summer, 2001)**

Ans. (b)

(v) An ideal gear-tooth profile is

(a) an involute of a circle (b) parabolic (c) cycloidal. **(AMIE—Summer, 2001)**

Ans. (a)

(vi) A cone is cut by a vertical plane parallel to the axis, but not passing through the apex. The true shape of the section is

(a) an ellipse (b) a parabola (c) a rectangular hyperbola.

(AMIE—Summer, 2001)

Ans. (c)

(vii) If a point moves along a straight line, which simultaneously rotates about one of its ends, the curve traced by the point is called

(a) spiral (b) helix (c) cycloid (d) involute. **(AMIE—Winter, 2001)**

Ans. (a)

11. State true or false by writing T or F :

(i) The curve, traced by a point, moving uniformly in one direction along a rotating line is called a spiral. **(AMIE—Summer, 2002)**

Ans. (T)

(ii) The difference of focal distances from any point on an ellipse is constant. **(AMIE—Summer, 2002)**

Ans. (F)

(iii) When a small circular disc rolls outside another bigger disc, the locus of a point on the circumference of the smaller disc is hypocycloid. **(AMIE—Summer, 2002)**

Ans. (F)

(iv) Pressure-volume diagram for the isothermal expansion of a gas is represented by the curve Rectangular Hyperbola. **(AMIE—Summer, 2002)**

Ans. (T)

(v) The curve, traced by the falling water from the open horizontal pipe, placed at a certain height from the ground, is a parabola.

Ans. (T)

12. A stone, thrown from the top of a building fell on the ground. Name the curved-path of the falling stone.

Ans. Parabola.

13. Choose the correct answer :

(i) Locus of a point, which moves around a centre and simultaneously moves either towards or away from the centre, is called

(a) (cycloid) (b) helix (c) spiral. **(Summer, 2005)**

Ans. (c)

(ii) A curve, traced out by an end of a rope unwound from a circle, is

(a) Cycloid (b) Involute (c) Epicycloid. **(Winter, 2006)**

Ans. (b)

(iii) The curve generated by the locus of a point, which moves such that the ratio of its distance from the fixed point to its distance from the fixed line is constant and is always greater than 1, is called

(a) Hyperbola (b) Ellipse (c) Parabola. **(Summer, 2006)**

Ans. (a)

Chapter 7

Orthographic Projections

7.1 Introduction

With the help of orthographic projection, we can exactly represent a point, a line, a plane, or a solid, situated in space. This is done with the help of three planes, known as Vertical plane (V.P.), horizontal plane (H.P.) and profile plane (P.P.), as will be explained later.

7.2 Preparation of a Simple Solid

(i) Draw Fig. 7-1 on a stiff drawing sheet.

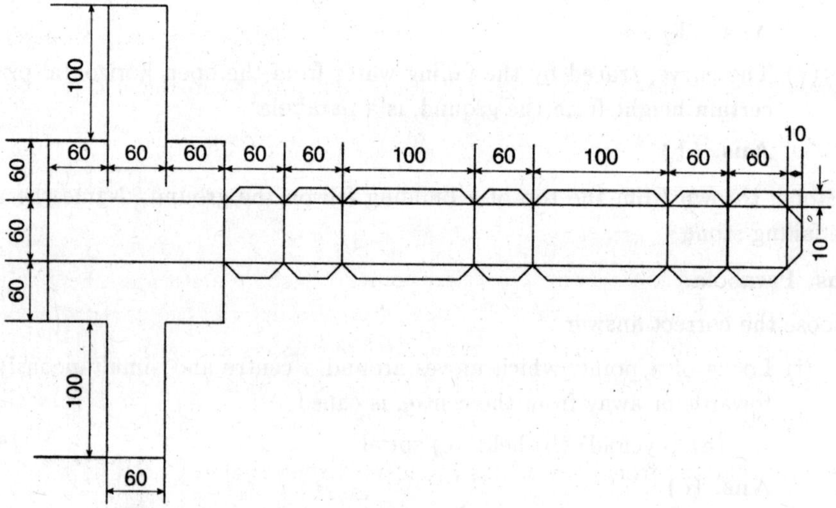

Fig. 7-1

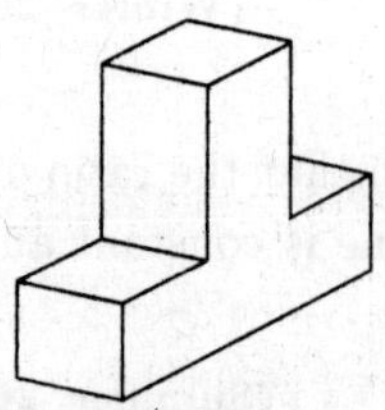

Fig. 7-2

(ii) Cut the outline of complete drawing with a sharp blade.

(iii) Fold the same along the lines drawn. Make the model as shown in Fig. 7-2.

(iv) The edges are glued and pasted together with the adjacent planes.

(v) We will get a solid like Fig. 7-2.

(vi) Now we take two paste boards (24″ × 12″). Cut the board width-wise and paste as shown in Fig. 7-3.

From our school-experience, we know that this will represent four quadrants, i.e., the 1st, 2nd, 3rd and 4th quadrants, moving in anti-clockwise direction. Each plane is at right angles with its adjacent plane.

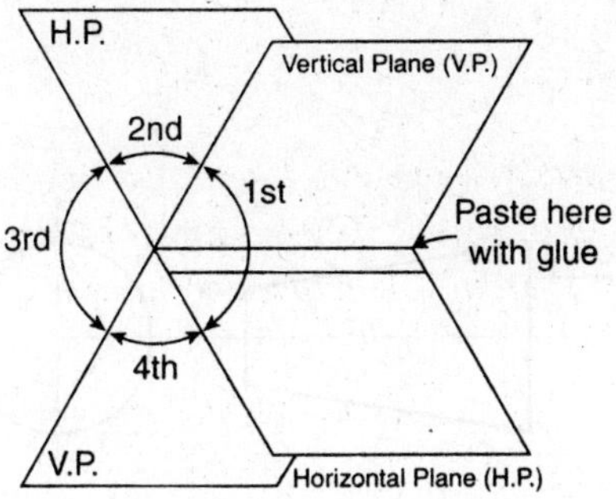

Fig. 7-3

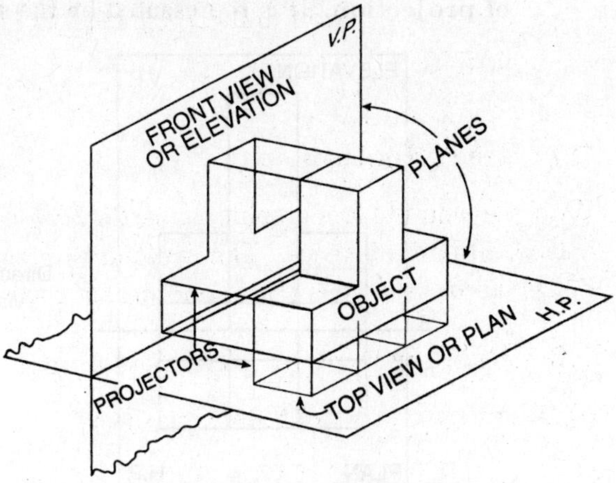

Fig. 7-4 (First-Angle Projection)

(vii) Now, place the model of Fig. 7-2 in the first quadrant as in Fig. 7-4.

(viii) Draw imaginary horizontal lines from every corner of the model.

(ix) All the lines are perpendicular to the vertical plane. These lines are known as projections or projection lines.

(x) Draw deep lines joining the projected points, which create a figure as shown in Fig. 7-4. This figure is known as front view or elevation. In drawing the front view or elevation, we are to see the object in such a manner that our vision will be at right angles to the front face of the object only. Thus, we shall not be in a position to see the top or side faces of the object.

(xi) We note that the front view is drawn on the vertical plane (V.P.).

(xii) In order to draw the top view or plan, we draw imaginary vertical lines from every corner of the object.

(xiii) These projection lines are perpendicular to the horizontal plane (H.P.).

(xiv) All the four corner points on H.P. are joined together. We shall get a rectangular figure. This is known as top view or plan.

(xv) Thus, we observe that, the top view or plan is drawn on the horizontal plane (H.P.).

(xvi) If the sheet of paper consisting of the first quadrant is made flat (i.e., spreading on a table), the elevation and plan can now be seen, as shown in Fig. 7-5.

(xvii) The horizontal plane (H.P.) and the vertical plane (V.P.) meet in a line, known as the ground line. This is denoted by G.L. in Fig. 7-5. We also note that the elevation is drawn above the G.L. and the plan is drawn below the G.L. "G.L." is the Ground Line.

(xviii) When the object is placed on the first quadrant, we generally call the front view as ELEVATION and the top view as PLAN. This system is known as the **first angle method of projection.** It is represented by the symbol shown in Fig. 7-5A.

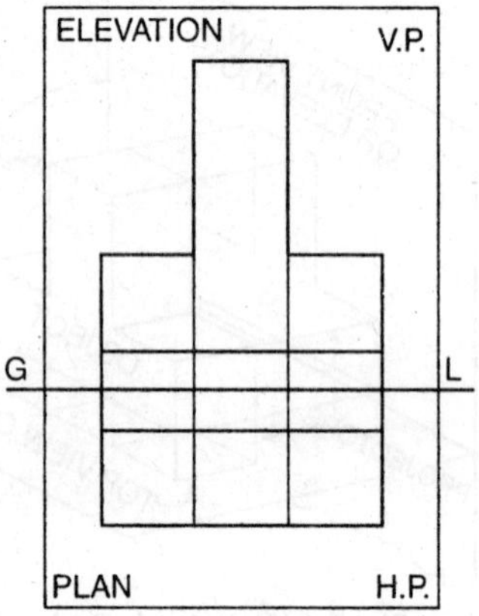

Fig. 7-5

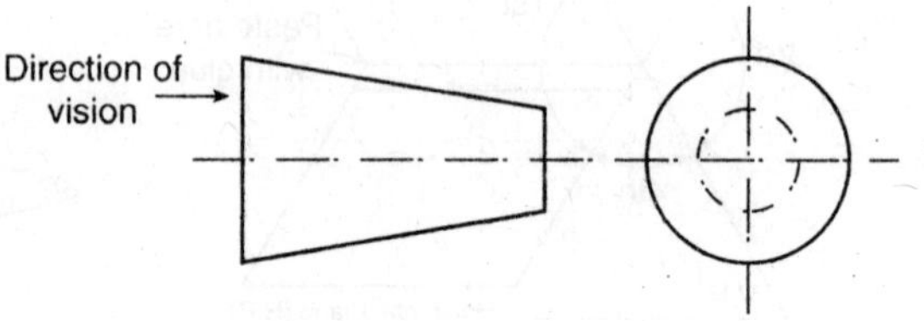

Fig. 7-5A

(xix) When the object is placed in the first quadrant, the placing of the object is expressed as follows :

The object is placed X mm in front of vertical plane (V.P.) and Y mm above horizontal plane (H.P.), X and Y may be stated in the problem. If X and Y are not stated in the problem, these should be taken according to suitability, say $X = Y = 25$ mm.

(xx) We also note from Fig. 7-5 that, **in the first angle method**, we draw the projection (elevation or plan) on the opposite side of viewing the object.

7.3 Third-Angle Projection

(i) We now imagine that two transparent plastic sheets are joined at right angles to each other. These represent H.P and V.P. The object is placed in space in such a manner that it is X mm behind V.P. and Y mm below H.P., as shown in Fig. 7-6.

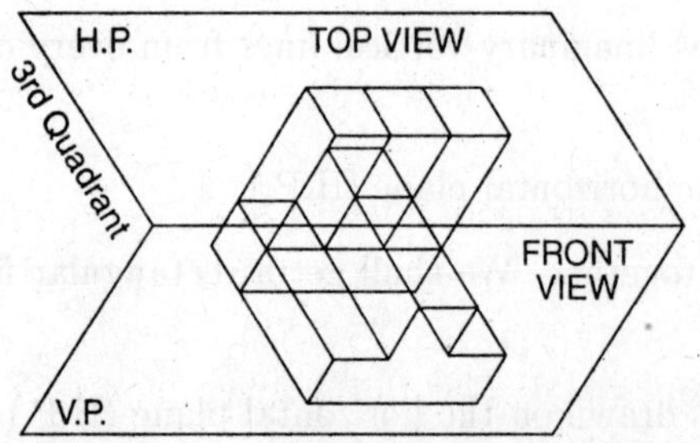

Fig. 7-6 (Third-Angle Projection)

(ii) We observe that the object is placed in the 3rd quadrant.

(iii) Since the two planes (V.P. and H.P.) are transparent, we can see the object through the planes.

(iv) As in the case of 1st angle projection, the projection lines are drawn from the object to get the front view and the top view.

(v) Again we observe that, the front view is drawn on the V.P. and the top view on H.P.

(vi) Now, the plastic sheet is made flat, and spread on a table. The top view and the front view may now be seen, as shown in Fig. 7-7. The meeting line of H.P. and V.P. is called the reference line, and is denoted by XY as shown in Fig. 7-7.

The ground line is denoted by G.L.

(vii) When the object is placed on the 3rd quadrant, system is called the 3rd angle projection. It is represented by the symbol shown in Fig. 7-7A.

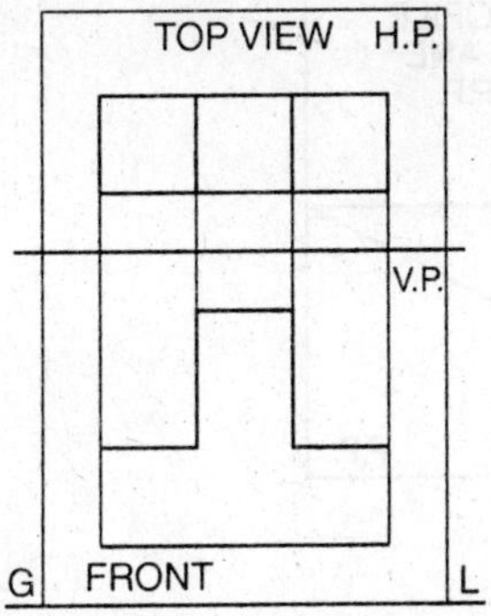

Fig. 7-7 (Third-Angle Projection)

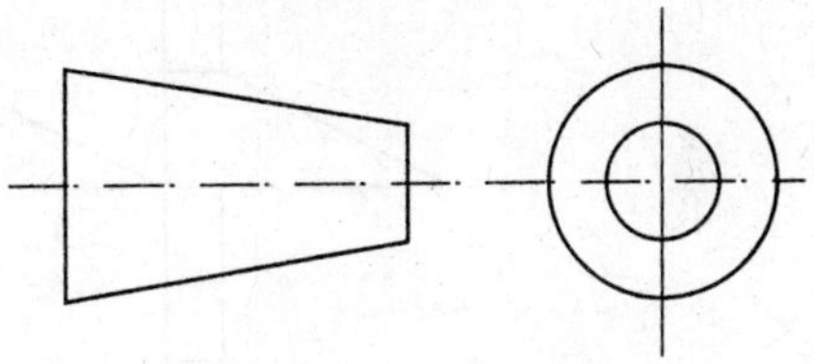

Symbol for third-angle Projection

Fig. 7-7A

(viii) Here we observe that the top view is drawn above the reference line XY, and the front view is drawn below the line XY.

(ix) We also note that, in the 3rd angle projection, we draw the projection (front view or top view) on the same side from which we see the object.

(x) comparing the projections of Fig. 7-7 and Fig. 7-5, we observe that, the front view and top view in both these systems (3rd & 1st angle) are exactly same.

7.4 Second and Fourth-Angle Method of Projection

If we place the object in the second quadrant (Fig. 7-3), the front view will be on V.P. and the top view will be on H.P. Now, by rotating the V.P. in an anti-clockwise (as in other systems) direction, the front view and top view will overlap each other. Thus, projections cannot be drawn. Similarly, by placing the object in fourth quadrant (Fig. 7-3), the front view and top view will also overlap each other.

For this difficulty, we do not use second and fourth angle projection methods.

7.5 Orthographic Projection

When all the projectors (or projection lines) are drawn perpendicular to the given planes (V.P. and H.P.), the projection are called orthographic projections.

Thus, we note that 1st and 3rd angle projection methods are orthographic projections.

7.6 Bureau of Indian Standards (B.I.S.) Code of Practice

B.I.S. recommended first angle method of projection in our country. In this book, we shall concentrate mainly on first angle method. But third angle method of projection will be utilized through problems and exercise.

7.7 Profile Plane

When we see the object either from left-hand or from right-hand side, we draw the projection on the profile plane (PP). This plane is at right angles both to H.P. and V.P. as shown in the Fig. 7-8.

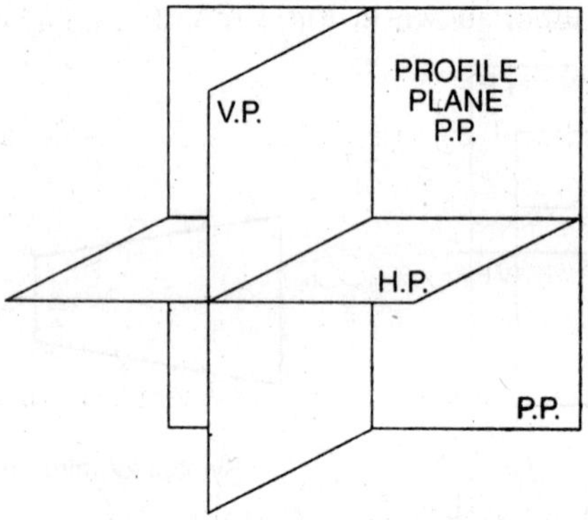

Fig. 7-8

Problems

1. What do you mean by a projection?
2. How do you get the projection of an object?
3. Define orthographic projection.
4. What is the difference between the first-angle and the third-angle methods of projection?
5. Fill in the blanks :

   (a) In the first-angle projection, the ______ is situated between the ______ and the ______ of projection.

   (b) In the third-angle projection, the ______ of projection is situated between the ______ and the ______.

   (c) Orthographic projection requires minimum ______ views to represent an object.

(AMIE—Winter, 2000)

Ans. (a) Object, Observer, Plane, (b) Plane, Object, Observer, (c) Two.

6. State whether the following statements are true or false, by indicating T or F against each :

   (i) The front view of an object is drawn on the V.P.

(ii) In the 3rd-angle method of projection, the front view is drawn above the reference line.

(iii) The first and the third-angle methods of projections are orthographic projections.

(iv) In the first-angle method, the projections are drawn on the same side from which the object is seen.

(v) The second and fourth-angle methods of projection, are not used.

(vi) In both first and third-angle methods of projection, the side view is drawn on the profile plane.

(vii) When an object is placed 10 mm in front of V.P. and 15 mm above H.P., the object must be placed in the first quadrant.

(viii) When an object is placed 20 mm behind V.P. and 25 mm below H.P., the object must be placed in the fourth quadrant.

Ans. (i) T, (ii) F, (iii) T, (iv) F, (v) T, (vi) T, (vii) T, (viii) F.

7. Choose the correct answer :

When the object lies between the observer and the plane of projection, the method of projection is

(a) first angle projection (b) third angle projection (c) neither of them.

(AMIE—Summer, 2000, Summer, 2001)

Ans. (a)

8. Choose the correct answer :

(i) The plan and elevation of a point are 40 mm above and 60 mm below the XY line, respectively. the point lies in the

(a) second quadrant (b) third quadrant (c) fourth quadrant.

Ans. (b)

(ii) When a point lies in the H.P. its view from the front will lie

(a) on XY (b) below XY (c) above XY. **(Winter, 2005)**

Ans. (a)

(iii) The distance of a point from the H.P. is marked from XY to the

(a) view from above (b) view from the front (c) side view. **(Winter, 2005)**

Ans. (b), in both first-angle and third-angle projections.

Chapter 8

Projection of Points

8.1 Introduction

This chapter deals with the projection of points only.

Now, a point may be situated in any one of the four quadrants, and any one place in that quadrant.

8.2 Point situated on the 1st Quadrant

Use paste board quadrant as explained in the chapter 7.

(i) Given point A, situated in the first angle or 1st quadrant (Ref. Fig. 8-1).

(ii) The point is situated above H.P. and in front of V.P.

(iii) Points 1 and 2 are the top and front views of A.

(iv) Now, place the paste board on the table, when we get Fig. 8-2.

(v) The Front view of A is 2 and the top view 1, above and below X-Y respectively.

(vi) Join 1 and 2 with thin lines. This is known as projector, which intersects XY at right angles at a point O. (Fig. 8-2)

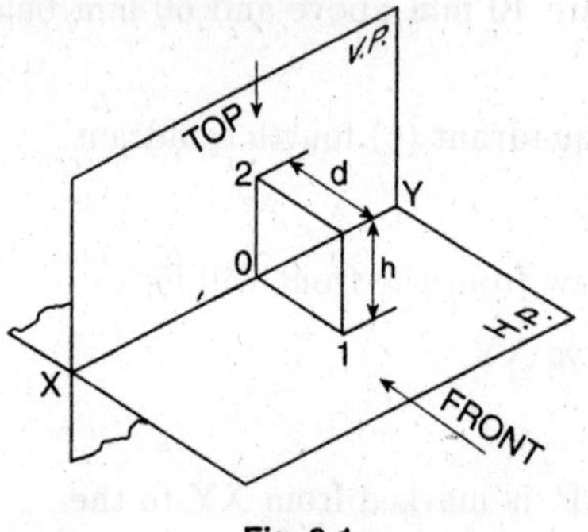

Fig. 8-1

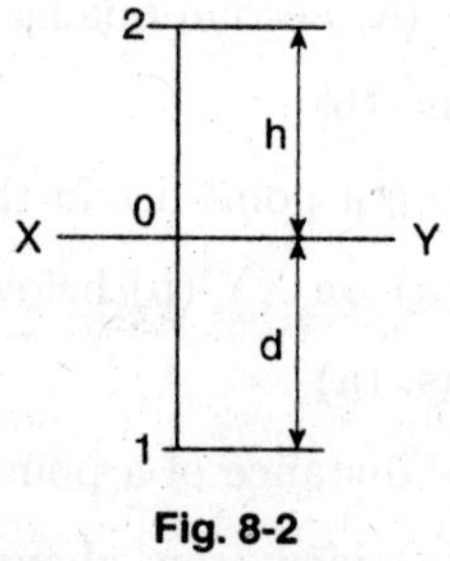

Fig. 8-2

(vii) We find that 2-0 = A-1 = height of the front view (Elevation) from XY = height of the front view from H.P. = h. XY is known as the reference line.

(viii) Similarly, we get 1-0 = A-2 = distance of the top view (plan) from XY = distance of the top view from V.P. = d.

(ix) if h or d is zero, then the point will be on XY line, i.e., the point lies on both H.P. and V.P..

8.3 Point on the 2nd Quadrant

Practise with quadrant board.

With reference to Fig. 8-3,

(i) Point B is on the Second Quadrant.

(ii) It is above H.P. and behind V.P..

(iii) 4 is the front view of B and 3 the top view.

(iv) We know that, in the second quadrant top and front views overlap each other. So, after rotating the plane and placing on the table, we get the Fig. 8-4.

(v) Both the views are above XY line.

(vi) Then, B-3 = 4-0 and B-4 = 3-0.

(vii) Join 3 and 4 with thin line projector, which intersects XY at O at right angles.

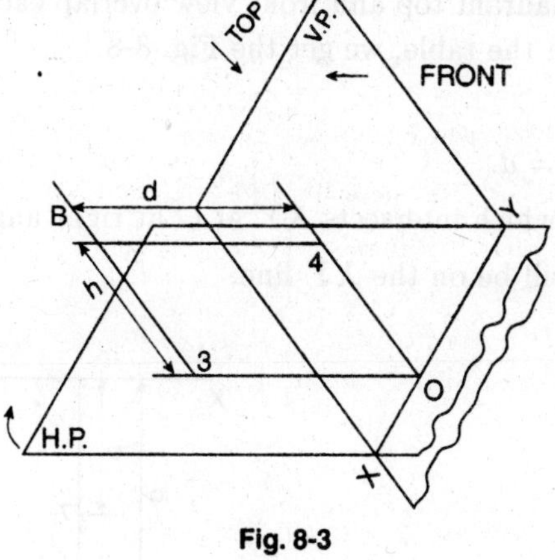

Fig. 8-3

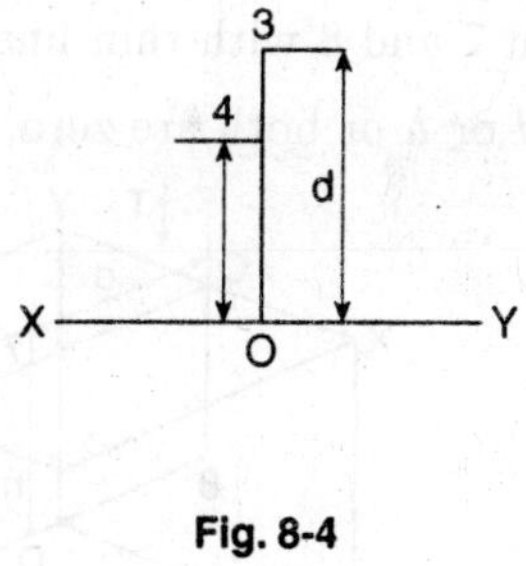

Fig. 8-4

8.4 Point on the 3rd Quadrant

With reference to Fig. 8-5, and transparent plastic sheet as quadrant, we get,

(i) Point C is on the third Quadrant.

(ii) It is below H.P. and behind V.P..

(iii) 6 is the front view of C and 5 the top view.

(iv) Now, rotating the plane and placing on the table, we get the Fig. 8-6.

(v) Join 5 and 6 with thin line projector, which intersects XY at O at right angles.

(vi) Its front view 6 is below XY line and top view 5 is above XY line.

(vii) When d, h or both are zero, the point will be on the X-Y line.

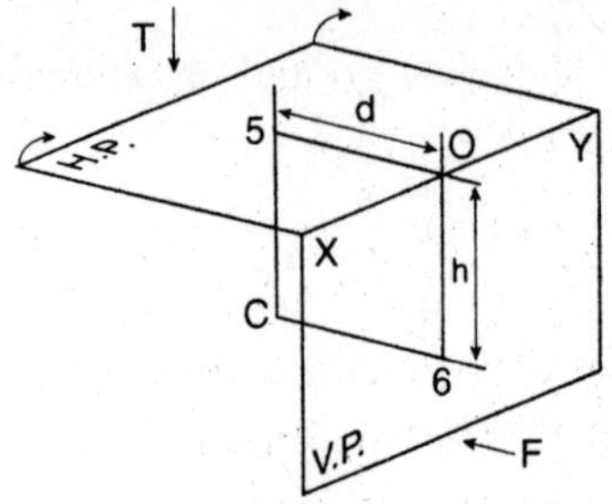

Fig. 8-5

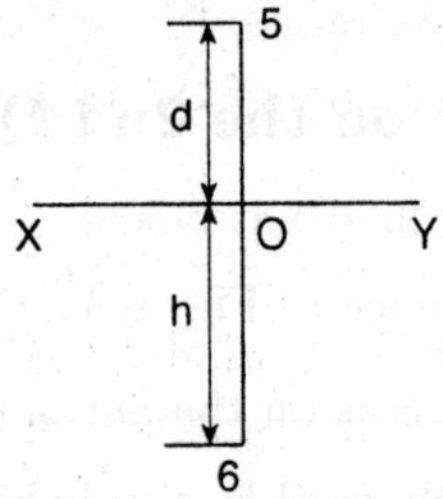

Fig. 8-6

8.5 Point on the 4th Quadrant

With reference to Fig. 8-7, we get,

(i) Point D is on the fourth Quadrant.

(ii) It is below H.P. and in front of V.P..

(iii) 8 is the front view of D and 7 top view.

(iv) Again, we know that, in the fourth Quadrant top and front view overlap each other. So, after rotating the plane and placing on the table, we get the Fig. 8-8.

(v) Both the views are below XY line.

(vi) Then D-7 = O-8 = h and D-8 = O-7 = d.

(vii) Join 7 and 8 with thin line projector, which intersects XY at O at right angles.

(viii) If d or h or both are zero, the point will be on the XY line.

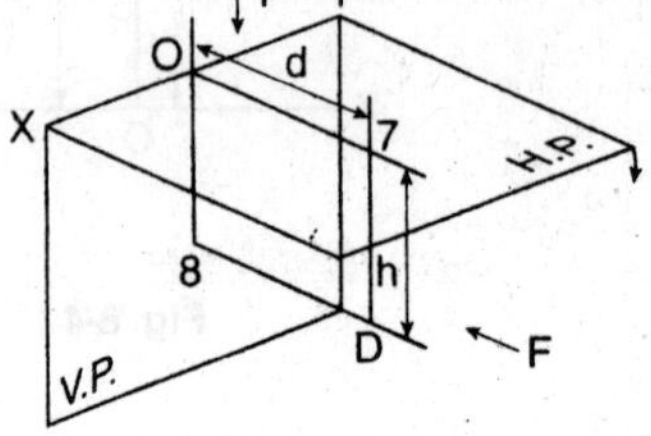

Fig. 8-7

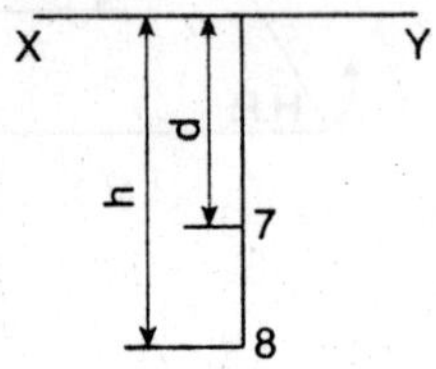

Fig. 8-8

8.6 Parameters

(i) The **projector** lines will always be **perpendicular** to XY line.

(ii) When the point is **above** H.P. the front view will be **above** XY line.

(iii) When the point is **below** H.P., the front view will be **below** XY line.

(iv) When the point is **in front** of V.P., its top view will be **below** XY line.

(v) When the point is **behind** V.P., its top view will be **above** XY line.

Problems

1. Draw the projection of the following points on the same ground or reference line, keeping the projectors of the two points 15 mm apart :

The point—

(i) *A*, on the H.P. and 20 mm behind the V.P.. (Third-Angle)

(ii) *B*, 45 mm above H.P. and 20 mm in front of V.P.. (First-Angle)

(iii) *C*, 35 mm below H.P. and 20 mm behind V.P.. (Third-Angle)

(iv) *D*, 50 mm in front of V.P. and 25 mm above H.P.. (First-Angle)

(v) *E*, on the V.P. and 40 mm below H.P.. (Third-Angle)

(vi) *F*, on the H.P. and 35 mm in front of V.P.. (First-Angle)

(vii) *G*, on both H.P. and V.P.. (First-Angle or Third-Angle)

(viii) *H*, 45 mm behind V.P. and 15 mm below H.P.. (Third-Angle)

(ix) *I*, on the V.P. and 35 mm above H.P.. (First-Angle)

Ans. In the case of First-Angle projection, the **elevation** is above the reference line *XY* and the **plan** is below the line *XY*.

In the Third-Angle projection, the front-view is below the line *XY* and the top-view is above the line *XY*.

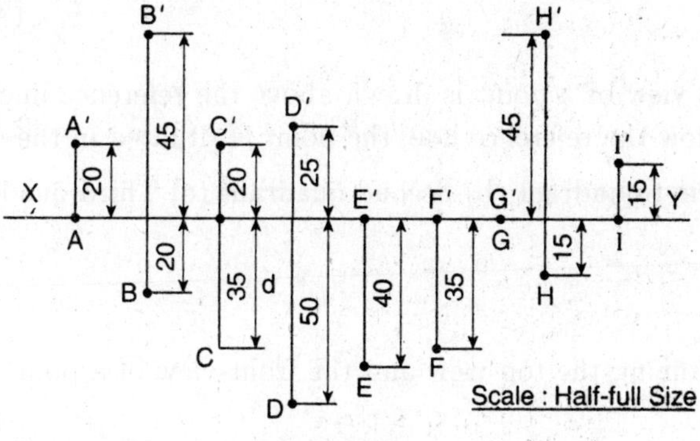

Fig. 8-9

Note : To solve the problem, draw the figure on the cardboard quadrant, as shown in Fig. 8-10.

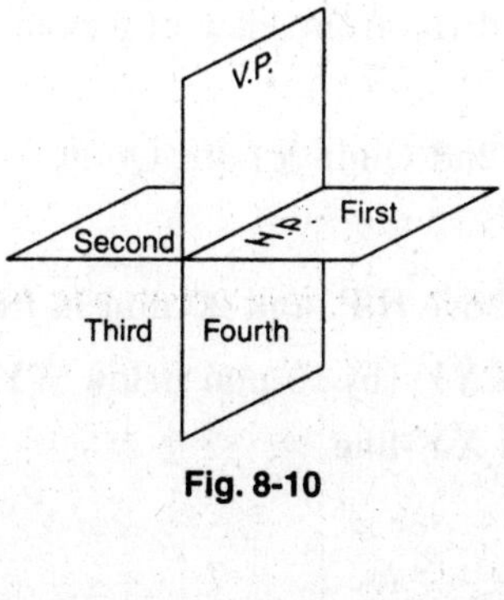

Fig. 8-10

2. Point '*A*' is 50 mm away from both the reference planes. Draw the projections in first and third-angles.

3. Describe the positions of points, as shown in the figure 8-11 :

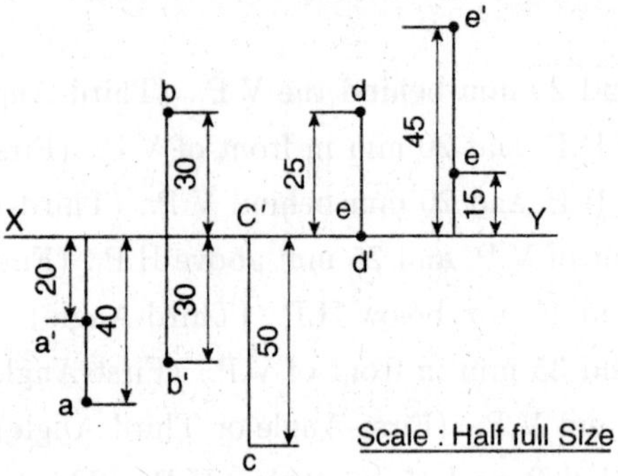

Fig. 8-11

4. Choose the correct answer :

(i) When the elevation of a point is drawn above the ground line and the plan is drawn below the ground line, the point is situated in the

(a) First quadrant (b) Second quadrant (c) Third quadrant.

(AMIE—Summer, 2000)

Ans. (a)

(ii) If the top-view of a point is drawn above the reference line and the point-view is drawn below the reference line, the point is situated in the

(a) First quadrant (b) Second quadrant (c) Third quadrant.

Ans. (c)

5. Fill in the blank :

(i) The line joining the top-view and the front-view of a point is called the ______ of the point. **(AMIE—Winter, 2000)**

Ans. projector

6. Choose the correct answer :

(i) Both the top view and the front view of a point lie above *X-Y* line. The point is in the

(a) 1st Qudt (b) 2nd Qudt (c) 3rd Qudt. **(AMIE—Summer, 2001)**

Ans. (b)

(ii) If a point is 10 mm above H.P. and 20 mm in front of V.P., then its top-view is

(a) 10 mm above *XY* (b) 10 mm below *XY* (c) 20 mm above *XY*

(d) 20 mm below *XY*-line.

Ans. (d)

7. State True or false by writing T or F :

(i) If a point is below H.P. and behind V.P., the top-view is below X-Y line

(AMIE—Summer, 2002)

Ans. (F)

8. Answer briefly :

(i) If a point is situated 20 mm above H.P. and 30 mm in front of V.P., what is the distance of its plan from XY line?

Ans. 30 mm below XY line (First-angle projection).

9. The elevation and plan of a point are at a distance of 30 mm above XY and 40 mm below XY-line respectively. In which quadrant the point is situated?

Ans. First quadrant.

10. The front-view and top-view of a point are at a distance of 30 mm below XY and 50 mm above XY-line respectively. State in which quadrant the point is situated.

Ans. Third quadrant.

Chapter 9

Projection of Straight Lines

9.1 Introduction

In chapter 7, it has been explained that, a straight line may be considered as the continuous movement of a point, without changing the direction. When the direction changes, it becomes a curved line.

Therefore, a straight line may be defined in two ways :

(a) a straight line is generated by a point moving in one direction only,

(b) a straight line is the shortest distance between any two points. These two points may lie situated on a plane or in the space.

A straight line may be specified in two ways :

(a) locating the two points on a plane by knowing their coordinates (i.e., X and Y coordinates); and joining them.

If the line is in space, then we should know the three coordinates (x, y, z) of the two points to make the straight line.

(b) locating one end of the line and fixing in a particular direction by knowing its angle of inclination.

9.2 Projection of Straight Lines

The main focal point of this chapter is to show how the plan and elevation of straight lines (in first angle projection) may be drawn. We can place a straight line mainly in six positions within two principal planes, i.e., V.P. and H.P.

A straight line

(i) perpendicular to H.P. and parallel to V.P.,

(ii) perpendicular to V.P. and parallel to H.P.,

(iii) parallel to both V.P. and H.P.,

(iv) making angle with H.P. and parallel to V.P.,

(v) making angle with V.P. and parallel to H.P.,

(vi) making angles with both V.P. and H.P.

These will be explained one after another.

9.3.1 Straight Line Perpendicular to H.P. and Parallel to V.P.

All the projections will be drawn in first angle projection.

(i) The straight line AB is perpendicular to H.P. and parallel to V.P. It is situated in the 1st quadrant, i.e., 1st angle [Fig. 9-1(a)].

(ii) Looking from the front, we find that the view on the V.P. is $a'b'$, i.e., the straight line of given length AB. This is also called true length.

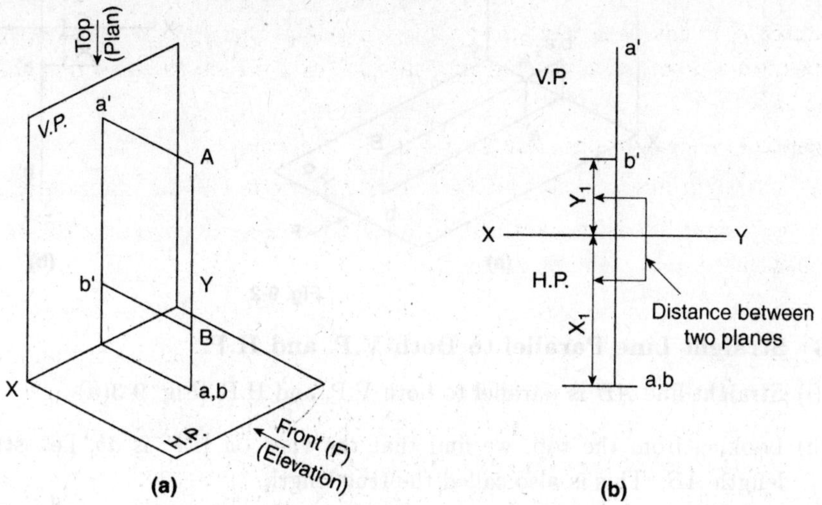

Fig. 9-1

(iii) Looking from the top, we find that the view on the H.P. is a point a, b (i.e., plan of the line AB is a point).

(iv) Now, turning the V.P. in an anti-clockwise direction and placing on the table, the elevation and plan may be obtained [See Fig. 9-1(b)].

(v) Keep the straight line at an equal distance from the two planes, if not mentioned specifically in a problem. The distances from V.P. and H.P. are shown by X_1 and Y_1, respectively [Fig. 9-1(b)].

Note : Use the pencil or a small stick as a straight line and paste board quadrant.

Hint : To solve a problem on a straight line, draw a figure [like Fig. 9-1(a)]. From this figure, the position and condition of the line will be clear. Lines AB and $a'b'$ will be dark and other lines faint.

9.3.2 Straight Line Perpendicular to V.P. and Parallel to H.P.

(i) The straight line AB is perpendicular to V.P. and parallel to H.P. [Fig. 9-2(a)].

(ii) Looking from the top, we find that the view on the H.P. is $a'b'$, i.e., the straight line of given length AB. This is also called the true length.

(iii) Looking from the front, we find that the view on the V.P. is a point (b, a). Thus, the elevation of the line AB is a point.

(iv) We get plan and elevation as in the Fig. 9-2(b). Draw the line AB and $a'b'$ dark and other projection lines faint.

(v) Other instructions are same as in previous conditions.

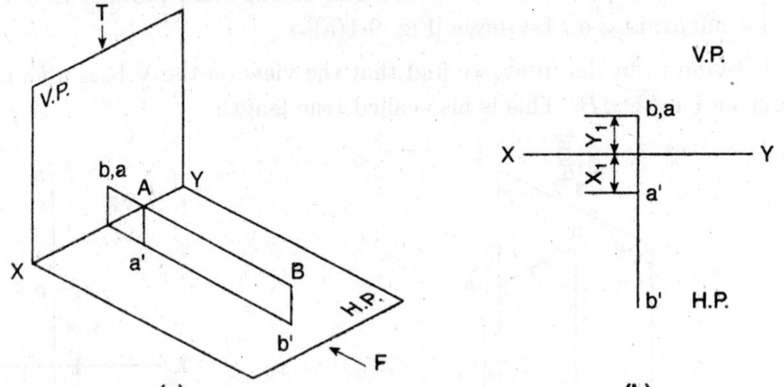

(a) (b)

Fig. 9-2

9.3.3 Straight Line Parallel to Both V.P. and H.P.

(i) Straight line AB is parallel to both V.P. and H.P. [Fig. 9-3(a)].

(ii) Looking from the top, we find that the view on H.P. is ab, i.e., straight line of given length AB. This is also called the true length.

(iii) Looking from the front [Fig. 9-3(a)], we find that the view on the V.P. is $a'b'$, i.e., straight line of the given length AB also. This is also the true length. Then we may conclude that a straight line parallel to any plane will give the true length, when projected on that plane. Thus, in order to get the true length of a line, we shall have to bring that line parallel to any plane.

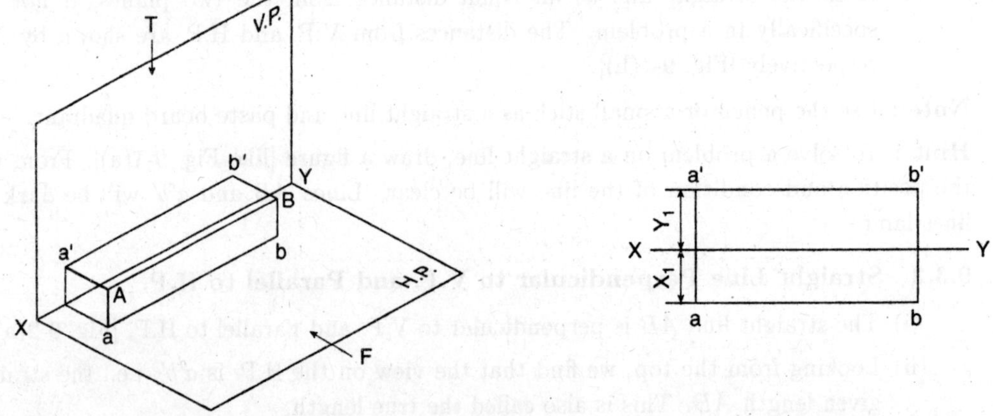

(a) (b)

Fig. 9-3

(iv) We get the plan and elevation of the line AB in Fig. 9.3(b) after rotating the V.P. and placing on the table.

(v) Other procedures are same as in the previous one.

9.3.4 Straight Line Making Angle with H.P. and Parallel to V.P.

(i) Place the straight line ascending to Fig. 9-4(a) and 9-4(c), i.e., parallel to both the planes V.P. and H.P.

(ii) Now rotate the stick (used as a straight line), i.e., the end point B keeping the end A fixed in its original position, i.e., inclined at an angle θ with H.P. [Fig. 9-4(b)].

(iii) b_1' is the new position of B.

(iv) Thus, the straight line is parallel to V.P. and making an angle with H.P. [Fig. 9-4(b)].

(v) Now looking from the front, we get elevation $a'b_1'$ [Fig. 9-4(d)] of line AB, on the V.P. $a'b_1'$, is also the true length of the line, as it is parallel to V.P.

(vi) Looking from the top, we get plane ab_1 [Fig. 9-4(d)] of line AB on the H.P., ab_1 is shortened from the original length.

(vii) Now turning the V.P. and placing on the table, we get the Fig. 9-4(e).

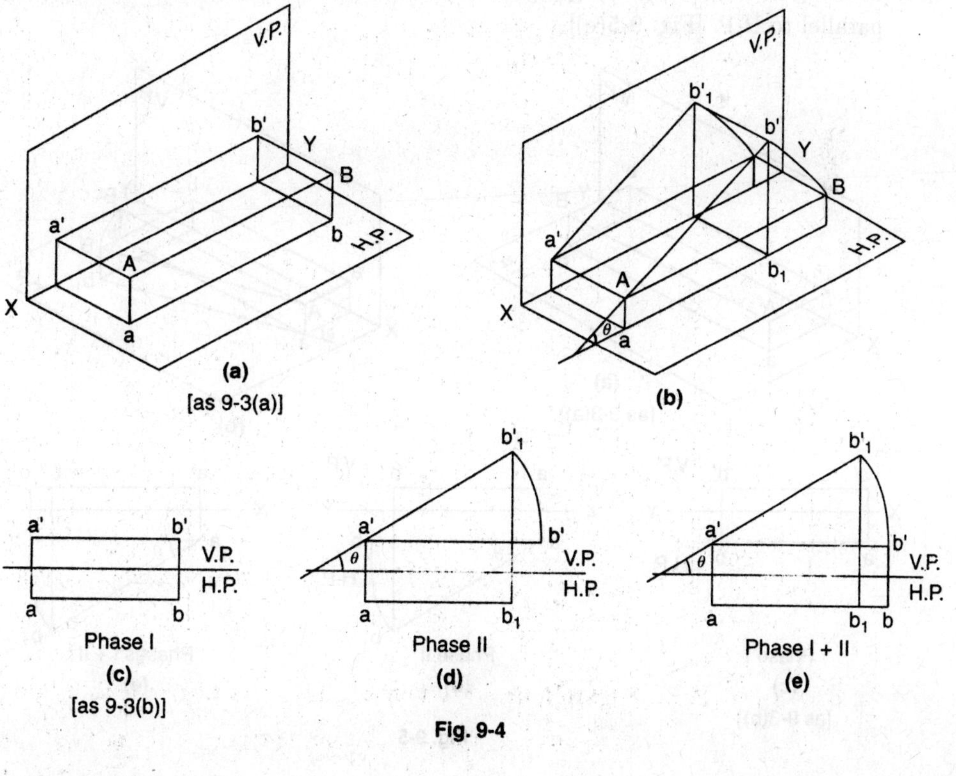

Fig. 9-4

(viii) We can draw the projections separately as Phase I and Phase II [Fig. 9-4(c) & (d)] or combine both Phases in one drawing as in Fig. 9-4(e). In case of problems on straight lines and planes, we will have to draw in one phase, i.e., combination drawing. But, in case of projection of solids, we will have to draw in two or three phases. To develop proper conception, the students should practice initially in two phases. After acquiring skill, the combination drawing may be drawn, and

(ix) final lines will be darker and other projection lines should be kept faint.

9.3.5 Straight Line Making Angle with V.P. and Parallel to H.P.

(i) Place the straight line according to Fig. 9-5(a), i.e., parallel to both H.P. and V.P.

(ii) All procedures are same as in the previous article, i.e., 9.3.4 except the nature of rotation.

(iii) Rotate the line AB, keeping the end A fixed and parallel to H.P., making angle θ with V.P.

(iv) It can be represented in two phases as in Fig. 9-5(c) and 9-5(d) or the combined one as in Fig. 9-5(e).

(v) Looking from the front, we get elevation as $a'b''$ on V.P. with shortened length as in Fig. 9-5(e).

(vi) Looking from the top, we get plan as ab on H.P. This length is the true length, as it is parallel to H.P. [Fig. 9-5(e)].

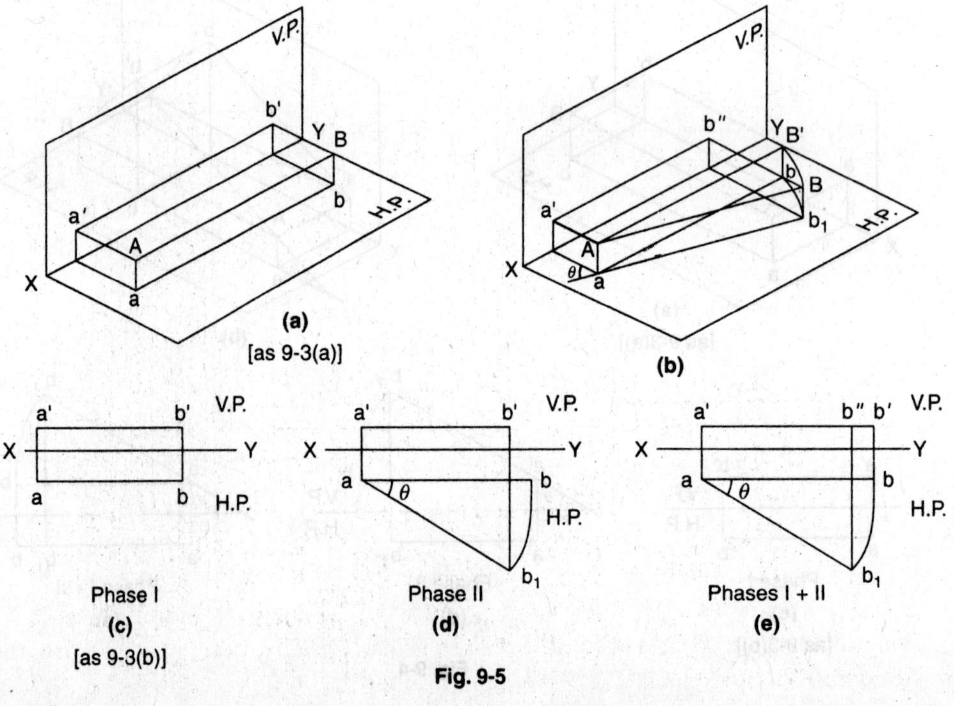

Fig. 9-5

9.3.6 Straight Line Making Angles with Both the Planes, i.e., H.P. and V.P.

(i) Place the straight line *AB* parallel to both the planes as in Fig. 9-6(a) or 9-6(d), i.e., Phase-I.

(ii) We have already acquired the knowledge of rotating the straight line, making angle with V.P. and H.P. separately. So, to make angle with both the planes, we first make the straight line parallel to both the planes. Then make angle with any one of the planes, and then with the other. We may make an angle with V.P. or H.P. first, then with V.P. or H.P., i.e., Phase-II.

(iii) Here Phase-I is parallel to both the planes V.P. and H.P. [Fig. 9-6(a) and 6(d)].

(iv) Phase-II making angle with V.P. and parallel to H.P. [Fig. 9-6(b) and 6(e)].

(v) Phase-III making angle with H.P. and also making angle with V.P. [Fig. 9-6(c) and 6(f)].

(vi) Combining the above three Phases, we get the plan and elevation in one drawing as in Fig. 9-6(g).

(vii) Final lines will be dark, others faint.

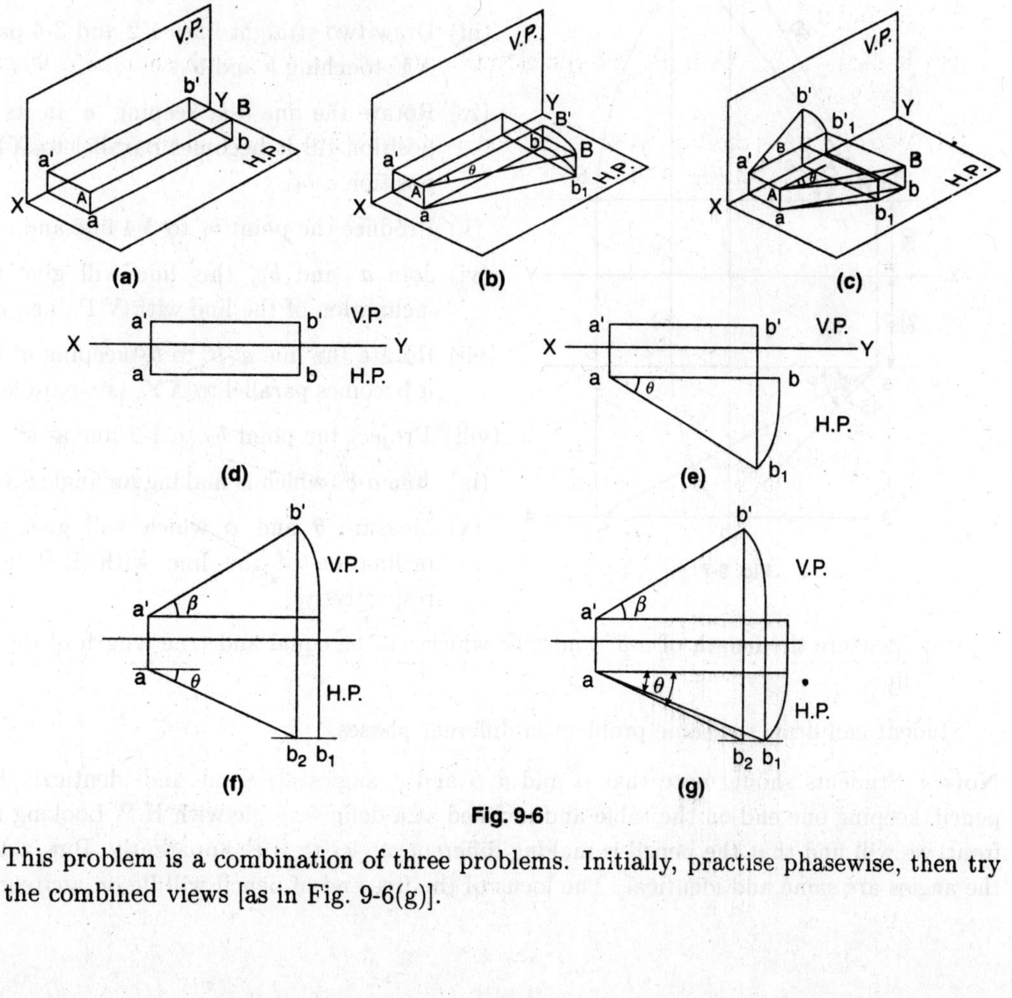

Fig. 9-6

Note : This problem is a combination of three problems. Initially, practise phasewise, then try to draw the combined views [as in Fig. 9-6(g)].

9.4 To Find the True Length of a Straight Line, when it is Inclined to Both the Planes (V.P. and H.P.)

Problem : The end A of a straight line AB is 20 mm above H.P. and 25 mm in front of V.P. The length of elevation of the line 50 mm and the elevation is inclined at 60° to H.P. The plan of the line is inclined at 45° to V.P. Find the true length of the line and its true inclinations with H.P. and V.P.

Solution : We know that, to get the true length of a straight line, it should be projected parallel to any plane.

According to Fig. 9-7,

(i) Draw $AB = ab = 50$ mm, 60° inclined to H.P. and end 'a' is 20 mm above H.P. (given), i.e., α.

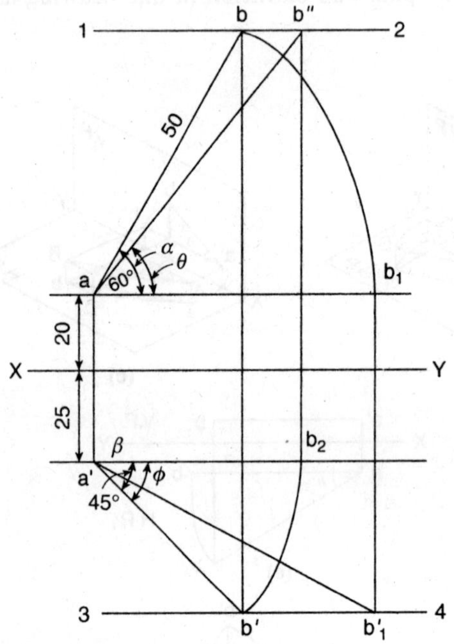

Fig. 9-7

(ii) Draw plan of ab as $a'b'$, 45° inclined to V.P. and end a' at a distance of 25 mm in front of V.P. (given), i.e., β.

(iii) Draw two straight lines 1-2 and 3-4 parallel to XY touching b and b'.

(iv) Rotate the line a-b, keeping 'a' in its original position till it becomes parallel to XY in the position a-b_1.

(v) Produce the point b_1 to 3-4 line and cut at b'_1.

(vi) Join a' and b'_1, this line will give the true inclination of the line with V.P., i.e., ϕ.

(vii) Rotate the line a'-b' to b_2 keeping a' fixed till it becomes parallel to XY, i.e., parallel to V.P.

(viii) Project the point b_2 to 1-2 line at b''.

(ix) Join a-b'' which is making an angle θ with H.P.

(x) Measure θ and ϕ which will give you true inclination of the line with H.P. and V.P. respectively.

(xi) Measure the length of a-b'' and $a'b'_1$ which will be equal and true length of the straight line.

Student can draw the same problem in different phases.

Note : Students should note that α and θ, β and ϕ angles are same and identical. Rotate a pencil, keeping one end on the table and inclined at a definite angle with H.P. Looking from the front, we will find that the pencil is making different angles to H.P. apparently. But, actually all the angles are same and identical. The locus of the free end of pencil will lie on a circumference of a circle. In the elevation, it will be a straight line parallel to H.P. We have solved the problem in this manner as we shifted b to b'' and b' to b'_1. See Fig. 9-8.

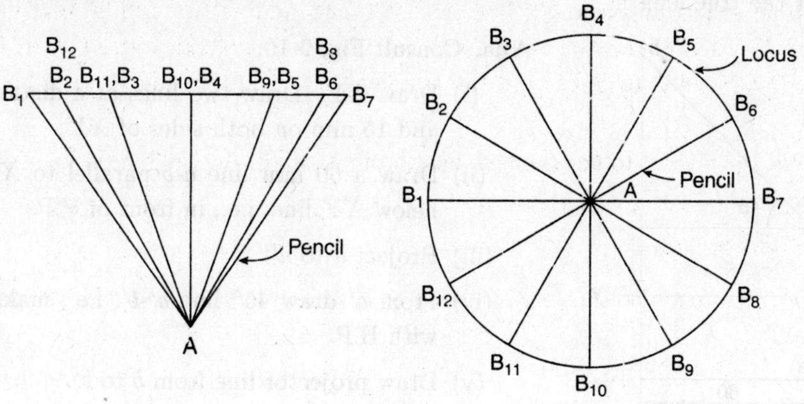

Fig. 9-8

Problems

For all problems, mention the angle (1st Angle or 3rd Angle) of projection and scale used (full size, half full size, etc.). Draw the final views in dark lines and the other faint. **Do not erase construction lines.**

1. One end A of a straight line AB, is 15 mm above H.P. and 20 mm in front of V.P. Length of the line is 70 mm. Draw elevation and plan for each of the following cases :

   (i) inclined at 30° to H.P. and parallel to V.P.,

   (ii) inclined at 45° to V.P. and parallel to H.P.,

   (iii) perpendicular to V.P. and parallel to H.P.,

   (iv) perpendicular to H.P. and parallel to V.P., and

   (v) parallel both to H.P. and V.P.

   Ans. Consult Fig. 9-9.

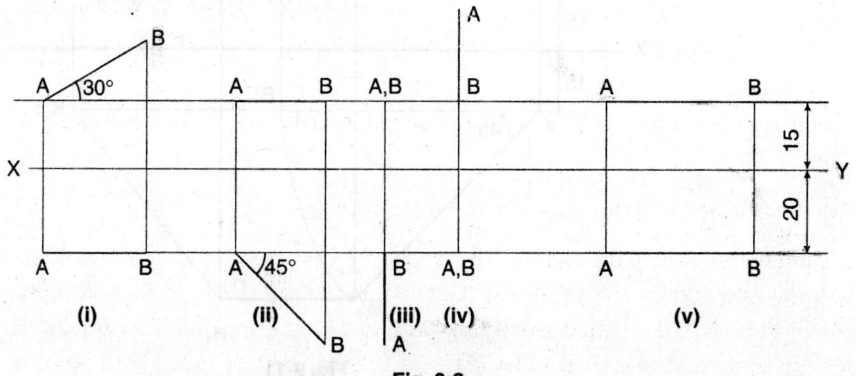

Fig. 9-9

2. The length of the top view of a line, parallel to V.P. and 45° inclined to the H.P., is 60 mm. One end of the line is 15 mm, above H.P. and 20 mm in front of V.P. Draw the line and find out the true length.

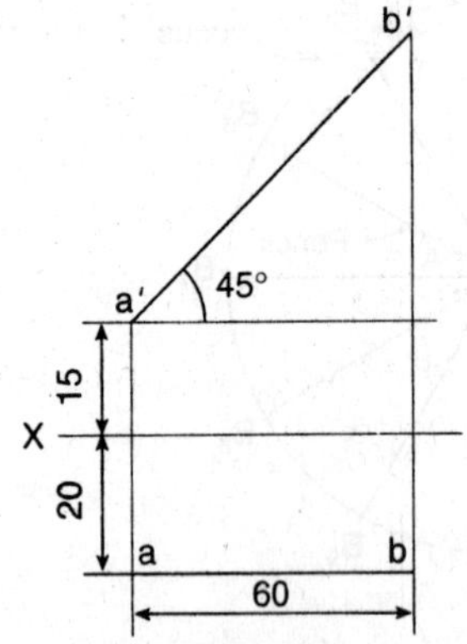

Fig. 9-10

Ans. Consult Fig. 9-10.

(i) Draw *XY*. Draw two lines at a distance of 20 mm and 15 mm on both sides of *XY*.

(ii) Draw a 60 mm line *a-b* parallel to *XY* at 20 mm below *XY* line, i.e., in front of V.P.

(iii) Project *a* to *a'*.

(iv) From *a'* draw 45° line *a'-b'*, i.e., making 45° **angle** with H.P.

(v) Draw projector line from *b* to *b'*.

(vi) *a'b'* is the true length of the line.

3. A straight line *AB*, 70 mm long, is placed with its end *A* is 15 mm above H.P. and 15 **mm** in front of V.P. It is inclined at 30° to H.P. and 45° to V.P. Draw elevation, plan and left side view.

Ans. Consult Fig. 9-11.

(i) Draw *XY* line.

(ii) Draw two faint lines at a distance of 15 mm on both sides of *XY*.

(iii) Take a point *A* and *A'*.

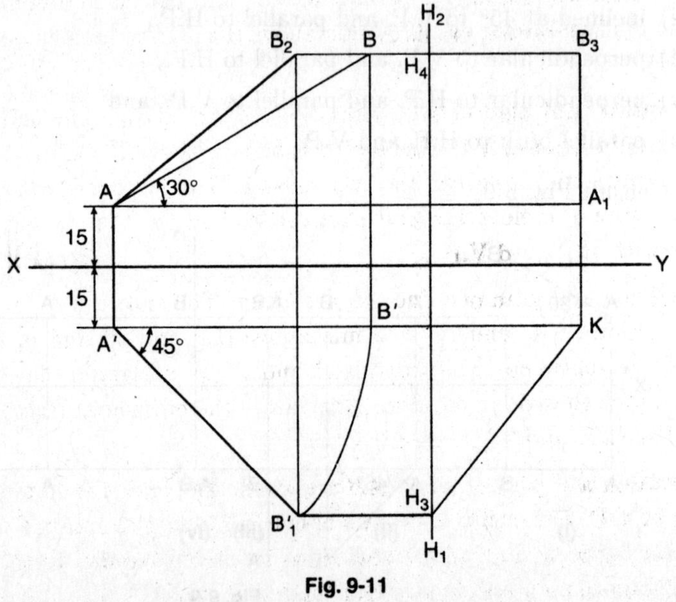

Fig. 9-11

(iv) From A, draw a 30° line of 70 mm length, i.e., AB.

(v) Project the line to $A'B'$.

(vi) Turn the line $A'B'$ through 45°, i.e., $A'B'_1$.

(vii) Through B, draw a line parallel to XY.

(viii) Draw projector from B'_1 to B_2.

(ix) Join AB_2.

(x) AB_2 and $A'B'_1$ lines are front and top views respectively of the 70 mm line.

(xi) To get left side view (in first angle projection), it will be at right side, i.e., looking from the left, but drawn at the right side.
(Note : Right end side-view will be reverse).

(xii) Draw a vertical line H_1H_2.

(xiii) From B'_1 and B, draw two lines parallel to XY which meet at H_3 and H_4 respectively on H_1H_2 line.

(xiv) From H_3 draw a 45° line which will meet the extended line $A'B'$ at K.

(xv) Projectors from A, B_2 and K meet at A_1 and B_3.

(xvi) Join A_1B_3 which is the left-side view of 70 mm line.

4. A 80 mm long straight line is parallel to H.P. and 30 mm above XY. Its two ends are 30 mm and 60 mm in front of V.P. respectively. Draw the projections and find the inclination with the V.P.

5. A straight line 70 mm long is parallel to V.P. and placed 20 mm in front of V.P. Its one end is on the H.P. and other end is 50 mm above H.P. Draw the line and find out the inclination with H.P.

6. The elevation and plan of a straight line AB are given. The end A is 15 mm above H.P. and 10 mm in front of V.P. The end B is 35 mm above H.P and 45 mm in front of V.P. The length of projection between the two ends A, B as measured parallel to the ground line, is 60 mm. Draw the elevation and plan. Also, find the true length and true inclinations of the line with H.P. and V.P.

7. The elevation and plan of a line AB are given. The end A is 15 mm above H.P. and 90 mm in front of V.P. The end B is 75 mm above H.P and 20 mm in front of V.P. The length of projection between the two ends A and B, as measured parallel to the ground line is 80 mm. Draw elevation and plan. Also locate the horizontal trace (H.T.) and vertical trace (V.T). For H.T. and V.T. consult Chap 10.

8. The elevation and plan of a line AB are given. The end A is 15 mm above H.P. and 30 mm in front of V.P. The end B is 80 mm above H.P. and 100 mm in front of V.P. The length of projection between the ends A and B, as measured parallel to the ground line, is 90 mm. Draw elevation, plan and locate H.T. and V.T.

9. The elevation and plan of a line AB are given. The end A is 20 mm above H.P. and 10 mm in front of V.P. The end B is 40 mm above H.P. and 40 mm in front of V.P. The length of projection between the ends A and B, as measured parallel to the ground line, is 60 mm. Draw elevation, plan and locate H.T. and V.T.

10. The length of a straight line AB is 70 mm. Its end A is on the ground line. The line is inclined at 60° to H.P. and 30° to V.P. The plane containing the line is perpendicular both to H.P. and V.P. Draw the elevation, plan and left-side view.

Ans. Here, the line is placed on a plane perpendicular to both H.P. and V.P. The sum of the inclinations 60° with H.P. and 30° with V.P., i.e., 60° + 30° = 90°. Isometric drawing of the line will help to solve this problem easily.

Refer to Fig. 9-12.

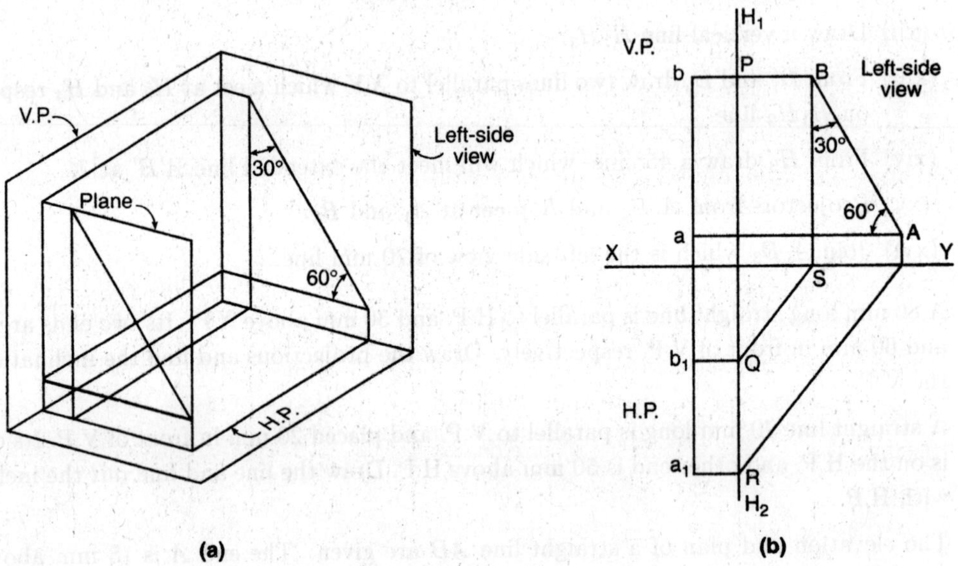

Fig. 9-12

To solve the problem, we will have to start from left-side view, (any side view will do) where we will get the true shape of the line.

(i) $AB = 70$ mm which is making 60° with H.P. and 30° with V.P.

(ii) Draw a vertical line H_1, H_2.

(iii) From B draw a vertical and a horizontal lines to meet XY at s and H_1H_2 at P respectively. Produce P to b in the vertical plane.

(iv) From A and s, draw 45° lines to meet H_1H_2 at Q and R respectively. These will be projectors to H.P. at a_1 and b_1.

(v) So, ab and a_1b_1, are the elevation and plan of the 70 mm line respectively.

11. A straight line CD is 70 mm long. Its end C is on the H.P. and 20 mm in front of V.P. The line is inclined at 30° to H.P. and parallel to V.P. Draw elevation, plan and left-side view.

12. A straight line AB is 65 mm long. Its end A is 25 mm above H.P. and 20 mm in front of V.P. The line is inclined at 45° to V.P. and parallel to H.P. Draw elevation, plan and left-side view.

13. A straight line PQ is 70 mm long. Its end P is 20 mm above H.P. and 20 mm in front of V.P. The line is inclined at 30° to H.P. and 45° to V.P. Draw elevation, plan and locate the traces.

14. A straight line AB is 75 mm long. Its end A is 25 mm behind V.P. and 20 mm below H.P. The line is inclined at 45° to H.P. and parallel to V.P. Draw front view, top view and locate the traces.

State Council Model Questions (West Bengal)

1. Draw the symbols used to represent 1st angle and 3rd angle projections.

2. What is meant by orthographic projection?

3. Differentiate between first-angle and third-angle projection.

4. Name the two systems of projection, which are in vogue.

5. Fill in the blanks with appropriate words :

(a) The projections of a point coincide and both fall below XY line. The point is situated in the ______ quadrant.

(b) The view of a point is 40 mm above XY and the front view, 20 mm below the top view : the point is situated in the ______ quadrant.

(c) When a straight line is perpendicular to one of the principal planes, it is ______ to other plane.

(d) When a line is parallel to the V.P., its ______ is a point (addition : when line is perpendicular to H.P.)

(e) When a straight line is inclined to ______ and parallel to ______ its front view represents the true length of the line.

(f) When a line is inclined to the V.P and parallel to the H.P., its front view is ______ to XY.

(g) When a line is inclined to V.P. and parallel to the H.P., the inclination of the top view with XY represents ______.

(h) The ______ view of a point is obtained as the intersection point between the ray of the sight and the V.P.

(i) The top view of a point is obtained as the intersection point between the ray of sight and the ______.

(j) When a line is parallel to both the H.P. and V.P., its side view is a ______.

(k) When a line is inclined to ______ and parallel to ______, the front view represents true length of the line. **(AMIE—Winter, 2001)**

Ans. (a) fourth, (b) second, (c) parallel, (d) top view, (e) horizontal plane, vertical plane, (f) parallel, (g) true inclination, (h) front, (i) horizontal plane, (j) point. (k) H.P., V.P.

6. State True or False by writing 'T' or 'F' respectively :

(a) In the first-angle projection, the object is positioned in between the observer and the plane of projection.

(b) In the third-angle projection, the object is positioned in between the observer and the plane of projection.

(c) When a point lies on the V.P., its top-view lies on XY line.

(d) The projection of a line on to a plane parallel to it, appears its true length.

(e) When a straight line is inclined to a plane, its projection on that plane is a line shorter than its true length.

(f) A profile plane is perpendicular to H.P. and inclined to the V.P.

(g) Front views of two lines are parallel. Hence the lines must be parallel.

(h) The plane, perpendicular to both the H.P. and the V.P., is called the profile plane.

(i) The straight line joining the projections of a point intersects the line XY at an angle of 90°.

(j) The distance of a point from the H.P. is marked from the XY line to the top view.

(k) When a point lies on the H.P., its front view lies above the XY line.

(l) In the third-angle projection, the view below the reference line is called top view and the view above the reference line is called front-view.

(m) When a straight line is inclined to a plane, its projection on that plane is a line shorter than its true length. **(AMIE—Winter, 2001)**

Ans. (a) T, (b) F, (c) T, (d) T, (e) T, (f) F, (g) T, (h) T, (i) T, (j) F, (k) F, (l) F (m) T.

7. A line measuring 75 mm has one of its ends 50 mm in front of V.P. and 15 mm above H.P. The other end of the line is 15 mm in front of V.P. and is above H.P. The top view of the line is 50 mm long. Draw and measure the length of the front view.

(AMIE—Summer, 2000)

8. One end A of a line AB is 20 mm above H.P. and 30 mm in front of V.P. The end B is 40 mm behind V.P. and 50 mm below H.P. The distance between the projectors of A and B is 60 mm. Draw the projection of the line. Find its true length, true inclinations with H.P. and V.P. and its traces. **(AMIE—Winter, 2000)**

9. A line AB, inclined at 30° to the H.P. has its ends A and B, 25 mm and 60 mm in front of V.P. respectively. The length of the plane is 65 mm and the V.T. is 15 mm above H.P. Project the elevation of the line and locate its H.T. **(AMIE—Summer, 2001)**

10. A line of 120 mm long makes an angle of 35° with H.P. and 45° with V.P. Its midpoint is 30 mm above H.P. and 20 mm in front of V.P. Draw the projections of the line.

(AMIE—Winter, 2001)

11. A line 100 mm long makes an angle of 35° with H.P. and 45° with V.P. Its mid-point is 20 mm above H.P. and 15 mm in front of V.P. Draw the projections of the line and locate its H.T. and V.T. **(AMIE—Summer, 2002)**

12. The end A of a line AB is 55 mm below H.P. and 55 mm behind V.P., while B is 100 mm above H.P. and 20 mm in front of V.P. The distance between their projectors is 120 mm. Draw the projections of the straight line and determine its true length, true inclinations with H.P. and V.P. and show its traces. **(AMIE—Winter, 2002)**

13. Pick-up the correct answer :

If a straight line is parallel to H.P. and inclined to V.P. then its true length is directly given by

(a) top view (b) front view (c) end view (d) none. **(AMIE—Winter, 2002)**

Ans. (a) [Because the line is parallel to H.P.]

14. Choose the correct answer for the following question :

When a line lies in the V.P., the true length will be on the

(a) Top view (b) Front view (c) End view. **(Winter, 2006)**

Ans. (b), in cases of both first-angle and third-angle projections.

Chapter 10

Projection of Planes

10.1 Introduction

Any flat or level-surface of a certain area may be called a **plane or lamina**. A 'lamina' is an enclosed figure, bounded by straight lines or a curved line. A good example of a plane is the flat surface of any **sheet-metal**. Any two points on the plane may be joined by a straight line. Then, this straight line will touch the plane at all points in the path of the straight line.

We also find that, **a plane has two dimensions only, i.e., length and breadth, but no thickness.**

10.2 Classification of Planes

For simplicity, seven classes of planes are explained below. These are shown in Fig. 10-1 on page 77.

(i) **Triangle :**

A closed figure, having three sides, is known as a triangle. Triangles may be—

- (a) **Equilateral**—All sides equal (each angle 60°).
- (b) **Isosceles**—Two sides equal and the two base-angles are equal.
- (c) **Scalene**—No two sides are equal, i.e., with unequal sides.
- (d) **Right-angled**—One of the three angles is a right angle (90°).

(ii) **Square :**

A square is a closed figure, having four equal sides and four right angles. The opposite sides are parallel.

(iii) **Rhombus :**

A four-sided closed figure with four equal sides. The opposite angles are equal, but not right angles. The opposite sides are parallel and equal.

(iv) **Rectangle :**

A four-sided closed figure with four right angles. The adjacent sides are unequal but the opposite sides are equal and parallel.

(v) **Pentagon :**

A five-sided closed figure. When all the sides are equal, it is called a regular pentagon. In this case, each internal angle is 108°. An irregular pentagon has unequal sides.

(vi) **Hexagon :**

A six-sided closed figure. Each internal angle of a regular hexagon is 120°. The **Seven, eight, nine and ten-sided polygons are known as heptagon, octagon, nonagon and decagon respectively**. They may be regular or irregular.

(vii) **Circle :**

If we imagine that, the number of sides of a regular polygon is increased to infinity, then the regular polygon will become a circle. A circle, of any size, has a centre and a fixed radius, as shown in Fig. 10-1.

N.B. Another example of a closed plane figure is the **ellipse** (See Chapter 6).

10.3 Preparation of Different Plane Figures

The different plane figures are shown in Fig. 10-1. These figures may be obtained by cutting old post-cards or paste-boards of rectangular shape. (shown dotted in Fig. 10-1) At the time of drawing projections of planes, the students are advised to use these cut-out figures.

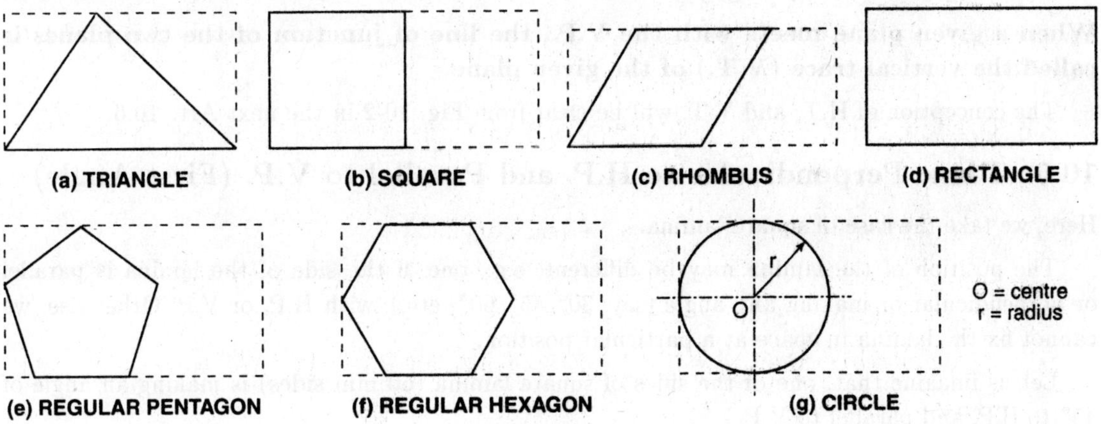

Fig. 10-1

10.4 Positions of the Plane

As in the case of a straight line, a plane may also be placed, with respect to the reference planes (H.P. and V.P.), in **six** different positions, as mentioned below :

The plane

(i) Perpendicular to H.P. and Parallel to V.P.

(ii) Perpendicular to V.P. and Parallel to H.P.

(iii) Perpendicular to both H.P. and V.P.

(iv) Perpendicular to V.P. and inclined to H.P.

(v) Perpendicular to H.P. and inclined to V.P.

(vi) Placed in space in an oblique manner such that any specific line on the plane is inclined both to H.P. and V.P. This specific line may be a diameter (in case of a circular plane), a major or minor axis (in case of an ellipse), a diagonal (in case of a rectangle, square or any regular polygon).

The above positions of a plane will be explained one after another.

10.5 Traces of Planes

When one given plane (supposing a rectangle or square) meets with any reference plane (V.P. or H.P.), we get a straight line at the junction of the two planes. **This straight line is known as the 'trace' of the given plane.**

10.5(a) Horizontal Trace (H.T.)

When a given plane meets with H.P., the line of junction of the two planes is called the horizontal trace (H.T.) of the given plane.

10.5(b) Vertical Trace (V.T.)

When a given plane meets with the V.P., the line of junction of the two planes is called the vertical trace (V.T.) of the given plane.

The conception of H.T. and V.T. will be clear from Fig. 10-2 in the next Art. 10.6.

10.6 Plane Perpendicular to H.P. and Parallel to V.P. (First Angle)

Here, we take the case of square lamina.

The position of the lamina may be different, e.g., one of the side of the lamina is parallel or perpendicular or making any angle (say, 30°, 45°, 60°, etc.) with H.P. or V.P. Otherwise, we cannot fix the lamina in space at a particular position.

Let us imagine that, one of the sides of square lamina (50 mm sides) is making an angle of 45° to H.P. and parallel to V.P.

Refer to Fig. 10-2(a) and (b).

(a) Since the plane is parallel to V.P., we get the true shape on the elevation.

(b) Place the square lamina of 50 mm sides, as shown in Fig. 10-2(b). The corner point '1' is at a distance d_1 above H.P., and at a distance d_2 in front of V.P.

(c) Now look first from the front. Since the lamina is parallel to V.P., we shall get the true shape (square) of the lamina in the elevation. Since the plane has no thickness, and it is perpendicular to H.P., we shall get a straight line in the plan [Fig. 10-2(b)]. This is shown as 4-2 in the plan.

(d) Join all the points (1,2,3,4) in the elevation, which is a square.

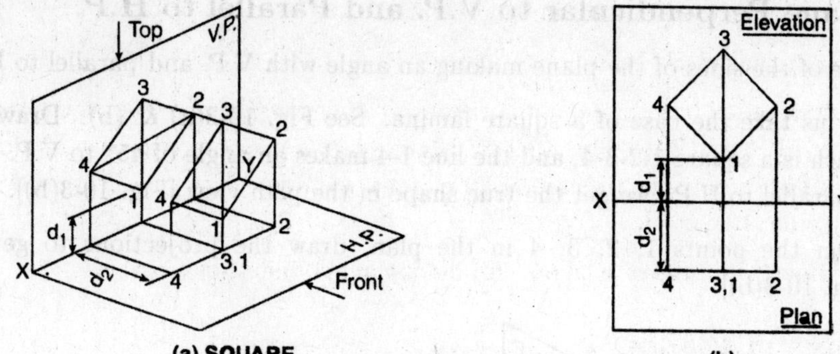

(a) SQUARE (b)

Fig. 10-2

(e) We shall get the elevation and plan above and below the reference line XY respectively, as shown in Fig. 10-2(b).

Examples with different types of lamina

Try to draw the projections with other laminas, as shown in Fig. 10-2(c) to (g).

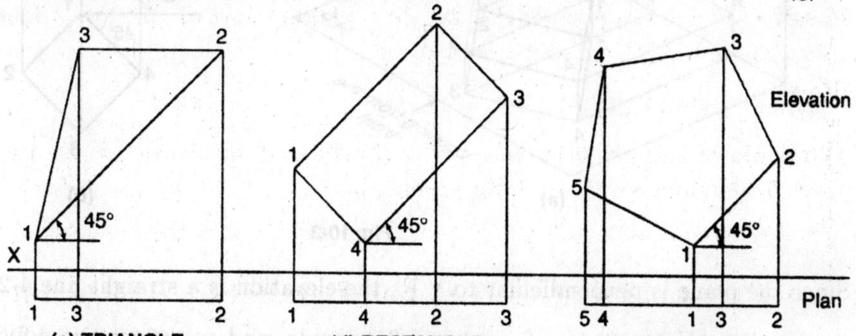

(c) TRIANGLE (d) RECTANGLE (e) PENTAGON

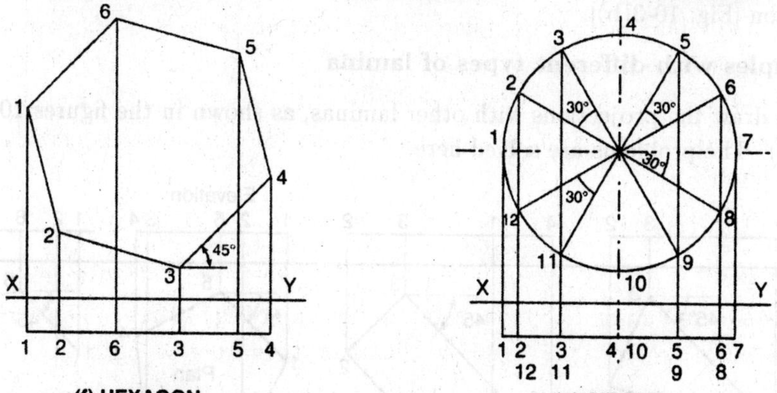

(f) HEXAGON (g) CIRCLE

Fig. 10-2

In all the cases, we get true shapes of the laminas in the elevation, and straight lines as the plan.

10.7 Plane Perpendicular to V.P. and Parallel to H.P.

(a) One of the sides of the plane making an angle with V.P. and parallel to H.P.

Let us take the case of a square lamina. See Fig. 10-3(a) & (b). Draw the plan first, which is a square 1-2-3-4, and the line 1-4 makes an angle of 45° to V.P. Since the plane is parallel to H.P., we get the true shape in the plan view [Fig. 10-3(b)].

(b) From the points 1, 2, 3, 4 in the plan, draw the projections to get the elevation [Fig. 10-3(b)].

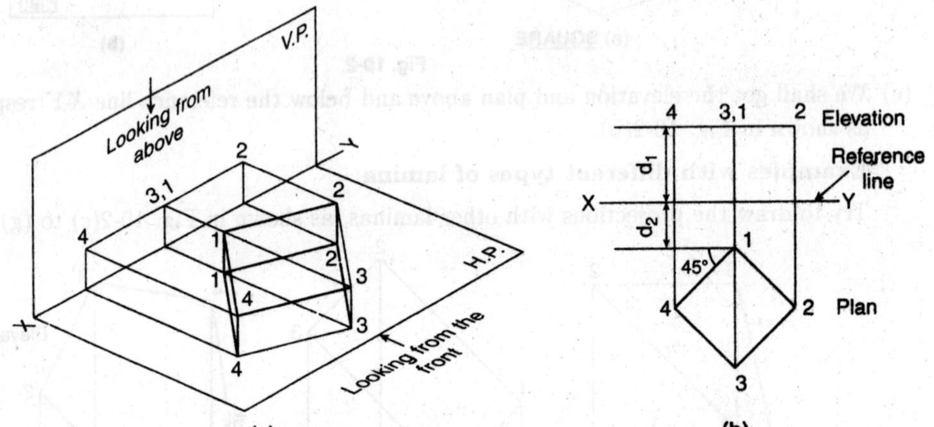

(a) (b)

Fig. 10-3

(c) Since the plane is perpendicular to V.P., the elevation is a straight line 4-2.

(d) In the plan, the projectors from 1 and 3 coincide, and so we get one point (1, 3) in the elevation [Fig. 10-3(b)].

Examples with different types of lamina

(e) Try to draw the projections with other laminas, as shown in the figures 10-3(c), (d), (e) and (f). The problems are solved here.

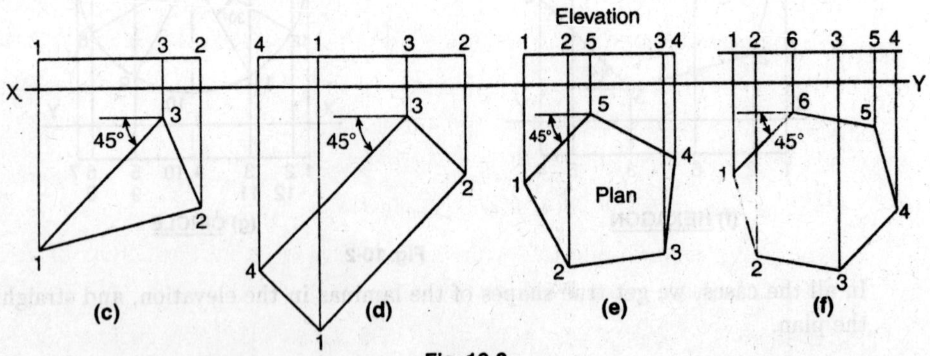

(c) (d) (e) (f)

Fig. 10-3

(f) Circular Lamina : Refer to Fig. 10-3(g).

In the case of a circle, we have noticed that there are no corners. So, we divide the circle into 12 equal parts with the help of a 30°-60° set-square. All division-points will meet the circumference, which should be numbered from 1,2,3,. . .,12. For simplicity, we may roughly consider a circle to be a 12-sided regular polygon. We draw the plan first, as the lamina is parallel to H.P. From the 12 points (1 to 12) in the plan, draw projections to get the elevation 1-7, which is a straight line, parallel to H.P. or X-Y line.

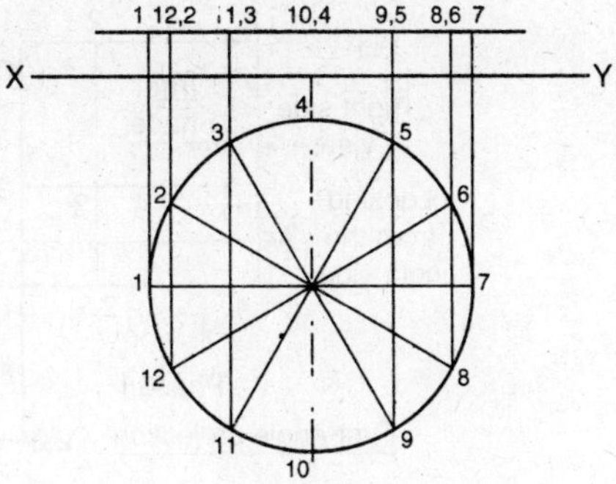

(g)

Fig. 10-3

10.8 Plane Perpendicular Both to H.P. and V.P.

Refer to Fig. 10-4(a) and (b).

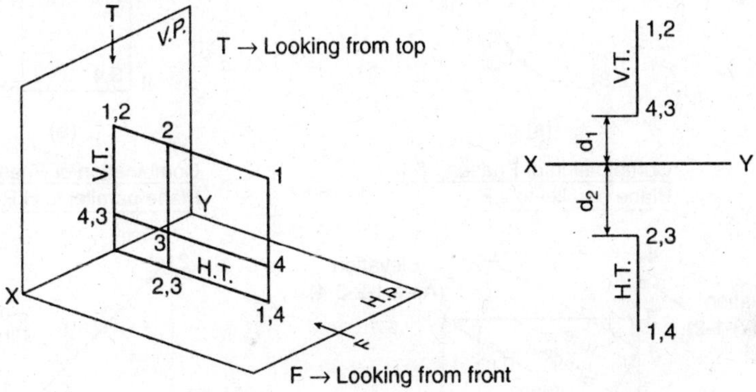

(a) (b)

Fig. 10-4

(a) Now, look first from the front and then from the top. In the elevation and plane, we shall get a straight line of same size, as that of the square. This is because the plane is a square and placed with one of the sides of the square parallel to H.P. or V.P.

(b) We cannot draw the top and front views, if the square is placed in other position. Because the true shape is not visible in top and front views. We shall have to start from the plane parallel to V.P. or H.P. and then we rotate the same.

This is shown in the figures 10-5(a) and (b).

(c) Study and practise the following figures. Rectangle perpendicular to V.P. and H.P. and one of the long side subtends and angle of 30° to H.P.

(d) Triangle is placed on the ground, and side of the base is parallel to V.P. first. Then front view is rotated through 90°. Now, we shall get the plane, perpendicular to V.P. and H.P.

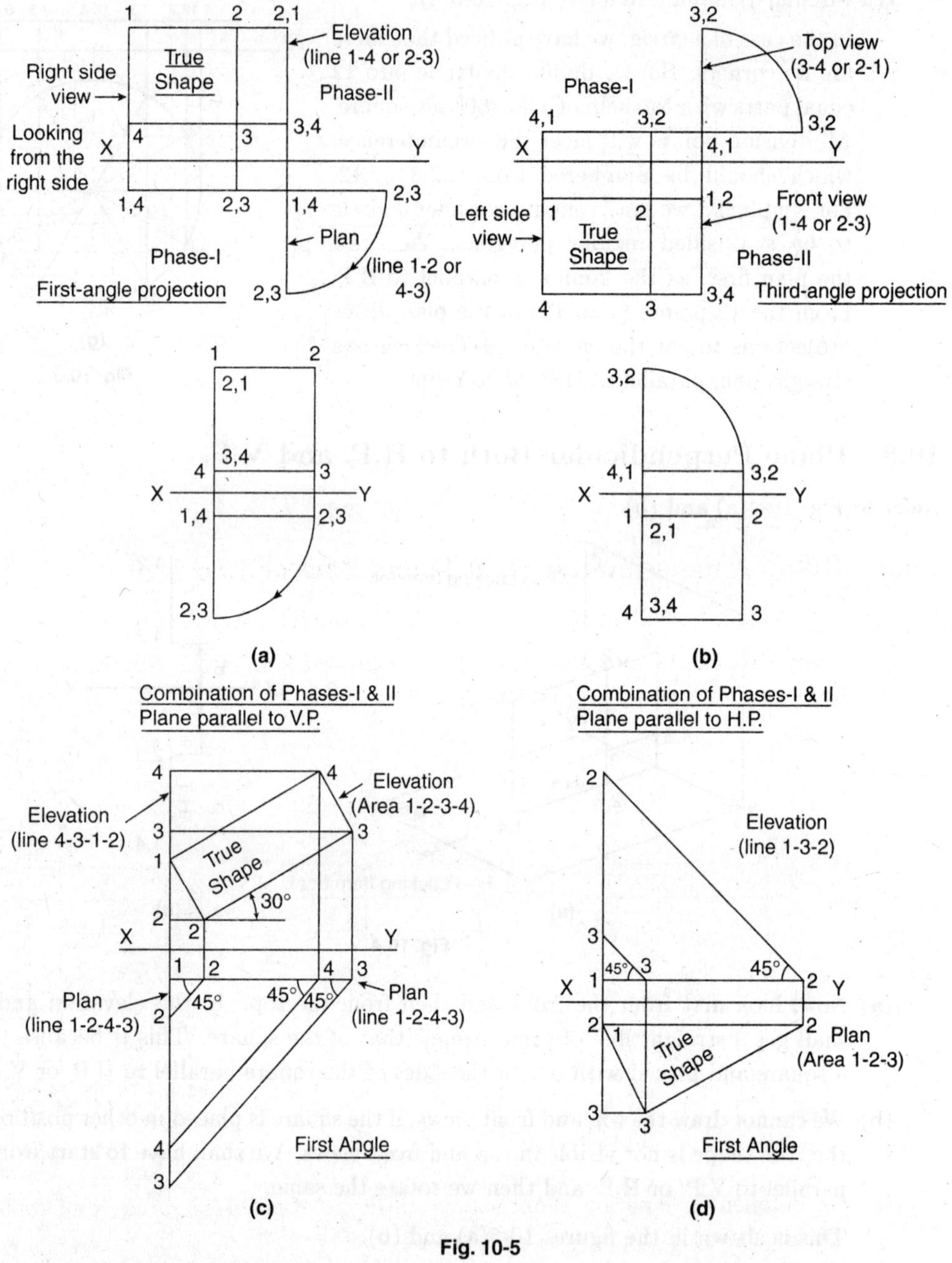

Fig. 10-5

(e) Regular pentagon, placed on the H.P. with one of its side and parallel to V.P. Then it is rotated through 90° in the vertical position of the pentagon. We get the plane perpendicular to both H.P. and V.P.

(f) For regular hexagon.

(g) For circle.

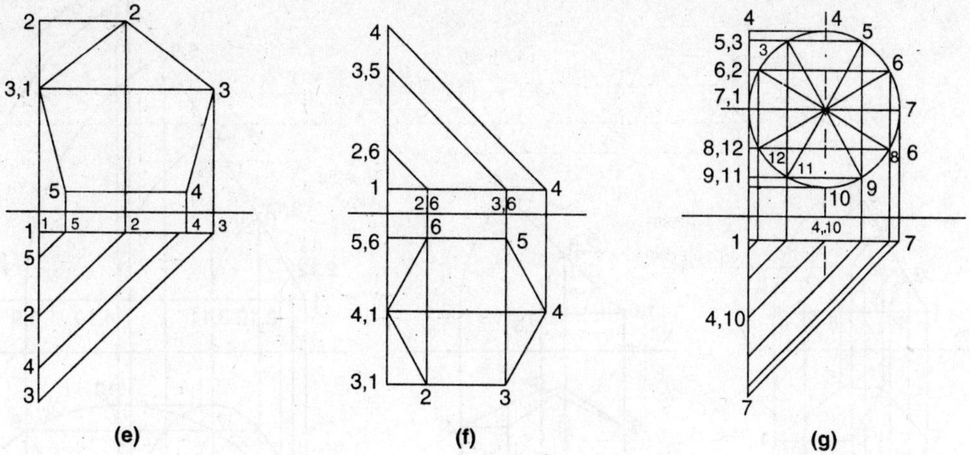

Fig. 10-5

10.9 Plane Perpendicular to V.P and Inclined to H.P.

(a) In this case, we will start initially by placing the plane parallel to H.P.

(b) Study with model and draw the following problems as in the Fig. 10-6(a) to (f). These figures may be drawn in two phases.

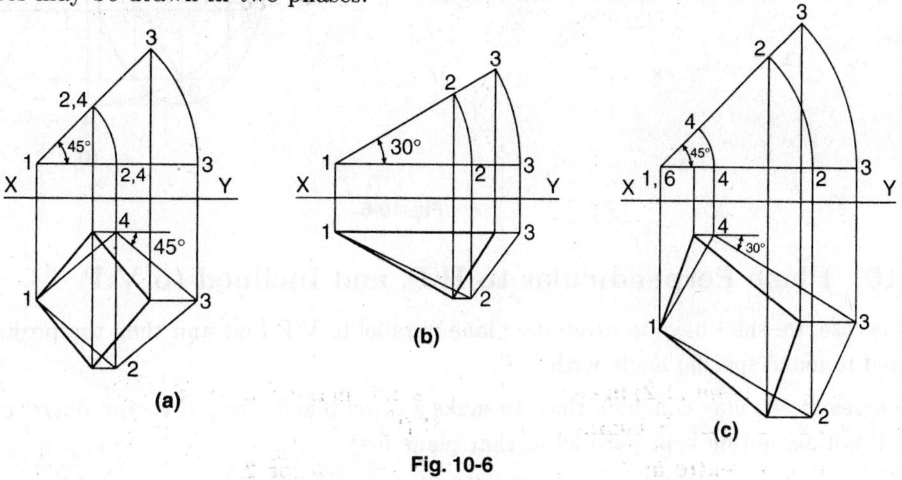

Fig. 10-6

(c) For detailed explanation about solving this type of problem, please see Problem 4 (page 89), Fig. 10-11.

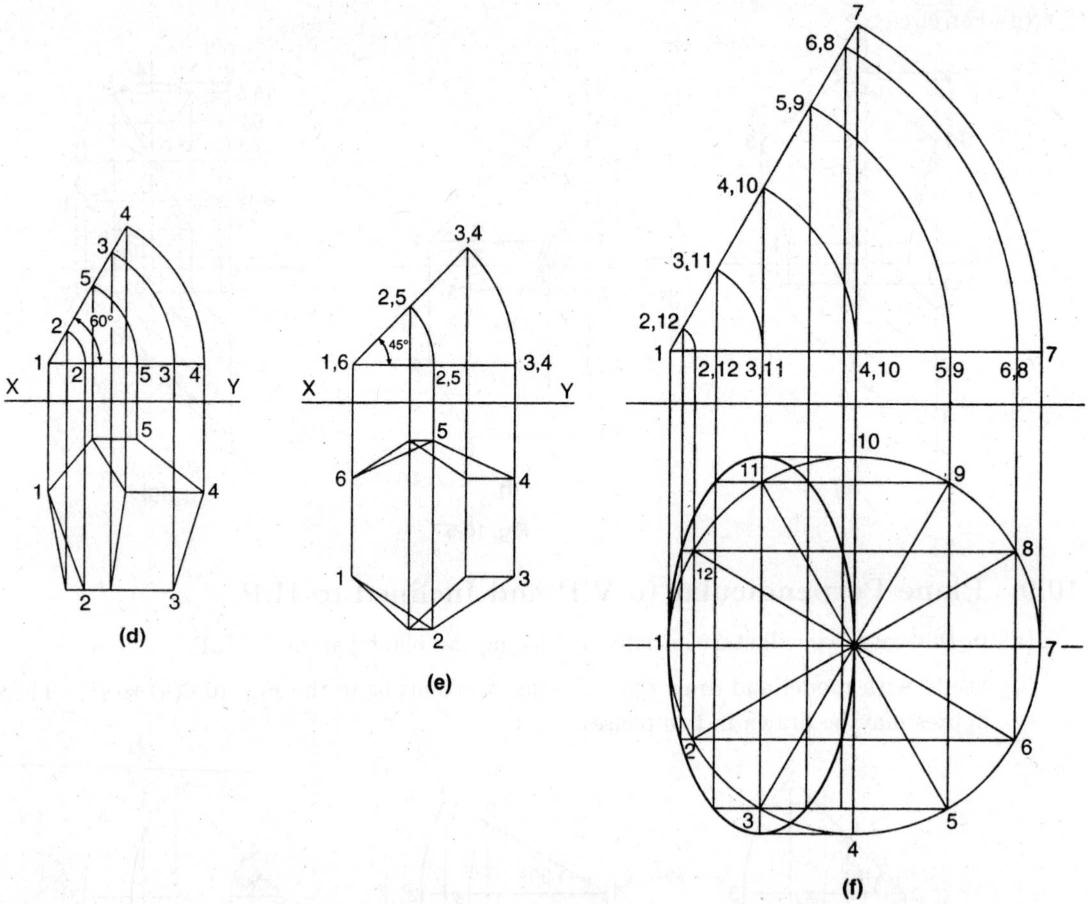

Fig. 10-6

10.10 Plane Perpendicular to H.P. and Inclined to V.P.

In this case, we shall have to draw the plane parallel to V.P first and then the projection is to be rotated to make specific angle with V.P.

Therefore, we may conclude that, to make a given plane inclined to any reference plane (H.P. or V.P.), it should be kept parallel to that plane first.

Six examples have been given on Fig. 10-7(a) to (f). Regarding solving this type of problem, please see problem 5, page 89.

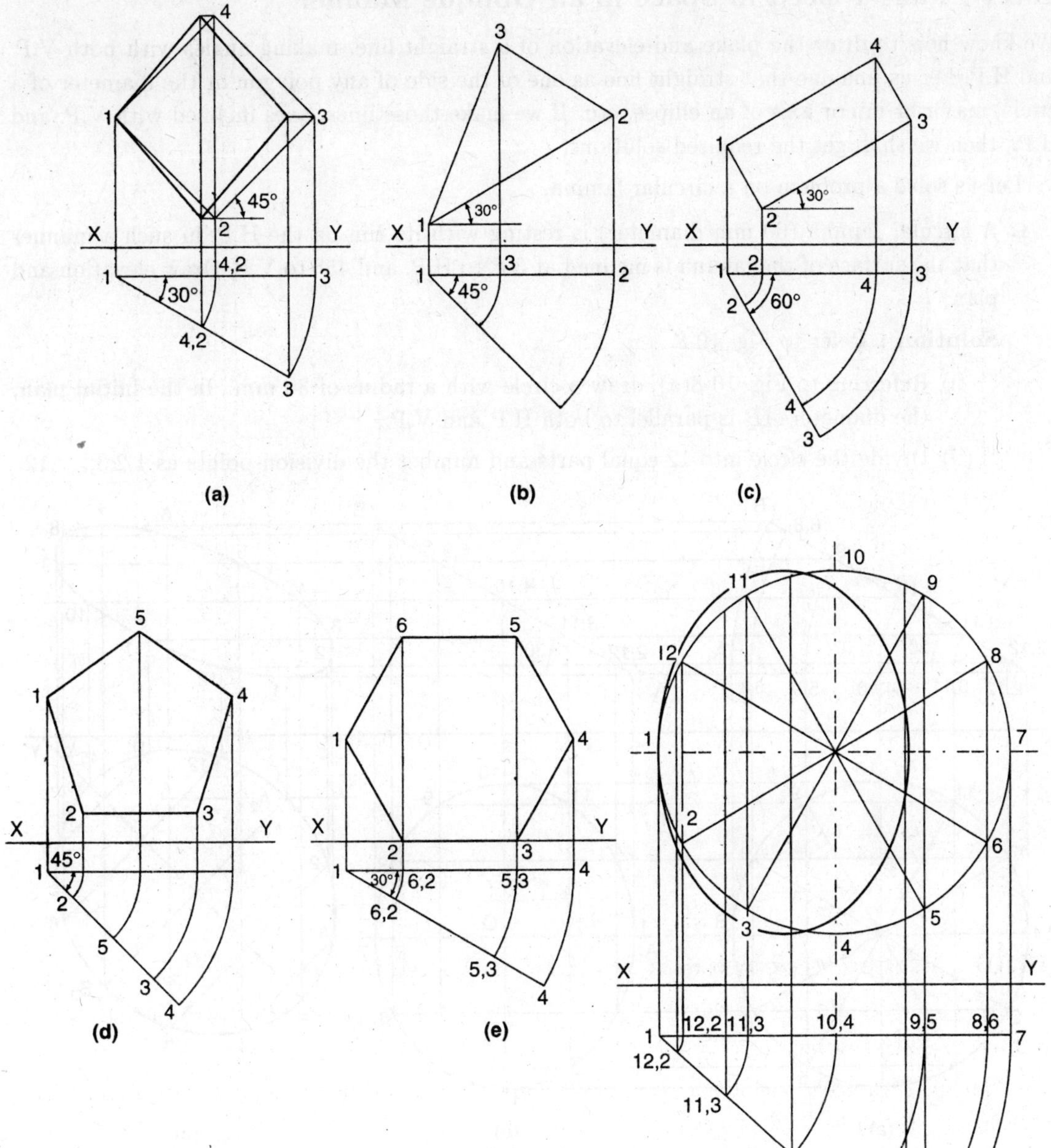

Fig. 10-7

10.11 Pane Placed in Space in an Oblique Manner

We know how to draw the plane and elevation of a straight line, making angles with both V.P. and H.P. Let us imagine that straight line as one of the side of any polygon or the diameter of a circle, major or minor axis of an ellipse, etc. If we make those lines, axes inclined with V.P. and H.P., then we shall get the required solutions.

Let us solve a problem on a circular lamina.

1. A circular lamina (60 mm diameter) is resting with its rim on the H.P. in such a manner that the surface of the lamina is inclined at 30° to H.P. and 45° to V.P. Draw elevation and plan.

   Solution : Refer to Fig. 10-8.

   (i) Referring to Fig. 10-8(a), draw a circle with a radius of 30 mm. In the initial plan, the diameter AB is parallel to both H.P and V.P.

   (ii) Divide the circle into 12 equal parts and number the division-points as 1,2,3,. . .,12.

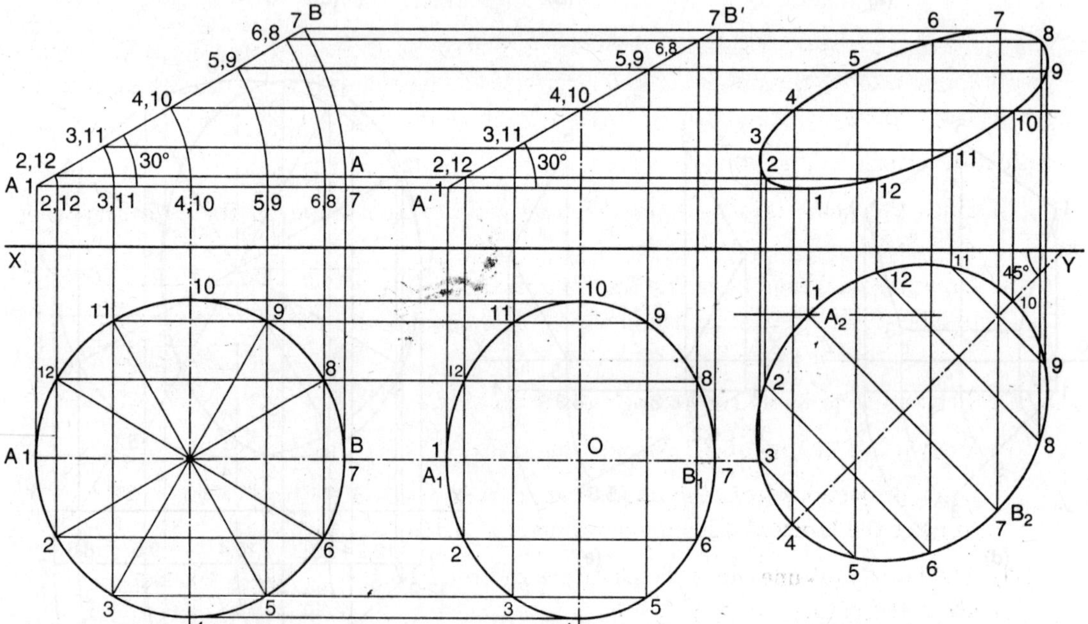

Fig. 10-8

   (iii) Draw the reference line XY and some distance above, draw another line A-A', parallel to XY. Project from all points (1–12) in plan to the line A-A'.
   On the line A-A', the points are 1, (2-12), (3-11), (4-10), (5-9), (6-8) and 7.

   (iv) Draw a line AB in elevation, making an angle of 30° to H.P. (i.e., A-A'). Transfer all the projected points 1, (2-12), etc. from line A-A' to AB. [Elevation of Fig. (a)]

(v) Now, to the right of the first elevation, draw a line $A'B'$ parallel to AB. Transfer all the projected points 1, 2-12, etc. on the line $A'B'$ [see Fig. 10-8 (b)].

(vi) Project all points in $A'B'$ and first plane (circle) to get the second plan view as shown in Fig. (b).

(vii) Since the circular lamina is inclined at 45° to V.P., rotate the axis A_1B_1 (in Fig. b) through 45°. This is shown by the line A_2B_2 in Fig. c, which is inclined at 45° to XY.

(viii) Now, bisect the line A_2B_2 to O to get the centre (Fig. c). Take the measurement O-4 and O-10 from second plane (Fig. b) and transfer the same to the third or final plane (Fig. c). In the same manner, obtain other points in the circumference of the first plane.

(ix) Project from all points of the third plan (Fig. c) and second elevation $A'B'$ (Fig. b) to get the final elevation (Fig. c).

(x) The final elevation and plane should be shown by dark lines.

2. An equilateral triangle, of sides 60 mm each, is placed on its plane parallel to H.P. so that one of the side is making an angle of 60° with V.P. The nearest corner from the principal planes is 20 mm away from both H.P. and V.P. Draw the plan and elevation.

Solution : Refer to Fig. 10-9.

(i) Since the plane is parallel to H.P., we get the true shape on H.P. We draw the plan first.

(ii) Draw the equilateral triangle abc, so that the base bc is parallel to XY. We take the point 'a' at a distance of 20 mm from the line XY.

(iii) The line ab or ac makes an angle of 60° to V.P.

(iv) Draw a line 20 mm above XY and parallel to XY.

(v) Draw projections from the points a, b, c from the plan to meet the horizontal line in elevation at a, b and c.

(vi) The horizontal line cab is the elevation and the triangle abc is the plan.

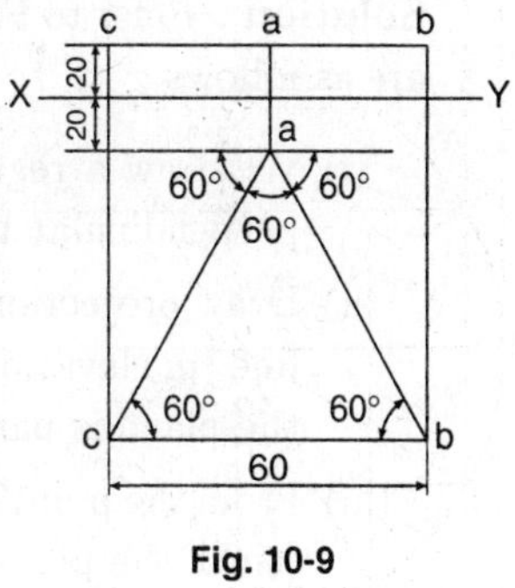

Fig. 10-9

3. A square, each side 50 mm, has one side parallel to H.P. and the adjacent side 25 mm in front of V.P. The plane of the lamina makes an angle of 30° to H.P. and perpendicular to V.P. Draw its elevation and plan.

Solution : (Plane perpendicular to V.P. and making an angle with H.P.)

(i) Draw a square (each side 50 mm) 1,2,3,4 on plan such that one of the sides (1-2 or 4-3) is parallel to V.P. The side 1-4 is at a distance of 25 mm from XY.

(ii) Project from plan to elevation, i.e., the line (2,1)-(3,4).

(iii) Turn the elevation through an angle of 30°, which is shown (1,2)-(3′,4′), where 2-3 = 2-3′.

(iv) Draw plan from the final elevation. This is 1-2-3′-4′, which is rectangular in shape. (See plan, Fig. 10-10)

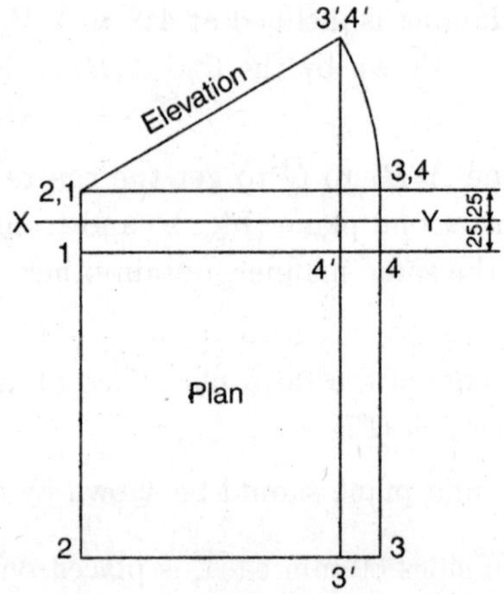

Fig. 10-10 (First-Angle)

4. A regular hexagon of 30 mm sides, is placed with its one side on the H.P., so that this side is perpendicular to V.P. The plane of the lamina is inclined at an angle of 60° to H.P. and perpendicular to V.P. Draw the elevation and plan.

Solution : Refer to Fig. 10-11. This problem is similar to the problem 3. The drawing-steps are as follows :

(i) We draw a regular hexagon of 30 mm sides. (See the plan), so that the side 1-2 is perpendicular to V.P. The hexagon is shown dotted.

(ii) Draw projections from all the points (1, 2, 3, 4, 5, 6 in the plan) to meet the horizontal line (in elevation) at the three points (1,2), (3,6) and (4,5). This is the position, when the plane is parallel to H.P.

(iii) From the point (1,2) in elevation, draw a line, making an angle of 60° to the horizontal (since the plane is inclined at 60° to H.P.).

(iv) With the centre at 1 (or 2-4) and radius = 1-5 (or 2-4), draw an arc to meet the inclined line 1-5′ at the point 4′ (or 5′). The **line 1-5′** (or 2-4′) is the **required elevation** (shown by firm line).

(v) Similarly, transfer the point 3 (or 6) to the inclined line 1-5′ at the point 3′ (or 6′). See the elevation.

(vi) On the plan, draw horizontal lines 3-3′, 4-4′, 5-5′ and 6-6′.

(vii) From the points (3′,6′) and (4′-5′) in elevation, draw projections to meet the horizontal lines (3-3′, 4-4′, 5-5′, 6-6′) on the plan at the points 3′, 4′, 5′ and 6′ respectively.

(viii) Join 2 with 3′, 3′ with 4′, 4′ with 5′, 5′ with 6′ in the plan. The enclosed figure (1-2-3′-4′-5′-6′-1) is the **required plan**, which is shown in firm lines.

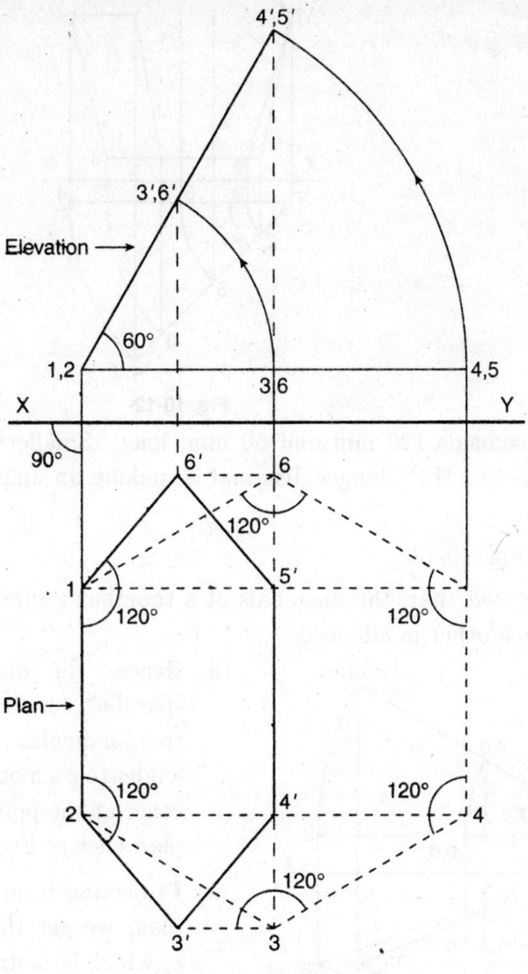

Fig. 10-11 (Problem 4)

5. A regular pentagon, of 25 mm sides, having its plane vertical and inclined at an angle of 45° to the V.P. One of its corners is 20 mm above H.P. and 15 mm in front of V.P. Draw elevation and plan.

Solution : Refer Fig. 10-12.

(i) Draw a regular pentagon on elevation as 1, 2, 3, 4, 5 of side 25 mm.

(ii) Project the points to plane parallel to XY. This is the line 1-2-5-3-4.

(iii) Turn the plane to 45°.

(iv) From plane, project the points to V.P. to get the final elevation 1-2-3′-4′-5′. This is shown by dark lines.

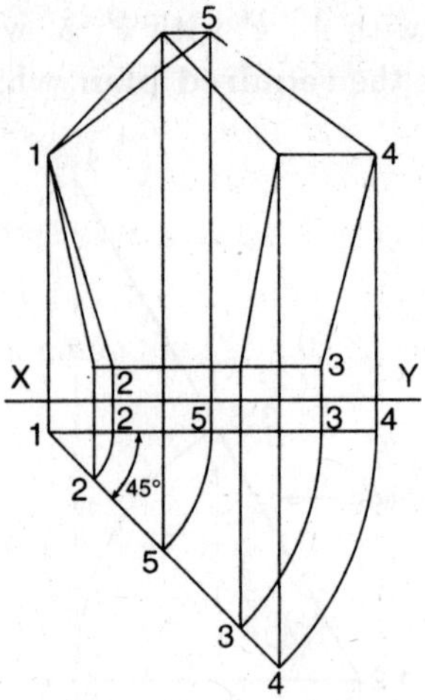

Fig. 10-12

6. A rhombus has diagonals 120 mm and 50 mm. long. Smaller diagonal is perpendicular to V.P., and parallel to H.P., longer diagonal is making an angle of 30° to H.P. Draw its elevation and plan.

Solution : Refer Fig. 10-13.

(i) It is to be noted that, the diagonals of a rhombus are at right angles (making 90° angle) to each other in all cases.

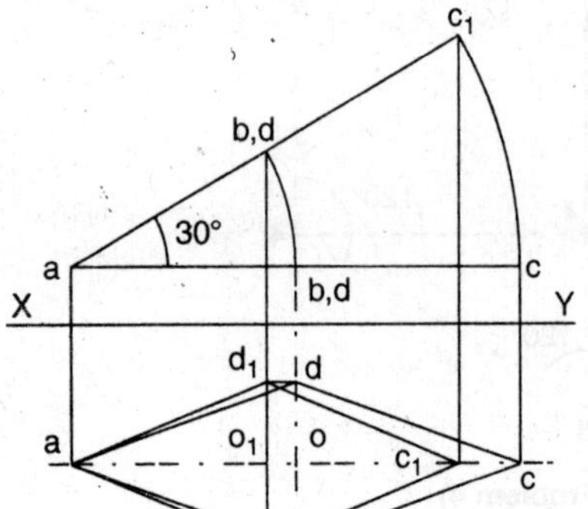

Fig. 10-13

(ii) Hence, the diagonals ac = 120 mm (parallel to V.P.) and bd = 50 mm (perpendicular to V.P.) are drawn at right angles to each other and also bisecting each other at the point O. This is shown in the plan view of Fig. 10-13.

(iii) Projecting from the points a, b, c, d in the plan, we get the elevation $a, b\text{-}d, c$, or $a\text{-}c$, which is a straight line, as the plane is parallel to H.P. [Fig. 10-13]

(iv) Draw a line ac_1 (in the elevation) making an angle of 30° to the horizontal.

(v) With centre at 'a' and radius ac, draw an arc to meet the line ac_1 at the point c_1.

(vi) With centre at 'a' and radius ab (or ad), draw an arc to meet the line ac_1 (in elevation) at the point b or d.

(vii) Projecting from the point b (or d) and c_1 (elevation), we get the points b_1, c_1 and d_1 in the plan.

(viii) Join 'a' with b_1, b_1 with c_1, c_1 with d_1 and d_1 with a to get the required plan $a\text{-}b_1\text{-}c_1\text{-}d_1$.

7. The diameter AB of a circle (70 mm diameter) is making angles of 30° and 60° with H.P. and V.P. respectively. Draw elevation and plan. (same as Prob No. 1).

8. A regular hexagonal lamina, each side 30 mm, rests with one of its side on H.P. in such a manner that this side is perpendicular to V.P. The surface of the lamina is inclined at 45° to H.P. and perpendicular to V.P. Draw elevation and plan.

9. A square lamina, each side 50 mm, is placed with one of its side on H.P., so that this side is inclined at 30° to V.P. The surface of the lamina is inclined at 45° to H.P. Draw elevation and plan.

Solution : Refer Fig. 10-14.

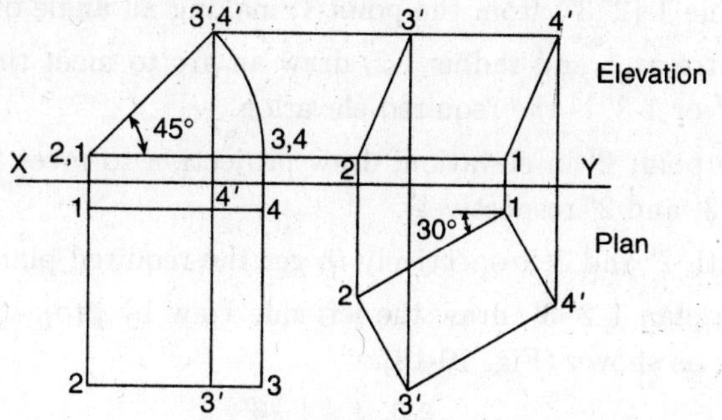

Fig. 10-14

10. A regular pentagonal lamina, each side 40 mm, is resting with one of its sides, on H.P., so that this side is perpendicular to V.P. The surface or plane of the lamina is inclined at 60° to H.P. and perpendicular to V.P. Draw elevation, plan and left-side view.

Solution : Refer Fig. 10-15.

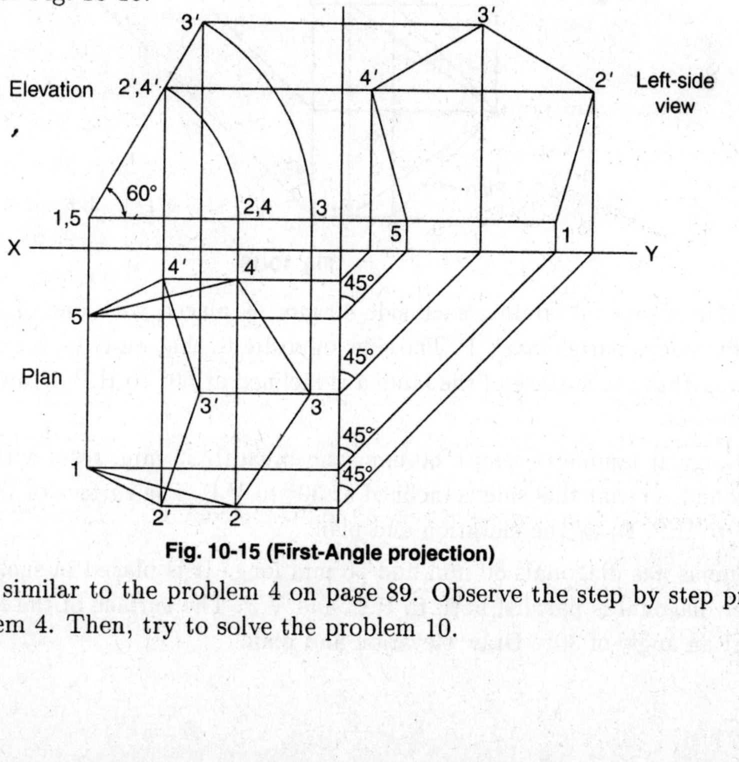

Fig. 10-15 (First-Angle projection)

This problem is similar to the problem 4 on page 89. Observe the step by step procedures in solving problem 4. Then, try to solve the problem 10.

11. An equilateral triangular lamina, each side 60 mm, is resting with one of its corners on H.P., so that the side, opposite to this corner, is parallel to H.P. and perpendicular to V.P. Draw elevation, plan and left-side view, when the surface or plane of the lamina is inclined at 45° to H.P. and perpendicular to V.P.

Solution : Refer Fig. 10-16. The steps are as follows :

(i) Draw the plan of the equilateral triangle (1-2-3), when it is parallel to H.P. and the side 2-3 is perpendicular to V.P.

(ii) Project from the points 2 and 3 in the plan to get the elevation 1-3 or 1-(2,3).

(iii) Draw a line 1-(2′,3′) from the point 1, making an angle of 45° to the horizontal.

(iv) With centre at 1 and radius 1-2, draw an arc to meet the line 1-2′ at the point 2′. Now, 1-2′ or 1-3′ is the required elevation.

(v) From the point 2′ in elevation, draw projection to meet the horizontal lines from 3 and 2 at 3′ and 2′ respectively.

(vi) Join 1 with 2′ and 3′ respectively to get the required plan 1-2′-3′.

(vii) From the plan 1-2′-3′, draw the left-side view by projecting from the plan and the elevation, as shown (Fig. 10-16).

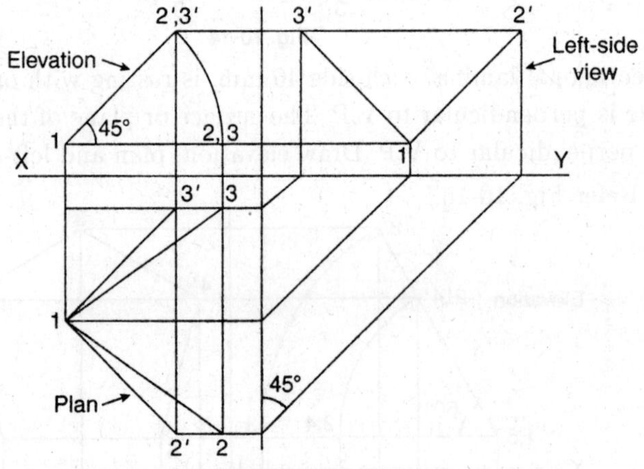

Fig. 10-16

12. A regular hexagonal lamina, each side 40 mm, is placed with one of its sides on H.P., so that this side is parallel to V.P. The side, opposite to that on H.P., touches the V.P. in such a manner that the surface of the lamina is inclined at 60° to H.P. Draw elevation, plan and left-side view.

13. A rectangular lamina of length 60 mm, and breadth 35 mm, rests with its 35 mm side on the ground, so that this side is inclined at 30° to V.P. The surface of the lamina is inclined at 45° to H.P. Draw the elevation and plan.

14. A rhombus has diagonals 80 mm and 50 mm long. It is placed in such a manner that the smaller diagonal is parallel both to H.P. and V.P. The surface of the lamina is inclined to H.P. at an angle of 30°. Draw elevation and plan.

15. Objective Questions :

Fill in the blanks with proper words :

(i) When a lamina is perpendicular to a principal plane (H.P. or V.P.), the projection of the lamina on that plane is a ______.

(ii) If a lamina is parallel to a principal plane (H.P or V.P.), the projection of the lamina on that plane is the ______ ______ of the lamina.

(iii) When a lamina is inclined at some angle, other than 90°, to a principal plane (H.P. or V.P.), the projection of the lamina on that plane is the ______ ______ of the lamina.

(iv) When a lamina is perpendicular to H.P., the lamina may or may not be perpendicular to ______.

(v) When a lamina is perpendicular to both H.P. and V.P., the true shape of the lamina will be obtained from ______.

(vi) When a lamina is inclined at some angle, other than 90°, to both H.P. and V.P. we cannot get ______ ______ of the lamina in elevation, plan or side-view.

Ans. (i) Straight line, (ii) True shape (or form), i.e., equal in size of the actual lamina, (iii) Apparent shape (or form), which is reduced in size of the actual lamina, (iv) V.P., (v) Side-view or end-elevation or end-view, (vi) True shape (or form).

16. A hexagonal plane *ABCDEF*, of side 30 mm each, is resting on the corner *A* in V.P. with its surface making an angle of 30° with V.P. The front view of the diagonal *AD* is inclined at 40° to *XY*. Draw top view and front view of the plane. **(AMIE—Winter, 2002)**

17. Answer briefly :

If the front-view of an equilateral triangle is an isosceles triangle, then whether the plane of the triangle is inclined to V.P. or to H.P.? **(AMIE—Winter, 2002)**

Ans. The plane of the equilateral triangle is inclined to V.P. with its base on the ground (i.e., H.P.). Only when the equilateral triangle is parallel to V.P., then the front-view is an equilateral triangle. In this case, the base is on the ground or H.P.

18. Choose the correct answer :

(i) A plane perpendicular to V.P. and parallel to H.P.

(a) It has horizontal trace (b) It has no horizontal trace

(c) Its horizontal trace coincides with vertical projection. **(Winter, 2006)**

Ans. (b)

19. A regular octagonal lamina (each side 30 mm) is placed with one of its side on H.P., so that the plane of the lamina is perpendicular both to H.P. and V.P. State in which plane (H.P., V.P., Profile plane) the true shape of the lamina can be obtained.

Ans. Profile plane.

Chapter 11

Projection of Regular Solids

11.1 Introduction

The idea of the projection of solid is based on the conception about the projection of points, lines and planes, taken together. We know that, a solid has length, breadth and thickness. Without going into the crystallographic classification of all kinds of solids, we may think of the following thirteen simple solids for our discussion here.

In this chapter, only regular or right solids will be discussed for simplicity.

11.2 Simple Classification of Solids

(i) **Tetrahedron**, i.e., equilateral triangular pyramid, bounded by four equilateral triangles,

(ii) **Cube**, which is a square prism of same length, breadth and thickness,

(iii) **Right Prism**—(a) triangular (b) square (c) pentagonal and (d) hexagonal prisms,

(iv) **Right Pyramid**—(a) triangular (b) square (c) pentagonal and (d) hexagonal pyramids,

(v) **Right circular cylinder**,

(vi) **Right circular cone**,

(vii) **Sphere**.

The cylinder, cone and sphere are also known as **solids of revolution**. Because these regular solids may be obtained by revolving a **generator**, which may be a straight line or a curved line. It should be noted that, for **right solids** (prism, pyramid, cylinder, cone), the axis is **perpendicular** to the base or end of the solid, and the axis will also pass through the centre of the base. When the axis is not perpendicular to the base (or end) of the solid, it is called an **irregular solid**.

11.3 Preparation of the Models of Solids

The following instructions will be helpful :

(a) Take a single stiff (not glossy) drawing sheet, which will be sufficient.

(b) Draw the following figures on the drawing sheet. [Fig. 11-1(i) to (vi)]

(c) Cut the paper-drawings with a sharp blade or knife and fold along the lines.

(d) Paste the 5 mm-edges of the models with glue or fevicol. You will get the shape of above solids, as discussed in chapter 10.

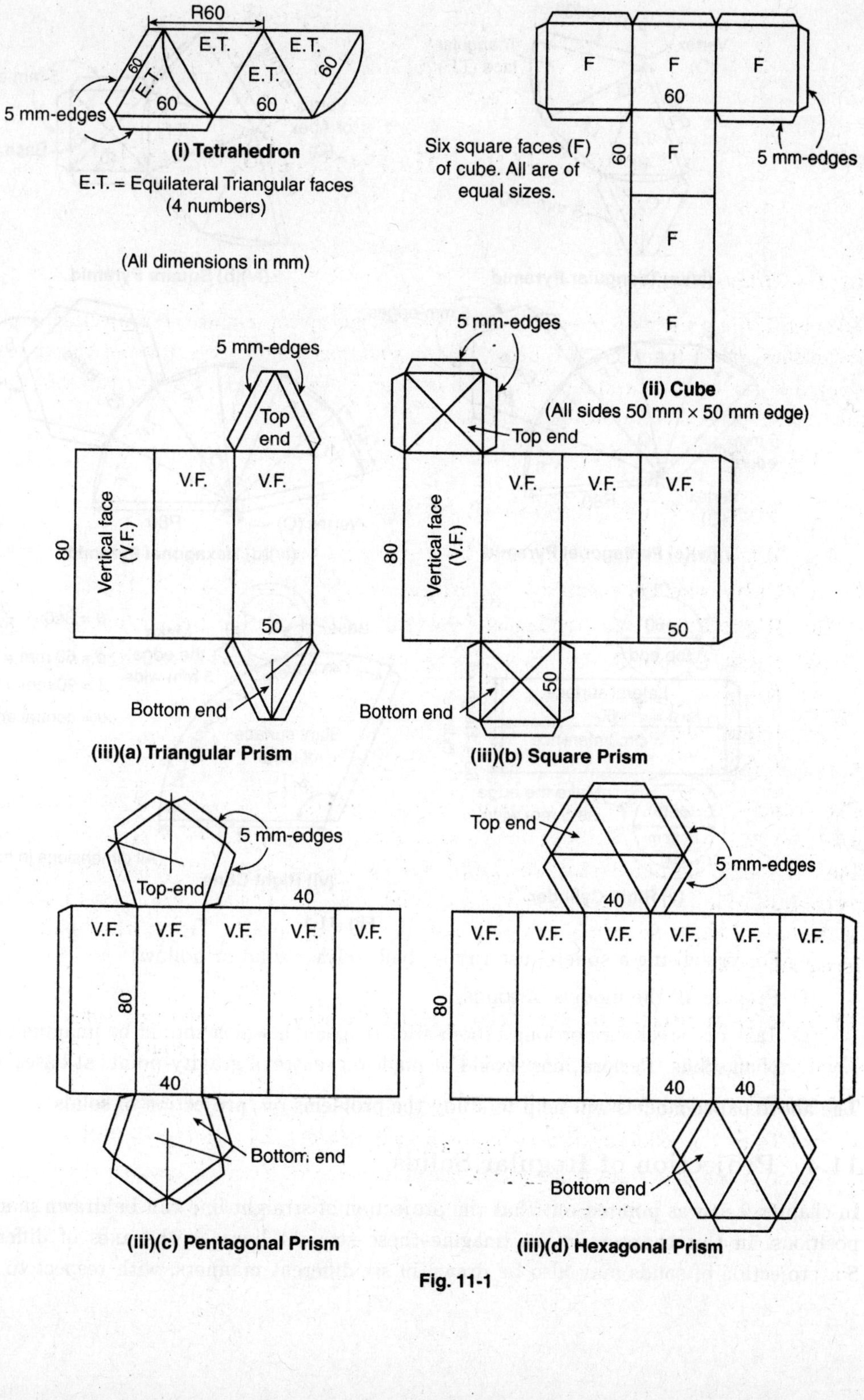
R60
E.T.
E.T.
E.T.
E.T.
60
60
60
60
5 mm-edges
(i) Tetrahedron
E.T. = Equilateral Triangular faces (4 numbers)
(All dimensions in mm)
F
F
F
60
Six square faces (F) of cube. All are of equal sizes.
60
F
5 mm-edges
F
F
(ii) Cube
(All sides 50 mm × 50 mm edge)
5 mm-edges
Top end
V.F.
V.F.
80
Vertical face (V.F.)
50
Bottom end
(iii)(a) Triangular Prism
5 mm-edges
Top end
V.F.
V.F.
V.F.
80
Vertical face (V.F.)
50
50
Bottom end
(iii)(b) Square Prism
5 mm-edges
Top-end
40
V.F.
V.F.
V.F.
V.F.
V.F.
80
40
Bottom end
(iii)(c) Pentagonal Prism
Top end
5 mm-edges
40
V.F.
V.F.
V.F.
V.F.
V.F.
V.F.
80
40
40
Bottom end
(iii)(d) Hexagonal Prism

Fig. 11-1

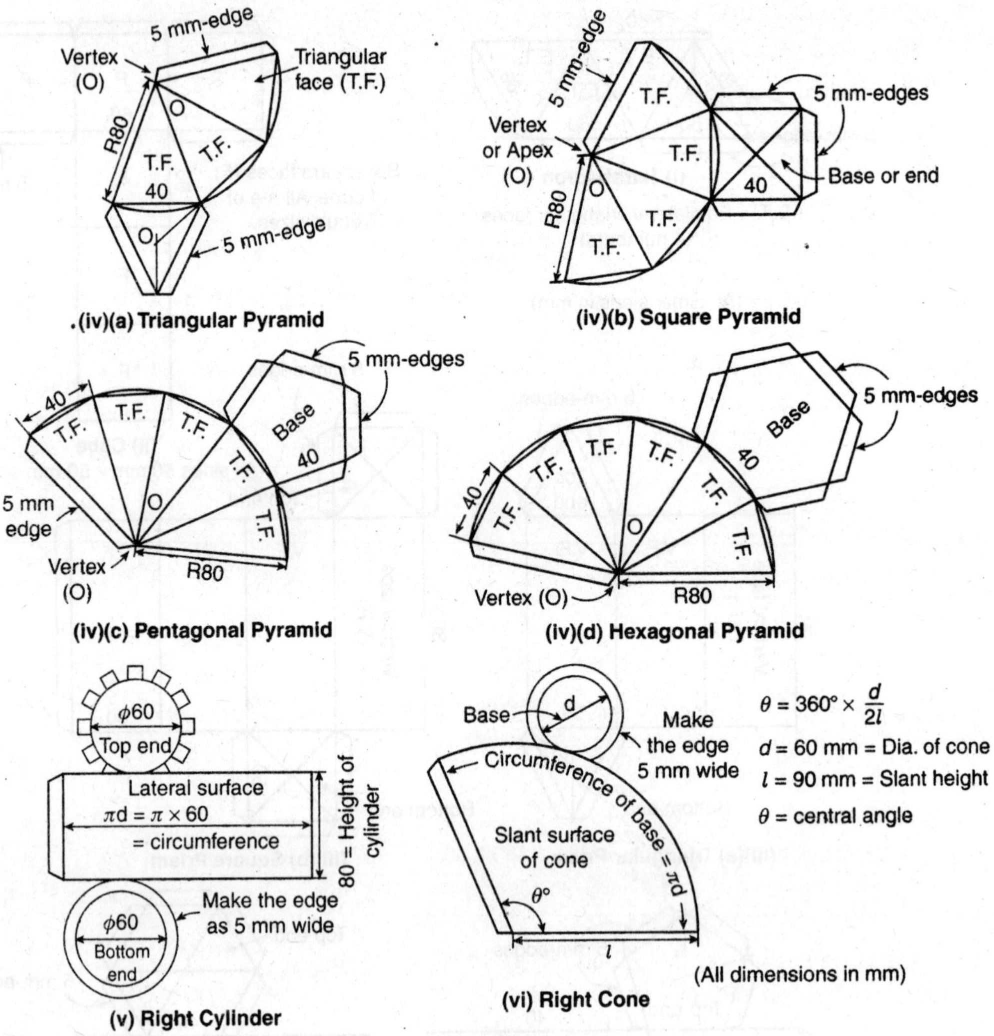

Fig. 11-1

(e) For visualising a sphere, use rubber balls (either solid or hollow)

(f) Prepare all the models of solids.

(g) Take one steel-wire or long broom-stick (coconut), which should be imagined as the axes of all solids. Perforations should be made on centre of gravity-points at bases of all solids.

The above paper-models will help to study the problems on 'projection of solids'.

11.4 Projection of Regular Solids

In chapter 9, it was pointed out that the projection of straight line can be drawn in six different positions. In this chapter, let us imagine those straight lines as the axes of different solids. So, projection of solids may also be drawn in six different manners, with respect to their axes.

Thus, the **position of axis** may be—

(i) Perpendicular to H.P. and parallel to V.P.

(ii) Perpendicular to V.P. and parallel to H.P.

(iii) Parallel to both H.P and V.P.

(iv) Parallel to V.P. and inclined to H.P. (Less then 90°)

(v) Parallel to H.P. and inclined to V.P. (Less then 90°)

(vi) Inclined to both V.P. and H.P. (Less then 90°)

In order to draw projection of solids in the above six positions, let us select only one solid, i.e., square prism. When problems with this solid will be clear to us, then we will be in a position to solve problems on other solids.

We have chosen the square prism, because the shape of all solids can be imagined from the shape of a square prism, excepting a sphere. With the help of model, quadrant board, conception of points, lines and planes, the projection of a square prism should be practised by the students several times in order to have a clear idea.

11.5 The Axis Perpendicular to H.P. and Parallel to V.P.

(a) Take the model of square prism, broom-stick and quadrant board.

(b) Insert the broom-stick through the centres of top and bottom square ends of the prism.

(c) Place the stick parallel to V.P. and perpendicular to H.P. as shown in Fig. 11-2. But in the above condition, the solid may be rotated about its axis, as any position of solid within 360° will satisfy the condition. So, another condition is required to fix the solid at a particular position.

(d) Some of the fixing conditions of the solids may be as follows :

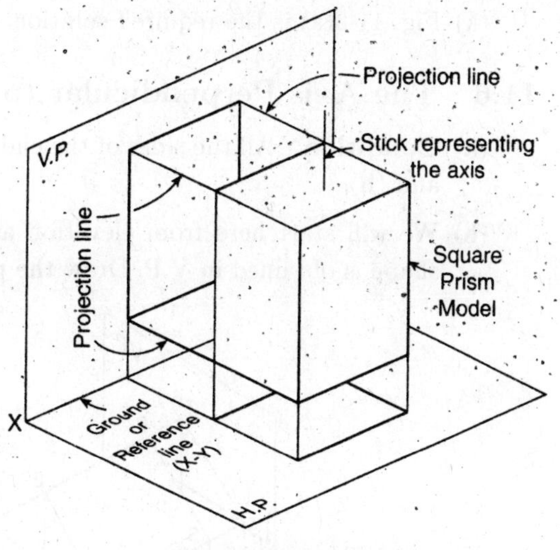

Fig. 11-2 (First-Angle)

(i) One of the side of base is parallel to V.P. or one of the rectangular face is parallel to V.P.

(ii) One of the side of base is perpendicular to V.P. or subtends 30°, 60°, 45° or any other angle with V.P.

(iii) All the sides of the base are **equally inclined** to V.P., i.e., all sides make 45° angle with V.P.

(e) Let us imagine this last condition as shown in Fig. 11-3(a) and (b).

(f) Draw *XY* line.

(g) Draw a square, one of the sides subtending 45° with XY. We are drawing the plan first, as we get the true shape in this view (for every problem, we shall start from the view (Elevation or Plan) where we will get the true shape. That will be clear from model and quadrant board). Complicated condition will be discussed in proper place.

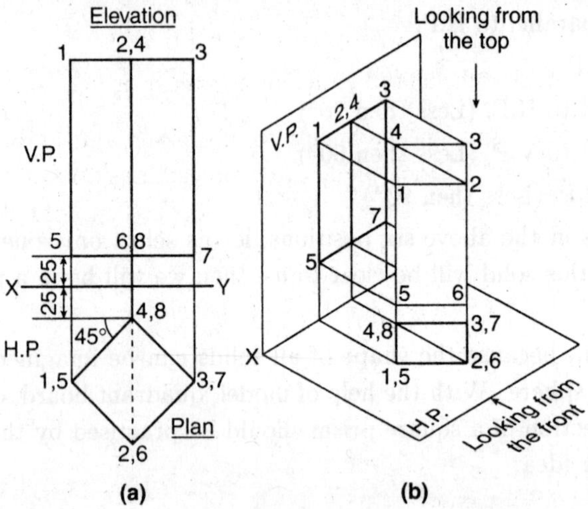

Fig. 11-3 (First-Angle)

(h) Number the corners as 1, 2, 3 and 4 (top end); 5, 6, 7 and 8 (bottom end).

(i) Draw all the projectors from plan to meet XY and extend to a certain distance in V.P. Complete the rectangle 1-3-7-5, which is the elevation.

(j) Number the respective corners in order (1, 2, 3, 4 and 5, 6, 7, 8).

(k) Fig. 11-3(a) is the required solution. (First-Angle projection)

11.6 The Axis Perpendicular to V.P. and Parallel to H.P.

(a) **Condition :** All the sides of the end are **equally inclined** to the H.P. as in Fig. 11-4(a) and (b).

(b) We will start here from elevation and complete it with previous instructions, as true shape is obtained in V.P. Draw the plan. See Fig. 11-4(b).

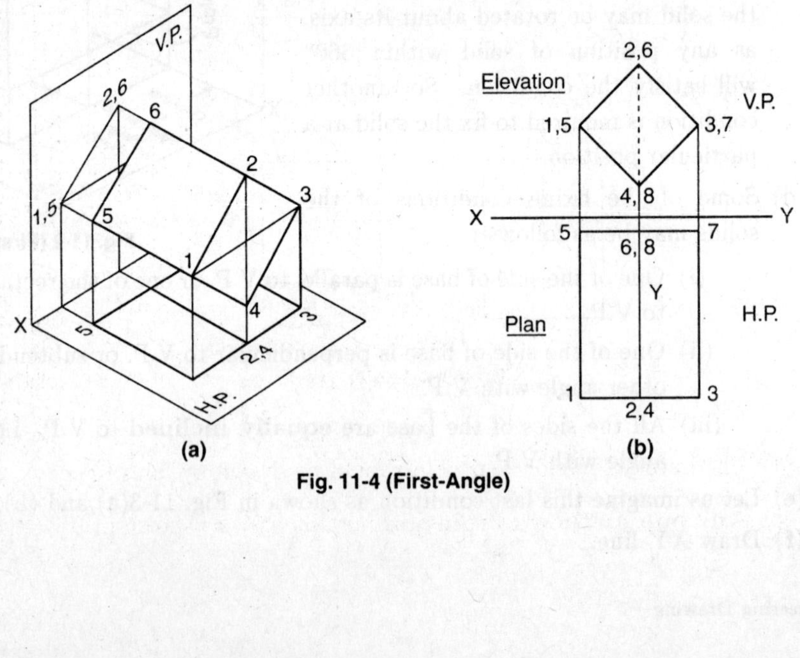

Fig. 11-4 (First-Angle)

11.7 Axis Parallel to Both H.P. and V.P.

(a) All the sides of the base are **equally inclined** to the V.P. and H.P., [Isometric view Fig. 11-5(a)].

(b) It is clear that we will not get the true shape either in elevation or in plan.

(c) So, at first we will have to draw axis perpendicular to any plane and parallel to other. Then rotate the plan through 90°. From plan, draw projections to get the required elevation (Fig. 11-5 b). Other method is shown in Fig. 11-5(c). The final elevation and plan should be drawn as dark lines.

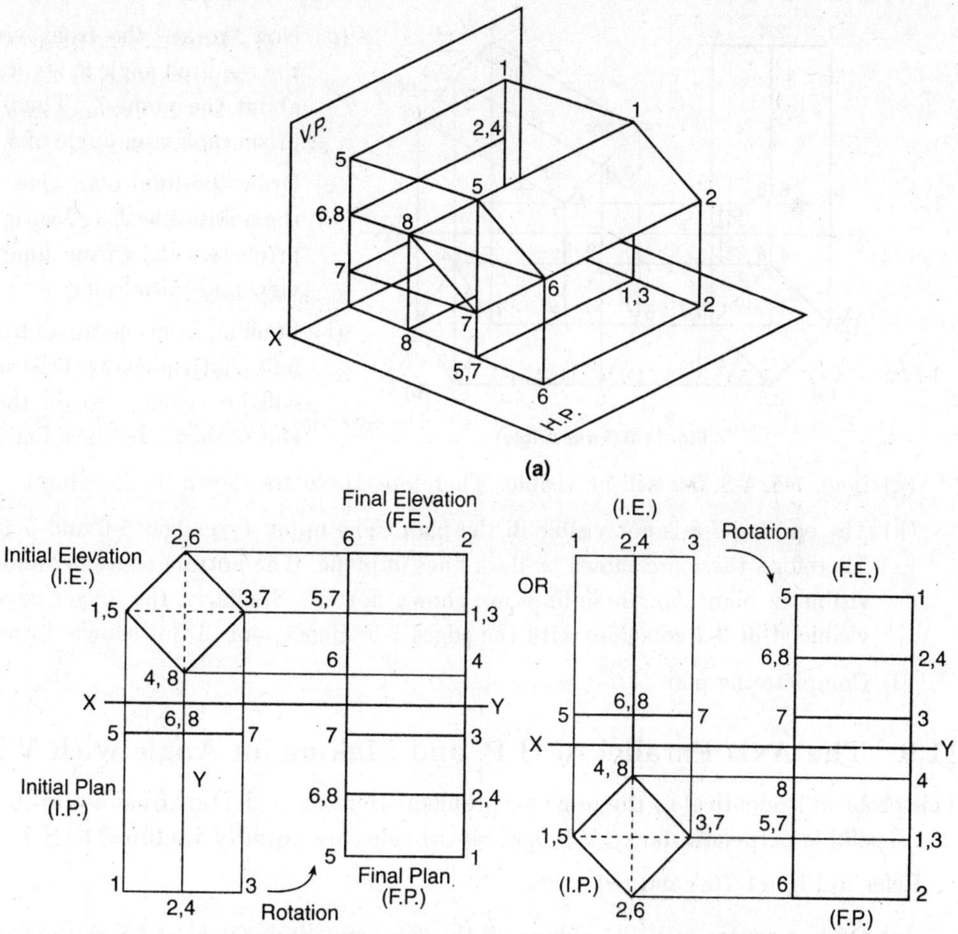

(a)

(b) (First-Angle)

(c) (First-Angle)

Fig. 11-5

(d) The initial elevation and plan should be drawn by faint lines.

(e) **Any one method of solution** may be used as shown in Fig. 11-5(b) and (c).

11.8 The Axis Parallel to V.P. and Making an Angle with H.P.

Condition : The prism is resting with one corner in space, so that this corner is 20 mm above H.P.

We will have to start by placing the axis perpendicular to H.P. and parallel to V.P. Thus, the plan and elevation of Fig. 11-5 (c) are drawn first with faint lines, as shown in Fig. 11-6 (a).

(a) In the plan, draw a square such that all the sides of the base are equally inclined to H.P. (i.e., at 45°). See Fig. 11-6(a).

(b) Draw projections from plan to get the elevation.

(c) Draw the front view and number all the corners as 1, 2, 3, 4, etc.

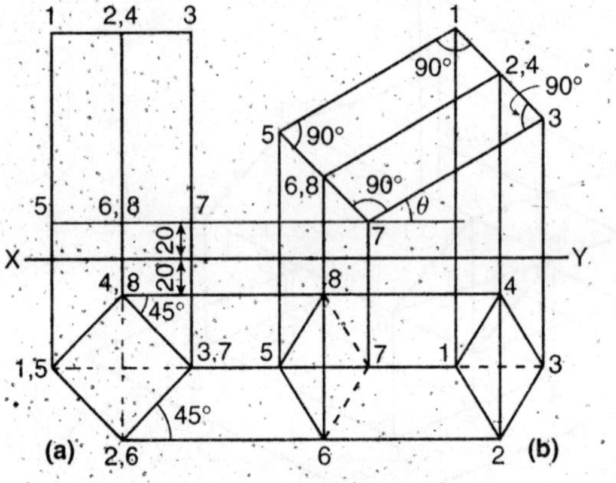

Fig. 11-6 (First-Angle)

(d) Now, rotate the front view through the required angle θ, i.e., to be rotated about the point 7. Thus, axis of the prism makes an angle of θ to the H.P.

(e) Draw the final plan view by meeting the points 1 with 1, 2 with 2, etc. with projection lines from final elevational view and initial plan.

(f) Looking from the top of front view, we find that the end consisting of 1,2,3,4 will be visible. So, in the plan, this end is shown by dark lines.

(g) Line, 1-5, 4-8, 6-2 will be visible. Therefore, these are shown by dark lines.

(h) The end 5-6-7-8 is not visible in the plan. The upper two edges 5-6 and 5-8 are visible. Therefore, these are shown by dark lines in plane. The bottom edges 6-7 and 7-8 are not visible in plan. So, these lines are shown dotted. Similarly, the longer edge 3-7 is not visible. But 3-7 coincides with the edges 1-5. Hence, only 1-3 is shown dotted.

(i) Complete the plan.

11.9 The Axis Parallel to H.P. and Making an Angle with V.P.

This problem is identical to the previous problem. Here we shall start from a position, where the axis of solid is perpendicular to V.P. and all the sides are **equally inclined** to H.P., i.e., at 45°.

Refer to Fig. 11-7(a) and (b).

(a) Draw a square in V.P., making all the sides equally inclined at 45° with *X-Y* line. This is the elevation. Fig. 11-7(a).

(b) Draw the top view, after projecting from elevation, Fig. 11-7(b).

(c) Mark all the corners 1, 2, 3, 4 and 5, 6, 7, 8 in elevation and plan Fig. 11-7(a) and (b).

(d) Now rotate the top-view about the edge 2-6 up to the specified angle, say $\theta°$, Fig. 11-7(c).

(e) From the final plan and initial elevation, project all the points as 1 with 1, 2 with 2 and so on. We will get the final elevation, Fig. 11-7(d).

(f) Looking from the front, the end (1, 2, 3, 4) is visible. So, the lines 1-2, 2-3, 3-4, 4-1 are shown as deep lines or firm lines. [Fig. 11-7(d)]

(g) Lines 1-5, 2-6, 4-8, 5-6, 5-8 will be visible. The end 5-6-7-8 will not be visible. The sides 6-7 and 7-8 are not visible and these are shown as dotted. Complete the final elevation. See Fig. 11-7(d).

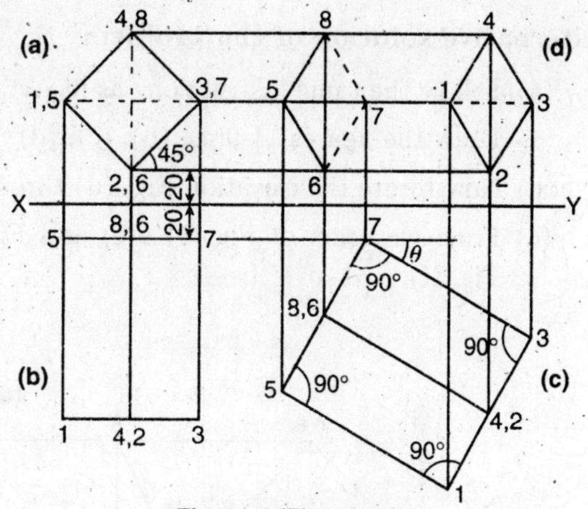

Fig. 11-7 (First-Angle)

11.10 Axis Making Angles with Both V.P. and H.P.

We will solve the problem with the help of both the previous examples. Art 11.5(iv) and (v).

(a) Follow the same instructions as in Art 11.5(iv).

Draw the figures 11-8(a), (b), (c) and (d).

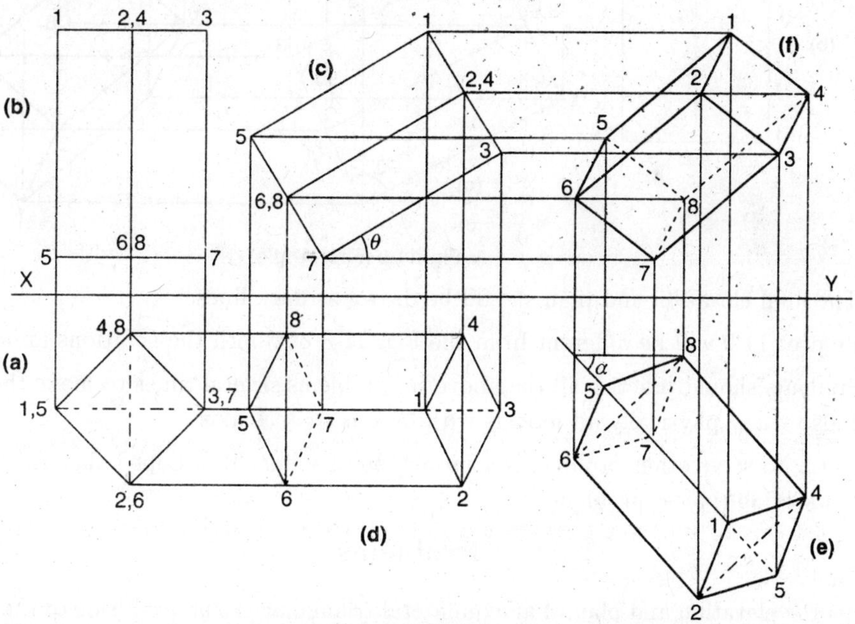

Fig. 11-8 (First-Angle)

(b) Now rotate the plan through the given angle, say α, with V.P. as in Fig. 11-8(e).

(c) From plan of Fig. 11-8(e), draw projections to get the final elevation Fig. 11-8(f), after projecting from previous elevation Fig. 11-8(c).

(d) The final elevation and plan should be drawn as dark lines. See Fig. 11-8(f) and (e).

Alternative solution of the problem

(a) Follow the same instructions as in Art. 11.5(v).

Draw the figures 11-9(a), (b), (c), (d).

(b) Now rotate the elevation Fig. (d) through the given angle, say θ, as in Fig. 11-9(e).

(c) From elevation of Fig. 11-9(e) and Fig. (c) draw projections to get the final plan at Fig. (f).

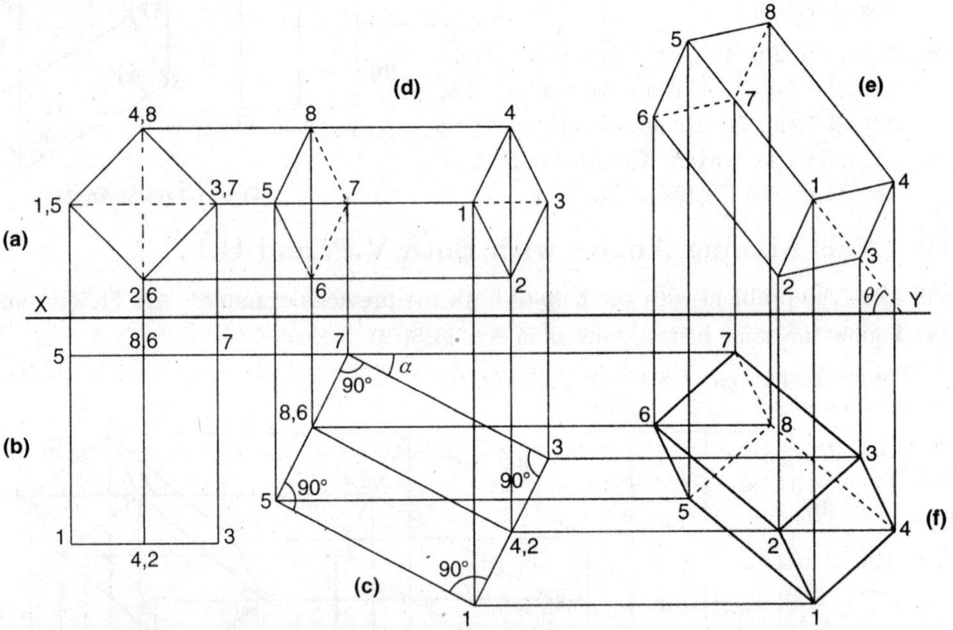

Fig. 11-9 (First-Angle)

(d) The final elevation and plan should be drawn as dark lines.

Figures 11-9 will be different from the Fig. 11-8 but both the solutions are correct.

Note : Students should practise all the above six problems several times to make the conception clear and also study physically all models with broom-stick as axis.

Then they try to solve other problems on projection of solids. It is hoped that they will find no difficulty in solving typical problems.

Problems

1. Draw the elevation and plan of an equilateral triangular prism, each side of the base 30 mm and axis 50 mm long, resting on one of its ends on the ground with a vertical face parallel to V.P.

Solution : Refer to Fig. 11-10(a) and (b).

Here, we can draw elevation or plan first, as we will get true shape in both views. Let us start with plan.

In Fig. 11-10(a), the longer edge points towards V.P. Thus the line 3-6 in elevation is shown dotted. Fig. 11-10(b).

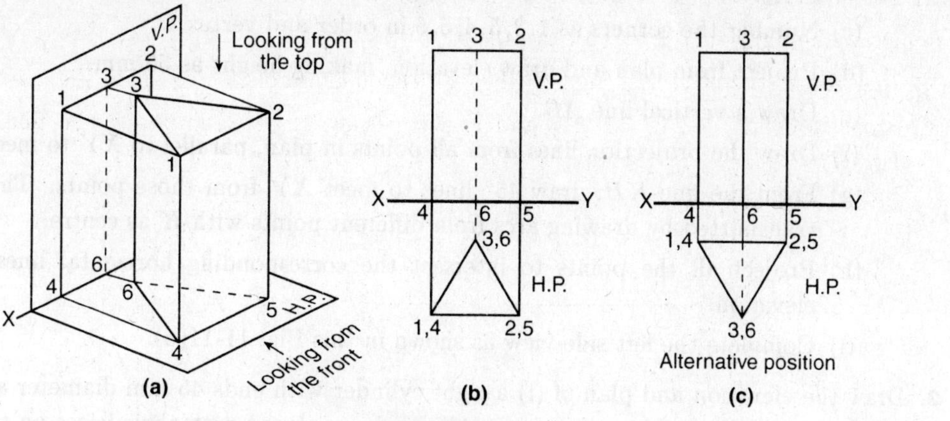

Fig. 11-10

The triangular prism may also be placed with a rectangular face facing V.P., as shown in Fig. 11-10(c). In this case, the line 3-6 in elevation is shown as firm line.

Note : As the solid is placed on the ground, so the elevation coincides with XY, which is the meeting line of V.P. and H.P.

2. Draw the projections (elevation, plan and left-side view) of a regular hexagonal pyramid, each side of the base-side 25 mm and axis 55 mm long, having its base 10 mm above the ground and one of the side of the base is perpendicular to V.P.

Solution : Refer to Fig. 11-11(a) and (b). The drawing steps are as follows :

(a) We will start from plan, as true shape will be available there. Draw X-Y line Fig. 11-11(b).

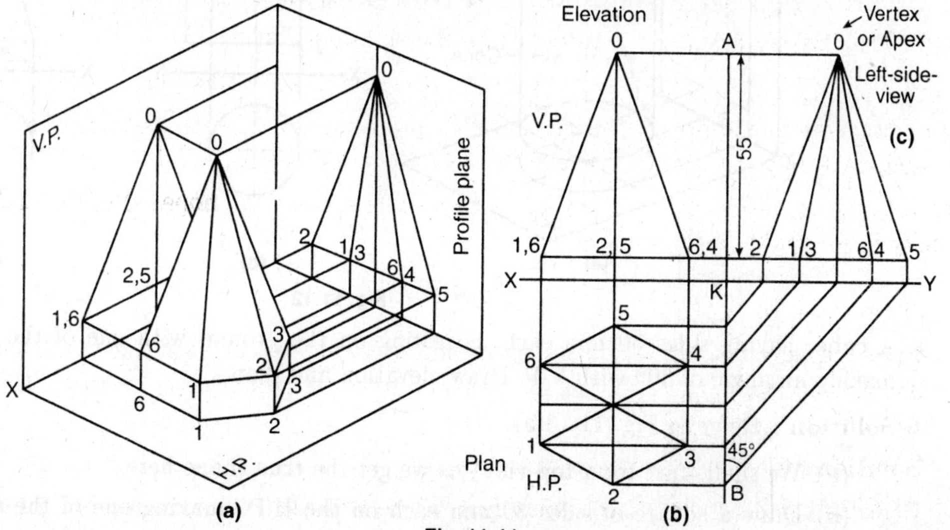

Fig. 11-11

(b) Draw a regular hexagon of sides 25 mm each making one of the sides of base perpendicular to XY [Fig. 11-11(b)].

(c) Number the corners as $1, 2, 3, 4, 5, 6$ in order and vertex O.

(d) Project from plan and draw elevation, making height as 55 mm.

(e) Draw a vertical line AB.

(f) Draw the projection lines from all points in plan, parallel to XY to meet at KB.

(g) From the line KB, draw 45° lines to meet XY from those points. These can be transmitted by drawing arcs from different points with K as centre.

(h) Project all the points to intersect the corresponding horizontal lines from the elevation.

(i) Complete the left side-view as shown in this Fig. 11-11(b).

3. Draw the elevation and plan of (i) a right cylinder with ends 45 mm diameter and axis 60 mm long and (ii) a right cone of same dimensions, placed with their bases on the ground, in both cases.

Solution :

(a) Here we can start from plan or elevation, as in both views we will get the true shape.

(b) Draw the views as shown in Fig. 11-12(b) and (c).

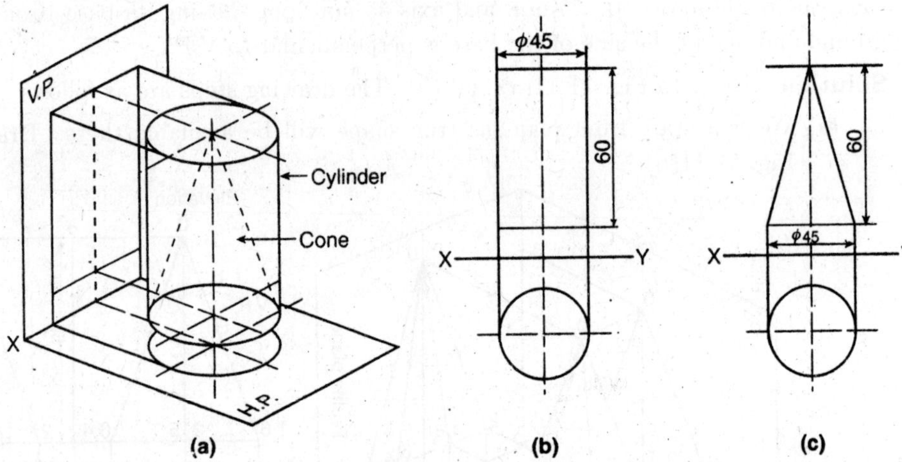

Fig. 11-12

4. A cube, having sides 50 mm each, is resting on the ground with one of the square-face making an angle of 30° with V.P. Draw elevation and plan.

Solution : Refer to Fig. 11-13(a).

(a) We shall start from top view, as we get the true shape here.

(b) Draw a square of sides 50 mm each on the H.P., making one of the sides at 30° with V.P. [Fig. 11-13(b)].

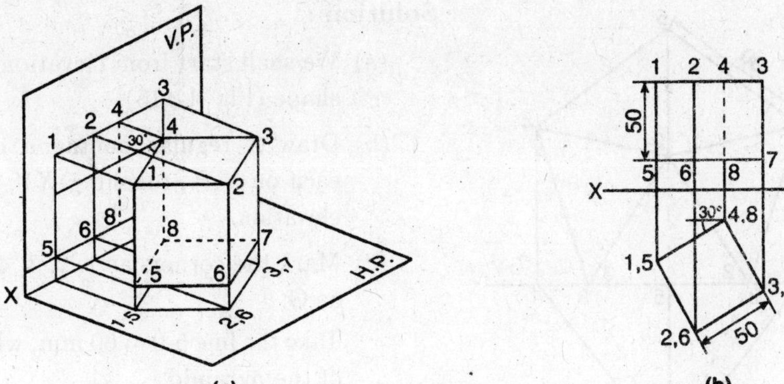

(a) (b)

Fig. 11-13

(c) Complete the front view, and mark the points. Note that the line 4-8 is shown dotted, indicating that the edge 4-8 is not visible when looking from front.

5. An equilateral triangular prism of sides 25 mm each and the length of axis 50 mm, has one of its rectangular face 10 mm above the ground. Axis is perpendicular to V.P. and nearer end 10 mm in front of V.P. Draw elevation and plan.

Solution : Refer to Figs. 11-14(a) and (b).

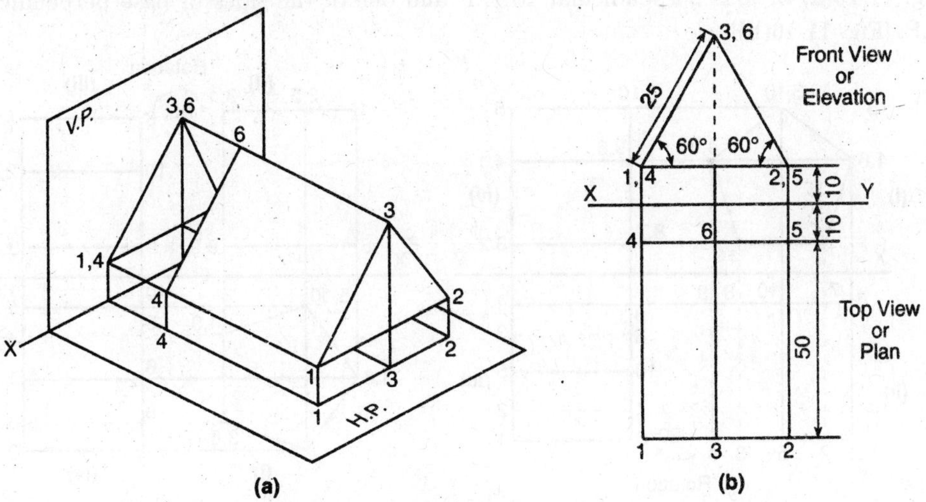

(a) (b)

Fig. 11-14

(a) We shall start from front view, as it is the true shape. Draw an equilateral triangle with sides 25 mm each, with the base parallel to X-Y line. Fig. 11-14(b)

(b) From the elevation, draw projections to get the plan. Mark all the points. See Fig. 11-14(b).

6. A right pentagonal pyramid, side of the base 30 mm each and axis 60 mm long, has its base in the V.P. One of the base-side is on the XY line. Draw elevation and plan.

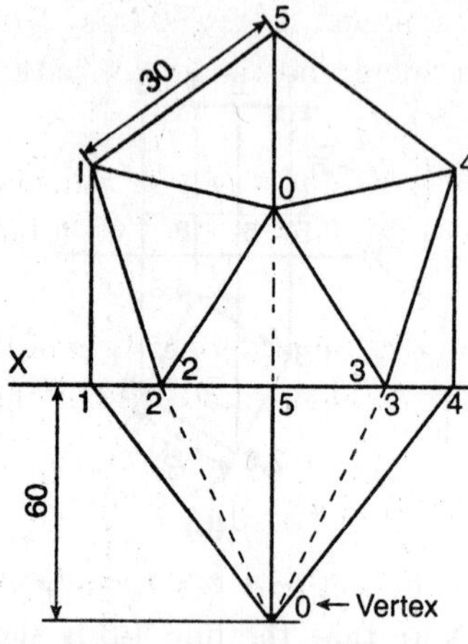

Fig. 11-15 (First-Angle)

Solution :

(a) We shall start from elevation, as it is the true shape (Fig. 11-15).

(b) Draw a regular pentagon of sides 30 mm each on V.P., touching *XY* line. This is the elevation.

(c) Mark the corners as 1, 2, 3, 4 and 5 and apex as O.

Take the line 5-0 = 60 mm, which is the height of the pyramid.

(d) Draw the projections and complete the plan. The edges 2-0 and 3-0 are shown dotted, because these are not visible from top.

7. A right pentagonal prism, base-sides 30 mm each and height 60 mm, is resting on the ground with one of its rectangular faces. The axis is parallel to both V.P. and H.P. Draw its projections.

Solution : Place the prism-axis perpendicular to V.P. and rectangular face on the ground Fig. 11-16(a) or axis perpendicular to H.P and one of the sides of base perpendicular to V.P. [Fig. 11-16(b)]

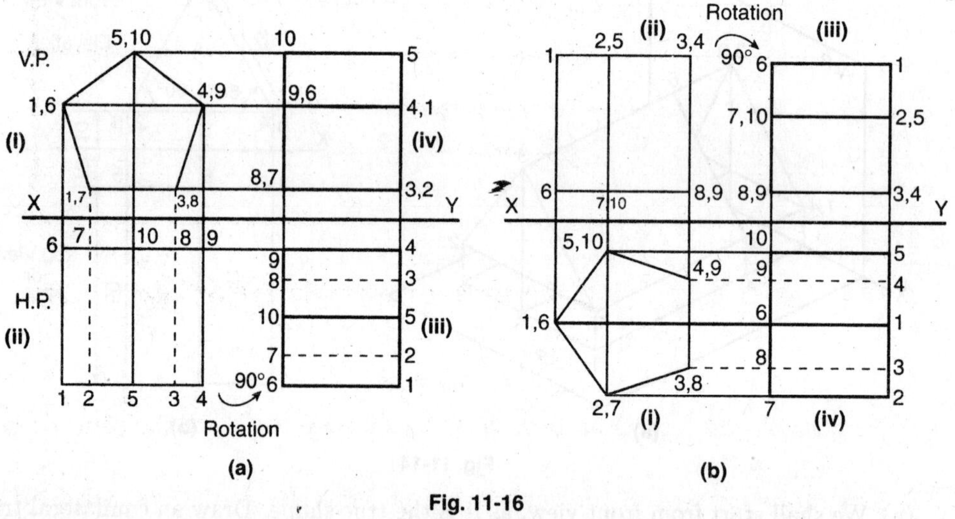

Fig. 11-16

In Fig. 11-16(a), draw the front view as in Fig. (i). Next, draw the plan as in Fig. (ii). Rotate the plan through 90°, as in Fig. (iii). Finally, draw the elevation as in Fig. (iv). For the position shown in Fig. 11-16(b), the axis is perpendicular to H.P. Draw the plan [Fig. (i)]. Next, draw the elevation [Fig. (ii)]. Rotate the elevation through 90° [Fig. (iii)]. Finally, draw the plan [Fig. (iv)].

8. A right hexagonal pyramid is having each side of the base 20 mm and height 60 mm. Draw elevation and plan, when one of the sides of base is on the ground and the axis is parallel to both V.P. and H.P.

9. A right square pyramid is having base of 45 mm each side and is 60 mm in height. All edges of the base are equally inclined to the H.P. and axis parallel to and 60 mm away from both H.P. and V.P. Draw elevation and plan.

10. A regular pentagonal pyramid has base-sides 20 mm each and axis 55 mm long. One of the edges of base is on H.P. The axis of the pyramid is inclined at an angle of 30° to H.P. and parallel to V.P. Draw elevation and plan.

Solution : Drawing steps are shown in the order of (i), (ii), (iii) and (iv).

(i) Draw a regular pentagon on H.P., so that one of the sides of base is perpendicular to V.P. This is the plan. Fig. 11-17(a)

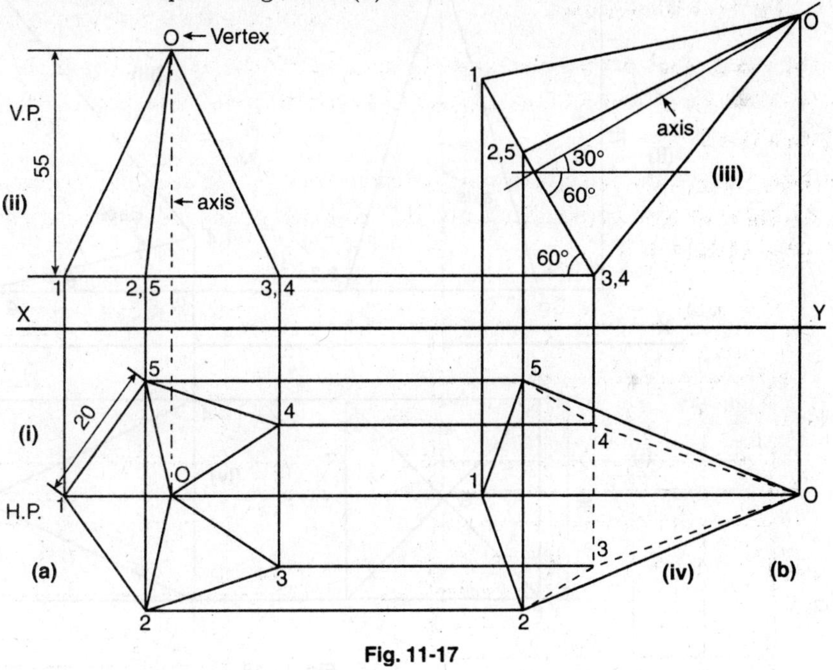

Fig. 11-17

(ii) Number all the corners as 1, 2, 3, 4, 5 and apex O. From the plan, draw the elevation.

(iii) Now rotate the base of front surface through 60° such that the axis makes 30° with H.P.

(iv) Complete the front view.

(v) Draw the plan, looking from the top of the elevation.

(vi) The base-edges 1-5, 1-2 and longer edges 5-0, 1-0, 2-0 are visible, and therefore, shown by firm lines. The base-edges 2-3, 3-4, 4-5 and longer edges 3-0 and 4-0 are shown dotted, as these are not visible from the top.

11. A right square pyramid, having sides of base 30 mm each and height 75 mm, is resting with one of the sides of base on H.P., so that this side is perpendicular to V.P. The triangular face, containing the side of base on H.P., is vertical. Draw elevation and plan. Measure the inclination of the base with H.P.

Solution : Drawing steps are shown in the order of (i), (ii), (iii) and (iv).

(i) Draw initially the plan and elevation of the pyramid with axis perpendicular to H.P., i.e., base on the ground and one of the sides of base is perpendicular to V.P., as shown in Fig. 11-18(i) and (ii).

(ii) A second elevation is drawn, so that the triangular face 0-2-3 is perpendicular to H.P. or XY line [See Fig. 11-18(iii)].

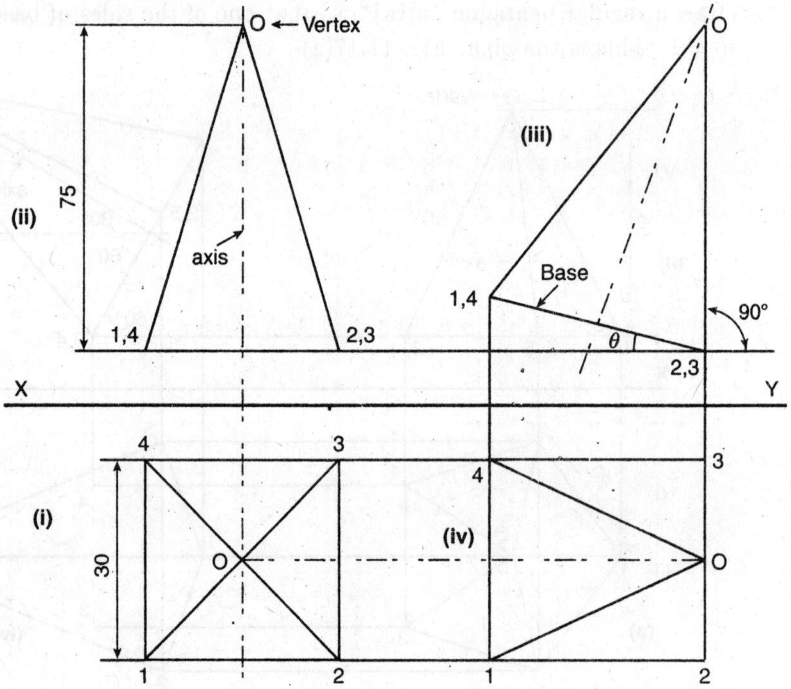

Fig. 11-18

(iii) Draw the top view from elevation. See Fig. 11-18(iv).

(iv) Measure the angle θ with the help of a diagonal scale or protractor.

(v) Although the base is not visible from top, all the base-edges are visible, and hence shown as firm lines.

12. A right circular cone, diameter of base 60 mm and height 80 mm, is resting with its base-rim on H.P., so that one of the slant edge in elevation is vertical. Draw the elevation and plan. Also measure the inclination of base with the H.P. Assume the axis of the cone is parallel to V.P. (This problem is identical to problem no. 11).

13. A cube of sides 60 mm each, is resting with one side on H.P., so that this side is perpendicular to V.P. The bottom surface of the cube is inclined at 30° to H.P. Draw elevation and plan.

Solution :

(i) Draw a square (plan view of cube), making one of the sides lying on the ground and perpendicular to V.P. as in Fig. 11-19(a), project and draw the elevation, Fig. 11-19(b).

(ii) Mark the corners as 1, 2, 3, 4 and 5, 6, 7, 8.

(iii) Rotate the 5-6-7-8 plane through 30° as in Fig. 11-19(c).

(iv) Draw the final plan as in Fig. 11-19(d). The edge 5-6 is not visible and shown as dotted.

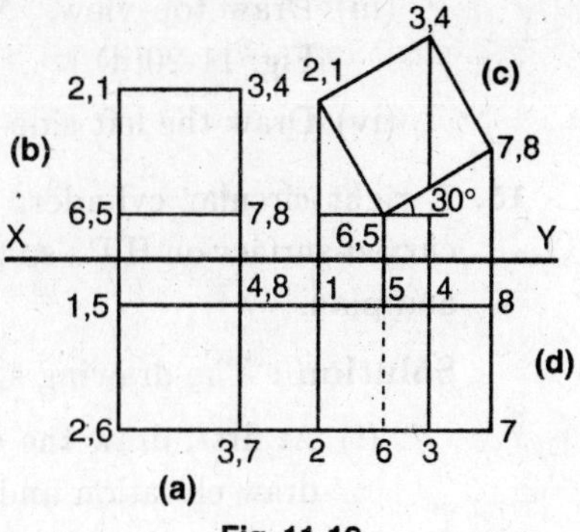

Fig. 11-19

14. A right hexagonal pyramid, each side of base 30 mm and height 75 mm, is resting with one of its triangular face on H.P., so that the axis of the pyramid is parallel to V.P. Draw elevation, plan and left side-view.

Solution : Drawing steps are shown in the order of (i), (ii), (iii), (iv) and (v).

(i) Draw plan and elevation of the hexagonal pyramid, placed on the ground and one of the side of base perpendicular to V.P., as in Fig. 11-20(a) and (b). It is the initial position and the projections are shown by faint lines.

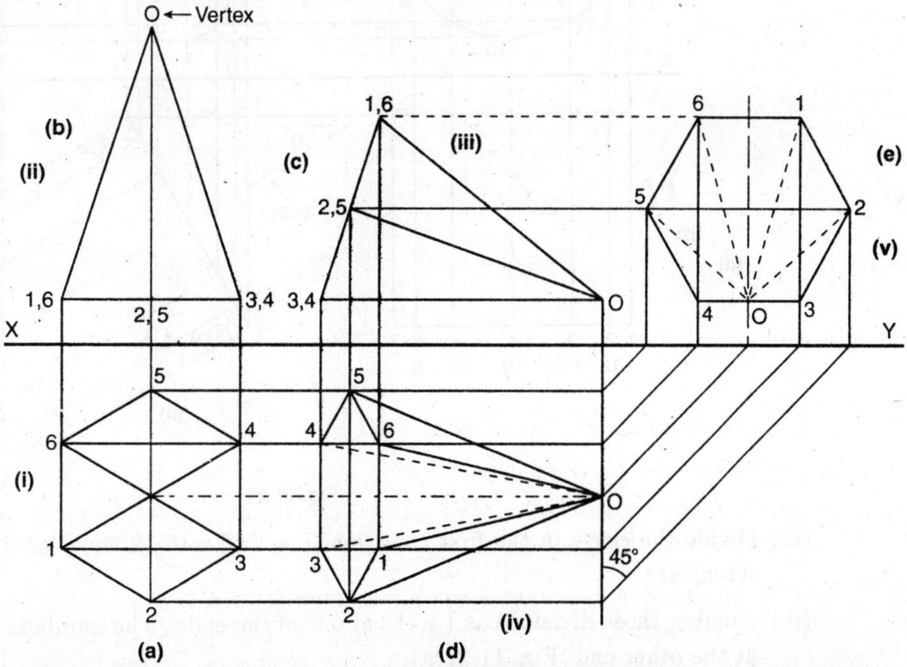

Fig. 11-20

(ii) Now rotate the elevation about 3-4 line and place the triangular plane 3-4-0 on H.P., i.e., parallel to XY line, as in Fig. 11-20(c).

(iii) Draw top view. Note that the base is visible in plan as shown by dark lines in Fig. 11-20(d).

(iv) Draw the left side-view. Fig. 11-20(e).

15. A right circular cylinder, diameter of base 60 mm and height 80 mm, is resting on its curved-surface on H.P., so that the axis is inclined at an angle of 30° to V.P. Draw elevation and plan.

Solution : The drawing steps are shown in the order of (i), (ii), (iii) and (iv).

(i) At first, draw the cylinder-axis perpendicular to V.P. and parallel to H.P. At first draw elevation and then plan [Fig. 11-21(a) and (b)].

(ii) Now, rotate the axis 60°, so that the axis makes an angle of 30° with V.P. [see plan of Fig. 21(c)].

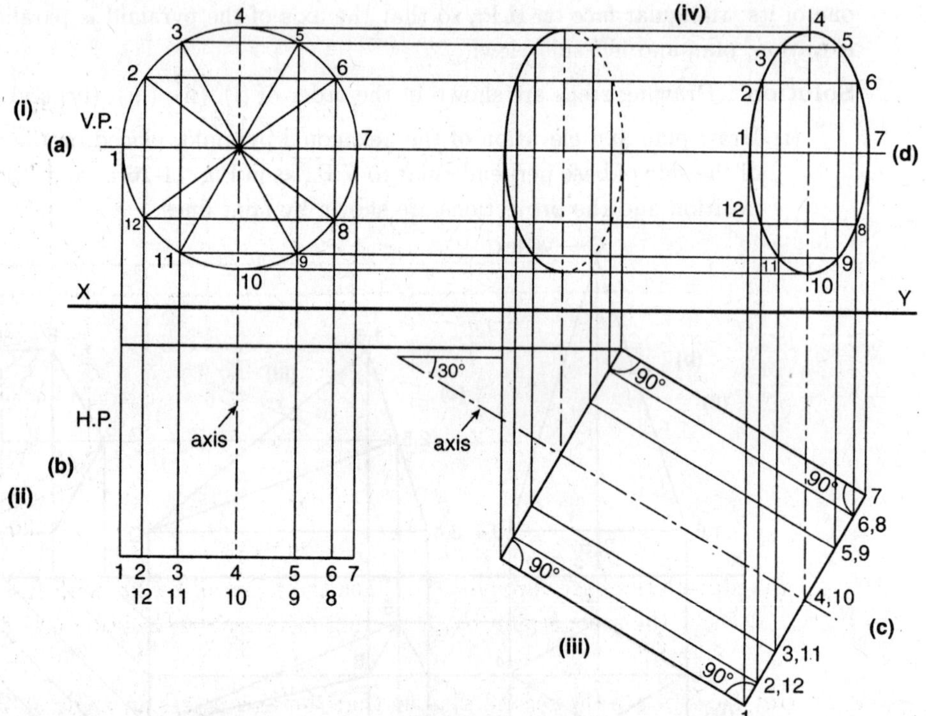

Fig. 11-21

(iii) Divide the circle in the first elevation [Fig. (a)] into 12 equal angles with 30°/60° set-square.

(iv) Number those divisions as 1 to 12 at one of the ends. The numbers will be identical at the other end, Fig. 11-21(a).

(v) Project all the numbers in the 1st and 2nd plan views, [Fig. (b) and (c)].

(vi) Now, imagine the cylinder to be a 12-sided prism.

(vii) Project from first elevation and second plan-views to get the second or final elevation [see Fig. 11-21(d)].

16. A regular hexagonal prism, side of ends 30 mm and height 80 mm, is resting with one of the rectangular faces on H.P., so that the axis of the prism is inclined at 30° with V.P. Draw the elevation and plan.

17. The dimensions of a right square pyramid are base-side 50 mm and height 80 mm. Its axis makes an angle of 30° with H.P. and plan of axis is inclined at an angle of 45° to V.P. It has one of its base-edges on H.P. Draw the elevation and plan.

Solution : We may start the problem keeping (a) the axis perpendicular to H.P. and parallel to V.P. or (b) axis perpendicular to V.P. and parallel to H.P. We are using the first method. The drawing steps are shown in the order of (i), (ii), (iii), (iv), (v) and (vi).

(i) Draw a square on H.P., so that one of the sides of base is parallel to V.P. as in Fig. 11-22(a). Draw the elevation.

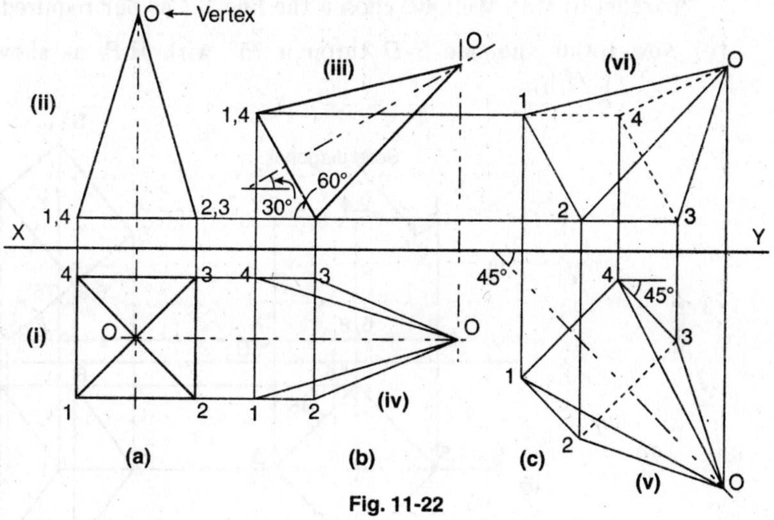

Fig. 11-22

(ii) Rotate the base through 60°. Then the axis will be 30° with H.P. in the elevation. Draw the second plan by projecting from second elevation and first plan [see Fig. 11-22(b)].

(iii) Now rotate the second plan so that the axis makes an angle of 45° with V.P. [See Fig. 11-22(c)].

(iv) Draw the third or final elevation by projecting from second elevation and third plan views [Fig. 11-22(c)].

(v) Note that in the final elevation, the base edges 3-4, 4-1 and the longer edge 4-0 are shown dotted. Similarly, in the final plan, the base-edge 2-3 is shown dotted as this is not visible from top.

18. The solid-diagonal of a cube (edges 60 mm) is inclined at an angle of 75° to H.P. and parallel to V.P. Draw elevation and plan.

Solution : Solid diagonal—The distance between two opposite corners, which are situated between two opposite planes of the cube, is known as solid-diagonal.

This is shown as SD in the elevation of Fig. 11-23(a). The drawing steps are shown in the order of (i), (ii), (iii) and (iv).

(i) Draw a square, making all the sides equally inclined to (i.e., 45°) V.P. which is the plane of the given cube [Fig. 11-23 (a)], for in that position we will get the solid-diagonal parallel to V.P.

(ii) Draw elevation of the cube.

(iii) Number all the corners as 1 to 8, Fig. 11-23(a).

(iv) Join 1 and 7 on elevation and plan with centre line as *S-D*, which is the solid-diagonal of the cube, placed parallel to V.P. (there may be 3-5, 2-8, 6-4, etc. solid-diagonals—these are all imaginary lines). But only 3-5 and 1-7 solid-diagonals are parallel to V.P. We have chosen the line 1-7 as our required solid-diagonal.

(v) Now rotate the line *S-D* through 75° with H.P. as shown in the elevation of Fig. 11-23(b).

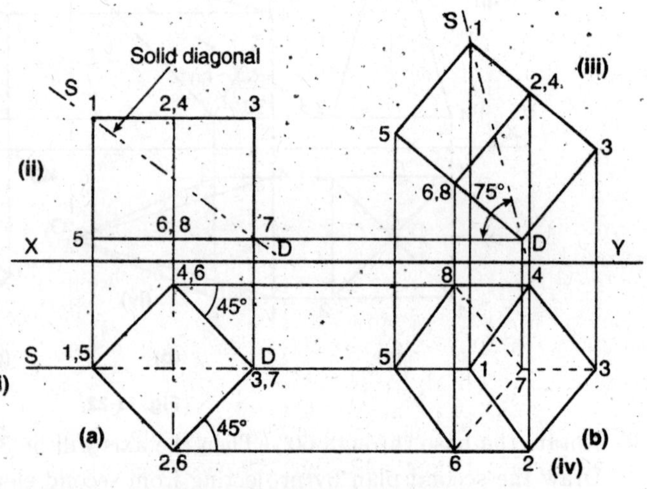

Fig. 11-23 (Problem No. 18)

(vi) Draw the plan view, projecting the points from elevation of Fig. 11-23(b) and plan of Fig. 11-23(a). That is the final plan of the problem.

(vii) The plan and elevation of Fig. 11-23(b) will be shown by dark lines. Line 1-7 is makes 75° with H.P. and parallel to V.P.

19. A regular hexagonal prism, side of end 30 mm and height 75 mm, is placed with one side of end on H.P., so that this side is perpendicular to V.P. The axis of the prism is inclined at 45° to H.P and parallel to V.P. Draw elevation and plane.

20. A regular hexagonal pyramid, side of base 30 mm and height 70 mm, is placed with one triangular face on the ground. The axis of the pyramid is parallel to V.P. Draw elevation, plan and left-side view.

21. The solid-diagonal of a cube (edges 50 mm) is inclined at an angle of 80° to H.P. and parallel to V.P. Draw elevation and plan.

22. A right circular cylinder (dia. of base 50 mm and height 80 mm) is placed with its rim on the ground. The axis of the cylinder is inclined at an angle of 60° to H.P. and parallel to V.P. Draw elevation, plan and left-side view.

23. A right circular cone (dia. of base 60 mm and height 80 mm) is resting with its curved surface on H.P. The plane of the axis makes an angle of 30° to V.P. Draw elevation and plan.

24. A right hexagonal prism (sides of end 25 mm each and height 75 mm), is placed with one rectangular face on the ground. The axis of the prism is inclined at an angle of 30° to V.P. Draw the elevation and plan.

25. A right pentagonal pyramid (sides of base 30 mm each and height 80 mm) is resting with one corner of its base on the ground. The axis of the pyramid is inclined at 30° to H.P. and parallel to V.P. Draw elevation and plan.

26. An equilateral triangular prism (sides of end 30 mm each and height 75 mm) is resting with one of its rectangular face on the ground. The axis of the prism is inclined at an angle of 45° to V.P. Draw elevation and plan.

27. A right circular cone (dia. of base 60 mm and height 80 mm) is placed with its vertex on the ground. The axis of the cone is inclined at an angle of 60° to H.P and parallel to V.P. Draw elevation, plane and left-side view, when the axis is inclined to the left of the observer.

28. A right hexagonal pyramid of base-edge 30 mm each and height 60 mm is resting on one of its base-edge on H.P. with a triangular face inclined to H.P. at an angle of 60° and the edge on which it is resting on H.P. is inclined to V.P. at an angle of 45°. Draw its projections in this position. **(W. B. State Council—1995)**

Solution : In this problem, the axis of the solid is inclined both to V.P. and H.P. So, solution of this problem will be a six-figure, i.e., three elevations and three plans.

First of all, we shall have to place the pyramid on the ground on its base, making one of the side of base perpendicular to V.P. Secondly, rotate the axis of the pyramid to make one slant face 60° with H.P. Finally, we will rotate the second top view through 45°. Fig. 11-24(c).

Refer to Fig. 11-24(a), (b) and (c). The drawing steps are shown in the order of (i), (ii), (iii), (iv), (v) and (vi).

(i) Draw a regular hexagon on H.P., i.e., plan and one of the sides will be perpendicular to V.P.

(ii) Number the corners as 1, 2, 3, 4, 5 and 6. Join the corners, which will meet at *O*, i.e., vertex of the hexagonal pyramid.

(iii) Project all the points to V.P. and draw the elevation of the pyramid as Fig. 11-24(a).

(iv) Draw another elevation, keeping triangular plane 0-3-4 inclined at 60° with H.P. as in Fig. 11-24(b).

(v) Draw the plan of the same projecting respective corners from the first plan and the second elevation.

All the four figures will be shown by faint lines.

(vi) Now, rotate the plan of Fig. 11-24(b) through 45° as plan of Fig. 11-24(c).

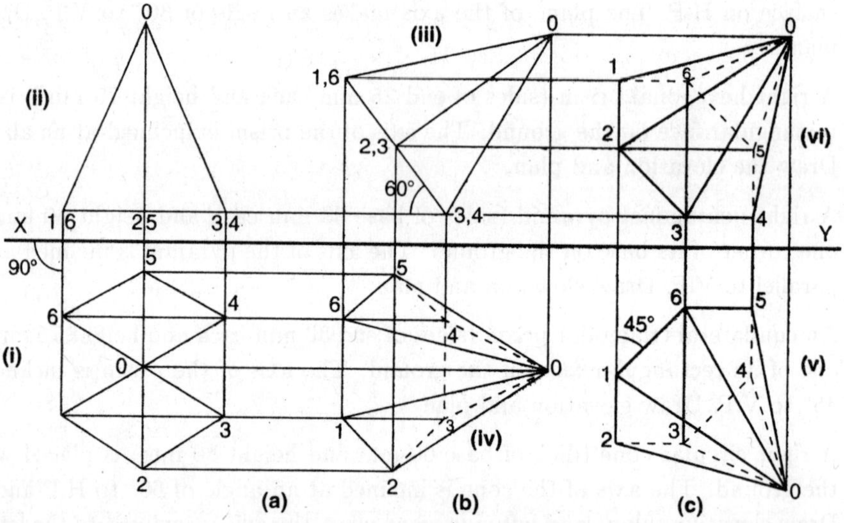

Fig. 11-24 (Problem No. 28)

(vii) Draw the final elevation of the same by projecting all the points from the final plan and the second elevation.

(viii) The lines 4-5, 5-6, 6-1, 6-0, 5-0 will not be visible and are shown as dotted line.

(ix) The plan and elevation of Fig. 11-24(c) is the required solution of the problem. These two figures will be in dark lines.

29. Draw three views (plan, elevation and left side-view) of a right pentagonal pyramid, base-edge 30 mm, height 60 mm, when one of its triangular faces is vertical. Measure the inclination of its base-surface with H.P. **(State Council—1996)**

Solution : Refer Fig. 11-25.

At first, place the pyramid on the ground and one of its side of the base perpendicular to V.P. as in problem no. 28. Solution will be of five figures. Last figure will be the left-side view. The drawing steps are shown in the order of (i), (ii), (iii), (iv) and (v).

(i) Place the pyramid on the ground as per direction. Draw the plan first, where we will get a regular pentagon. See Fig. 11-25(i).

(ii) Number the corners as 1, 2, 3, 4, 5 and vertex O. See Fig. (i).

(iii) Project the points above XY line and draw the elevation, Fig. 11-25(ii).

(iv) Draw another elevation of Fig. 11-25(ii), keeping the triangular plane 3-4-0, perpendicular to both the planes. See Fig. (iii).

(v) Complete the plan by projecting from previous plan and present elevation. Here all the lines projected to vertex will be visible and are shown as firm lines, Fig. 11-25(iv).

(vi) Now in elevation, measure the angle θ with diagonal scale (here it is 30°), Fig. 11-25(iii).

(vii) To draw the side-view, draw a vertical line H_1H_2 and project all the points from plan and elevation of Fig. 11-25(iii) and (iv) to that straight line (H_1H_2) parallel to XY and cut at points P, Q, R, S, T, U, V, M and N.

(viii) From T, U, V, M and N, draw 45° lines to meet XY line. (These can be transmitted by arcs 'A' as a centre). Draw vertical lines from those points to meet respective points from the second elevation. See Fig. 11-25(v).

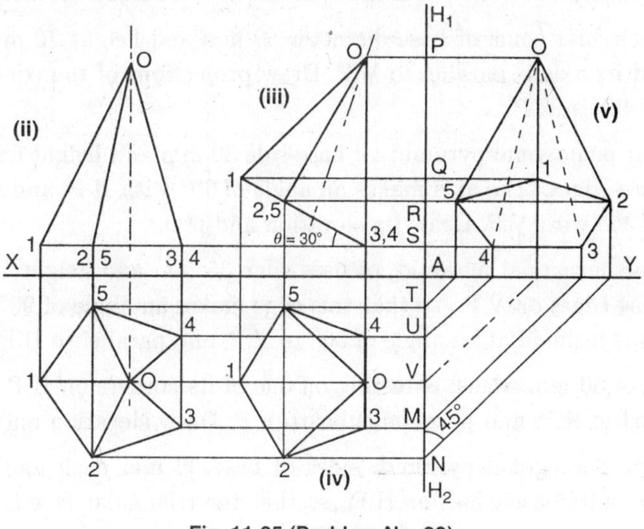

Fig. 11-25 (Problem No. 29)

(ix) Now join the points. We will get the side view as shown in Fig. 11-25(v). It is clear that, in the side-view the base 1-2-3-4-5 will be visible and lines 03 and 04 will not be visible, which are shown as dotted lines. Other lines will be shown dark with plan and elevation of Fig. 11-25(v).

State Council Model Questions (West Bengal)

1. A cube of 50 mm edges is resting with one edge on the H.P., so that this edge is perpendicular to V.P. and the face containing that edge makes an angle of 30° with H.P. Draw plan, elevation and left-side view.

2. A right square prism of side of base 40 mm and height 75 mm is resting with one of its rectangular faces on H.P. and axis inclined at 45° to V.P. Draw plan and elevation.

3. A right square pyramid-side of base 40 mm and height 70 mm, is resting with one of its triangular faces on the H.P., so that the axis of the pyramid is parallel to V.P. Draw elevation, plan and right side-view.

4. An equilateral triangular prism, of base-side 35 mm and 70 mm long, stands on one of its longer edges on H.P. such that the top face is parallel to H.P. and the touching edge makes an angle of 45° with V.P. Draw plan and elevation of the prism.

5. A right hexagonal pyramid of base-edge 35 mm and height 65 mm is resting on one of its corners on the H.P. and the diagonal passing through that corner is parallel to V.P. Draw the projections of the pyramid. (Elevation, plan and left side-view)

6. A right circular cylinder, diameter of ends 50 mm and height 70 mm, is resting on its rim on H.P. so that the axis makes an angle of 40° with H.P. and parallel to V.P. Draw projections of the cylinder. (Elevation, plan and left side-view)

7. A right circular cone, diameter of base 50 mm and height 70 mm, is resting with its curved surface on H.P., so that the axis is parallel to V.P. Draw elevation and plan.

8. A right circular cone, of base diameter 50 mm and height 70 mm, is resting on its rim on H.P. and its axis is parallel to V.P. Draw projections of the cone, when the axis makes an angle of 45° to H.P.

9. A regular pentagonal pyramid, of base-side 30 mm and height 65 mm, rests on H.P. on one of its base-edges. The axis makes an angle of 30° with H.P. and the plan of axis making an angle of 45° with V.P. Draw its elevation and plan.

10. A regular hexagonal pyramid, of base-edge 35 mm and height 70 mm, is resting with one of its base-edges on V.P., so that this edge makes an angle of 90° with H.P. The axis of the pyramid is inclined at an angle of 30° to V.P. and parallel to H.P. Draw elevation and plan.

11. A cube, of 50 mm edges, is resting on one of its corners on H.P. so that the solid-diagonal is parallel to H.P. and perpendicular to V.P. Draw elevation and plan.

12. A regular pentagonal pyramid, sides of base 30 mm each and height 65 mm, is resting with one of its base-edges on H.P., so that the triangular face containing that base-edge is perpendicular to V.P. and both to H.P. Draw plan and elevation of the pyramid.

13. A right circular cone, diameter of base 60 mm and height 80 mm, is resting with its curved surface on V.P. so that the plan of the axis makes an angle of 30° to V.P. and parallel to H.P. Draw elevation and plan.

14. A regular hexagonal prism, of side of end 40 mm and height 70 mm, has a drilled hole of diameter 30 mm along the axis of the prism. The solid is resting on one of its edges on H.P. so that the axis makes an angle 45° with H.P. and parallel to V.P. Draw plan and elevation.

15. A right square pyramid, side of base 40 mm and height 65 mm, is resting with one of its base-edges on the H.P. so that this base-edge is perpendicular to V.P. The axis of the

pyramid is parallel to V.P. and inclined to H.P. at an angle of 30°. Draw elevation and plan.

16. A right square pyramid, side of base 40 mm and height 65 mm, is resting with one triangular face on H.P., so that the axis is parallel to V.P. Draw elevation, plan and left side-view.

State Council Model Questions (West Bengal)

1. State True or False by writing T or F :

   (a) When the axis of a solid is perpendicular to the H.P., its relation with the V.P. is parallel.

   (b) When the axis of a solid is inclined to the H.P., it may be perpendicular to V.P.

   (c) When a square pyramid is resting on H.P. with all the edges making equal angles with V.P., the axis of the pyramid is 45° to the V.P.

   (d) If a solid rests on a corner of its base on the ground, the sides of base containing that corner should be kept equally inclined to V.P.

   (e) When a right circular cone is lying with its generator on H.P., its elevation is an ellipse.

   (f) A tetrahedron contains four equilateral triangles.

   (g) When a square prism is resting on its base on V.P., the elevation is a square.

   (h) The elevation of a sphere resting on H.P., is an ellipse.

   (i) The top view of the base of an oblique cone, resting on H.P. on its base, is an ellipse.

   (j) When the axis is not perpendicular to the base of a solid, it is called an irregular solid.

   Ans. (a) T, (b) F, (c) F, (d) F, (e) F, (f) T, (g) T, (h) F, and (i) F.

2. **(i)** Fill in the Blanks, Using Appropriate Words, Selected from those Given in the Brackets :

   (a) The projection of a ______ in any position on any plane is always a circle (circular lamina, sphere, cylinder).

   (b) When a cube is resting on H.P. with all the faces making equal angles with V.P., the solid-diagonal is parallel to ______ plane. (vertical, H.P., parallel to both V.P. and H.P.)

   (c) When the axis of a triangular prism is parallel to both the planes, the view in the profile plane is a ______. (rectangle, square, triangle).

   (d) When a right circular cylinder is resting on H.P., its elevation is a ______. (circle, rectangle, ellipse).

   (e) A cylindrical log of wood (Diameter 40 mm and height 60 mm) is cut by a plane along the axis and perpendicular to the end. One of the cut-piece rests with its rectangular face on H.P., so that the longer edge (60 mm) is parallel to V.P. The shape of the view on the profile-plane is ______. (a semicircle of radius 20 mm, a rectangle of size 60 × 20 mm, a rectangle of size 60 × 40 mm).

(f) When a hemisphere rests on V.P. with its flat surface, the plan appears to be an ______ (circle, semicircle, ellipse).

(g) When the axis of a right cylinder is parallel to both H.P. and V.P., the view in the profile-plane appears to be a ______ (rectangle, ellipse, circle).

(h) When a cylinder is resting on its curved surface on H.P., the top-face of which is inclined at 45° to V.P., the plan of the solid appears to be a ______ (ellipse, circle, rectangle).

(i) The solid which contains four equilateral triangles, is known as ______ (tetrahedron, hexahedron, dodecahedron).

2. (ii) Fill in the blanks with appropriate words :

(j) When the axis of the solid is parallel to both H.P. and V.P. ______ view reveals the true shape of the base. **(AMIE—Winter, 2001)**

(k) If a cone lies on the ground on one of its generators, with its axis parallel to V.P., its front view is a ______ **(AMIE—Winter, 2001)**

Ans. (a) sphere, (b) vertical, (c) triangle, (d) rectangle, (e) a semicircle of radius 20 mm, (f) semi-circle, (g) circle, (h) rectangle, (i) tetrahedron, (j) end or side-view, (k) triangle.

3. The isometric view of an insulator bracket is shown in Fig. 11-26. To a suitable scale, draw the

(i) Front view, looking along the arrow X;

(ii) End-view, looking along the arrow Y.

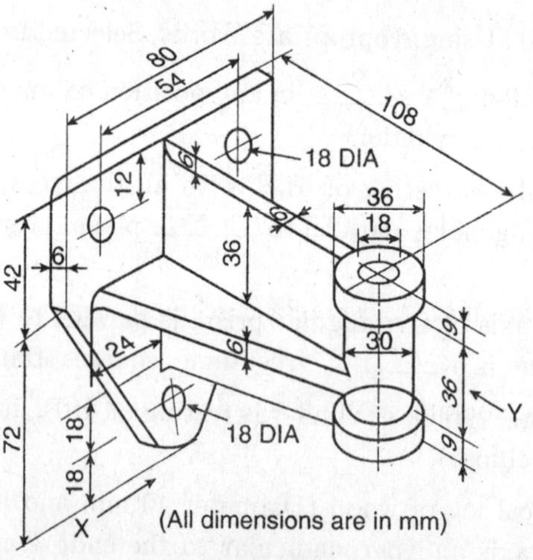

Fig. 11-26

4. A tetrahedron, of each side 40 mm, rests with one of its edges on H.P., which is inclined at 45° to V.P. The triangular face containing that edge is inclined at 30° to H.P. Draw the top and front views of the solid. (Use Third-Angle Projection) **(AMIE—Summer, 2001)**

5. A hexagonal pyramid, each side of base 30 mm and the axis 70 mm long, is resting in an edge of the base on H.P. Draw the projections of the solid, when the axis makes an angle of 45° with V.P. and the base of the solid is nearer to V.P. **(AMIE—Winter, 2001)**

6. Pick-up the correct answer :

   If the frustum of a pentagonal right pyramid stands on its base in H.P., with one of the sides of base parallel to V.P., its front view will be

   (a) triangle (b) rectangle (c) trapezium (d) pentagon. **(AMIE—Winter, 2001)**

   Ans. (c)

7. Draw the following views of the solid shown in Fig. 11-27 :

   (i) Front View (along x-direction);

   (ii) Top view (along z-direction);

   (iii) Side view (along y-direction). **(AMIE—Winter, 2000)**

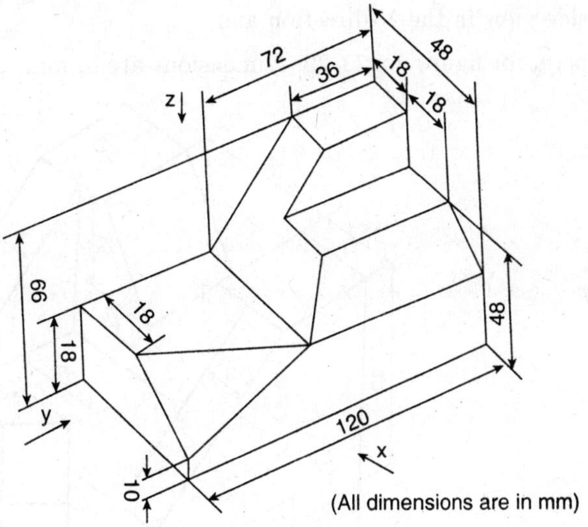

Fig. 11-27

8. Fig. 11-28 shows the isometric view of a machine-component. To a suitable scale, draw the following views :

   (i) A front view, looking along the arrow X.

   (ii) A top view. **(AMIE—Summer, 2001)**

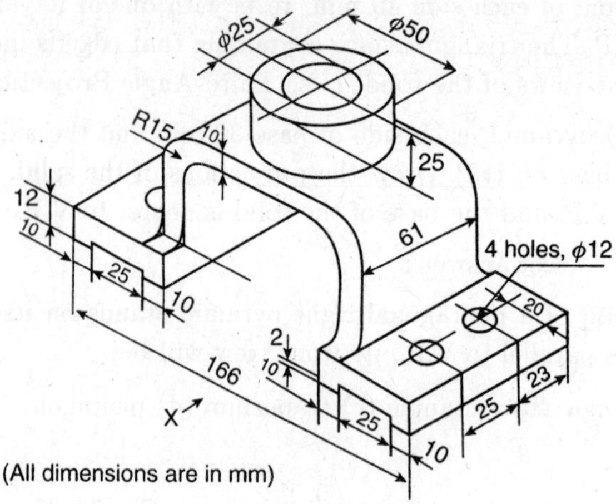

Fig. 11-28

9. Draw :

(a) a front view in the X-direction,

(b) a side view in the Y-direction and

(c) a plan, for figure no. 11-29. Dimensions are in mm. **(AMIE—Winter, 2001)**

Fig. 11-29

10. What is a tetrahedron? **(AMIE—Winter, 2002)**

Ans. A tetrahedron is a triangular pyramid, bounded by four equilateral triangles.

11. Draw the following orthographic projections of the machine-part shown in Fig. 11-30, using first-angle projection :

(i) Front-view, looking along the direction of arrow

(ii) Top-view and

(iii) Left side-view **(AMIE—Winter, 2006)**

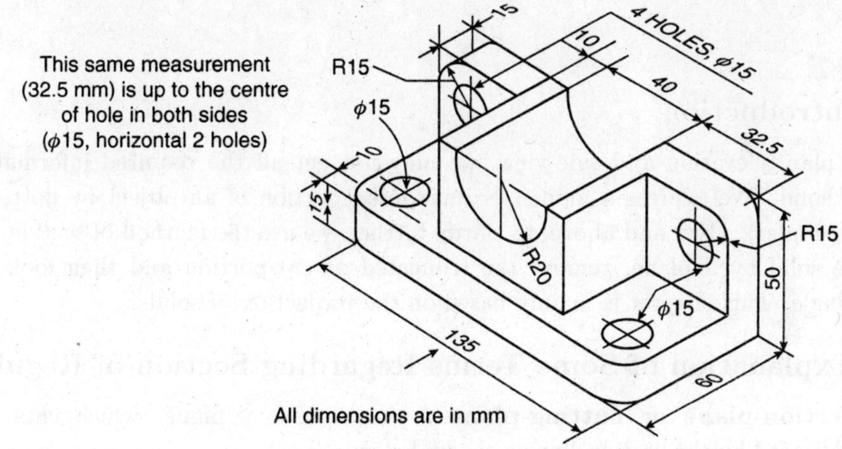

All dimensions are in mm

Fig. 11-30

Chapter 12

Section of Regular Solids

12.1 Introduction

From the plan, elevation and side-view, we may not get all the required information about a particular solid. We express a hidden or unexposed portion of an object by dotted lines (e.g., bore of a cylinder). Over and above, to clarify further, we use the method of section. We imagine to cut the solid by a plane, remove the truncated or cut-portion and then look through the required angle. This chapter is mainly based on the projection of solids.

12.2 Explanation of Some Terms Regarding Section of Regular Solids

(i) **Section-plane or cutting-plane :** An imaginary plane, which cuts the solid, is indicated by the section-line as shown below :

(ii) **Section :** The surface produced in a solid by a cutting plane is called section. This is indicated by equal-spaced thin lines, 45° inclined, covering the sectional area, known as **Section or hatch-lines**. Draw these lines with a 45°-set-square, 2 to 3 mm apart, making a pencil line as shown in Fig. 12-1 and moving the set-square (in the direction of arrowhead) along a T-square edge. The movement may be in the right direction also.

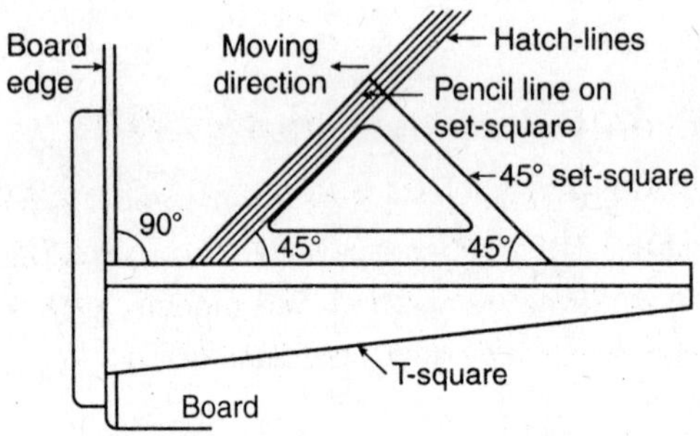

Fig. 12-1

The distance between the two hatch lines depends upon the area of the section. The students will take individual decision, keeping artistic sense, regarding coverage of space by the hatch-lines.

(iii) **Sectional view :** View of remaining portion of an object after cutting it by an imaginary plane is known as **sectional view or section**.

(iv) **Sectional plan :** If we look the cutting face of an object from the top, we get a sectional plan.

(v) **Sectional elevation :** When we look the cutting face of an object from the front, we get a **sectional elevation**.

(vi) **Sectional end-view** or **sectional side-view :** When we look the cutting face of an object, either from the left or from the right, we get a **sectional left-side view** or **sectional right-side view.**

(vii) **True-shape of section :** When our vision is perpendicular to the cutting plane, the sectional surface is known as **true shape of section.** This is because the size of the area of section is not reduced i.e. equal to the area after section.

(viii) **Positions of true-shape : (a)** When the cutting plane is horizontal, (i.e., parallel to H.P. and perpendicular to V.P.) the **sectional plan** becomes the **true-shape.**

(b) When the cutting plane is vertical (i.e., perpendicular to H.P. and parallel to V.P.), the **sectional elevation** becomes the **true-shape.**

(c) When the cutting plane is perpendicular both to H.P. and V.P., the **sectional side-view** becomes the **true-shape.**

12.3 Different Positions of Cutting Plane

Five different positions of the cutting plane are as follows :

The section plane may be

(i) Parallel to V.P. and perpendicular to H.P.

(ii) Parallel to H.P. and perpendicular to V.P.

(iii) Perpendicular to H.P. and making an angle with V.P.

(iv) Perpendicular to V.P. and making an angle with H.P.

(v) Perpendicular to both H.P. and V.P.

12.4 Section-Plane Parallel to V.P. and Perpendicular to H.P.

(a) We will solve problems in all the above five positions with the help of a square pyramid. The students should practise first with these five problems several times with the paper models, prepared by himself/herself and quadrants, and cut the models with a sharp blade or knife. Then he/she will be able to solve the other problems easily.

Example

(b) Draw plan and elevation of a right square pyramid, as shown in the Fig. 12-2. Side of the base 50 mm and height 100 mm. The pyramid is placed on the ground on its base and all the sides of the base are equally inclined to V.P.

(c) Number the (rners and apex with 1, 2, 3, 4 and 'O' respectively, in plan [Fig. 12-2(a)].

(d) Draw a cutting plane *PL*, perpendicular to H.P. and parallel to V.P. The cutting plane line *PL* is denoted by one long line with two dashes as —- —- —-. This is another convention as one long line with one dash and thick at ends as —. — — —. We can use any one of the above two conventions, (As per Bureau of Indian Standards).

(e) *PL* line cuts the 1-2 and 2-3 lines at *A* and *B* and line O-2 at *C*. *PL* line is passing through the mid-point of O2 line, i.e., $OC = C2$ in the plan [(Fig. 12-2(a)].

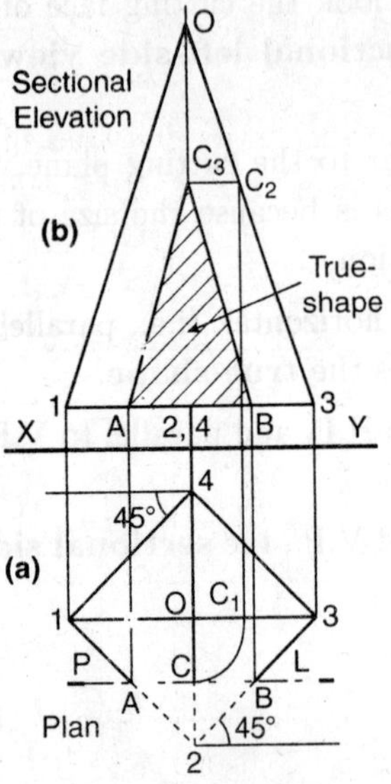

Fig. 12-2 (First-Angle)

(f) Project points A, B from plan to elevation [(Fig. 12-2(b)].

(g) To project the point C in plane to the elevation, project C to O-3 or O-1 line. With O as centre and OC as radius, draw an arc cutting C-3 line at C_1. As if we have turned the pyramid through 90° i.e., C coincides with C_1. This is done in order to make the longer edge O-2 parallel to V.P. [(Fig. 12-2(a)].

(h) Project the point C_1 to C_2 in elevation [(Fig. 12-2(b)].

(i) From C_2, draw a line parallel to XY to meet the line 0-2 at C_3, i.e., the point C at plan is projected to elevation at C_3 [(Fig. 12-2(b)].

(j) Join ABC_3 at elevation. Draw 45° lines (known as hatch lines) on triangle ABC_3, which is the sectional elevation. This view is also a true shape of section, as the section plane is parallel to V.P. [(Fig. 12-2(b)].

(k) In the plan view, the lines A2, B2 and C2 will be drawn faint or dotted and other lines dark [(Fig. 12-2(a)].

Note : (i) Students should note that removed-portion of the solid should be shown lighter than other portion or may be shown by dotted lines.

(ii) When cutting plane is perpendicular to H.P. or V.P., then cutting line will be on H.P. or V.P., i.e., if the cutting plane is perpendicular to V.P., we will get the cutting line in V.P. and the corresponding section in H.P.

12.4.1 Section-Plane Parallel to H.P. and Perpendicular to V.P.

Here, we will get the cutting line at V.P., as the cutting plane is perpendicular to V.P.

(a) Draw plan and elevation as shown in Fig. 12-3. The dimensions and conditions are as in Fig. 12-2.

(b) Numbering system is same as in the previous case.

(c) Cutting plane, parallel to H.P. and perpendicular to V.P., is denoted by PL which cut the line 01, 02, 03, 04 at A, B, C, D respectively in elevation [(Fig. 12-3(a)].

(d) Project A and C to plan. To get the points B and D in the plan, draw arcs with radius OC or OA and centre at O to cut O2 and O-4 at B and D respectively [(Fig. 12-3(b)].

(e) Join AB, BC, CD, and DA in plan and draw hatch lines on the enclosed area. This is also the true shape of section, as the cutting plane is parallel to H.P. [(Fig. 12-3(b)].

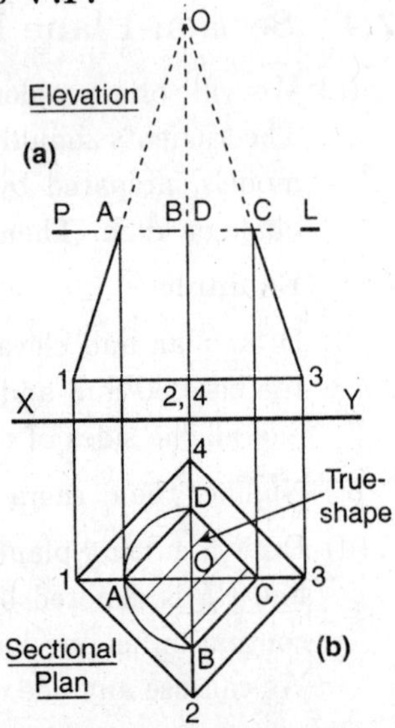

Fig. 12-3

(f) OA, OB, OC lines in elevation will be lighter or dotted, and other lines dark.

12.4.2 Section-Plane Perpendicular to H.P. and Making an Angle with V.P.

Example

Here, cutting line will be on H.P. as it is perpendicular to H.P. We will not get the true shape on plan or elevation. So, to get the true shape of the section, we will have to draw another view perpendicular to the cutting plane. Refer to Fig. 12-4.

(a) Condition remains the same as before.

(b) Cutting plane PL is perpendicular to H.P. and making angle θ with V.P.

(c) In the plan, cutting plane PL cuts the line 1-2 at A, O-2 at C and 2-3 at B [(Fig. 12-4(b)].

(d) Project those points to elevation as in Fig. 12-2(a).

(e) Join A, B & C in elevation and cover that area with hatch lines. That indicates sectional elevation [(Fig. 12-4(a)].

(f) Draw perpendicular lines from the cutting plane PL from points A, C and B [(Fig. 12-4(c)].

(g) Draw line AB parallel to PL at any distance from PL. Distance should be such that the true shape can be accommodated there [(Fig. 12-4(c)].

(h) Take height of C from elevation ($C'4$). Join A, B, C to get the true shape and cover the area with hatch lines. This is the true shape of section, because the vision is perpendicular to the cutting plane PL. Also the area or size of section is not reduced [(Fig. 12-4(c)].

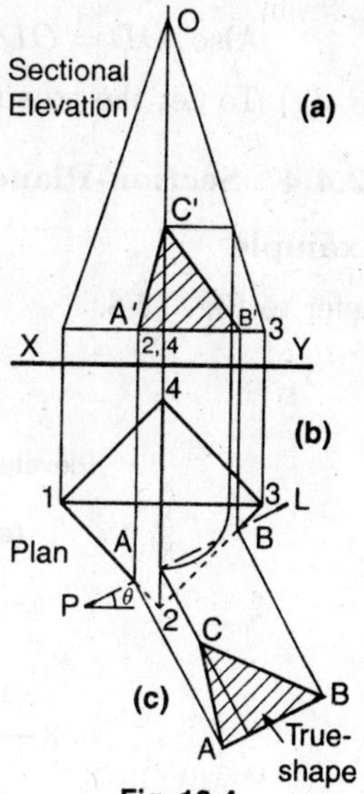

Fig. 12-4

12.4.3 Section-Plane Perpendicular to V.P. and Making an Angle with H.P.

Example

Refer to Fig. 12-5. Here, cutting line will be on V.P., i.e., in elevation. In this problem also, we shall get the true shape of section in another view other than plan or elevation.

Solution : In this case, the

(a) Condition of cutting plane is reversed as in previous example.

(b) The cutting plane PL is perpendicular to V.P. and making and angle θ with H.P. [(Fig. 12-5(a)].

(c) The cutting plane PL cuts the line O1, O2, O3 and O4 at A, B, C and D respectively in elevation. Draw a horizontal line Bh. From h draw a vertical line h-h' to meet the line O3 at h'. With centre at O and radius O-h', draw an arc to meet the lines O-2 and O-4 at $2'$ and $4'$ respectively.

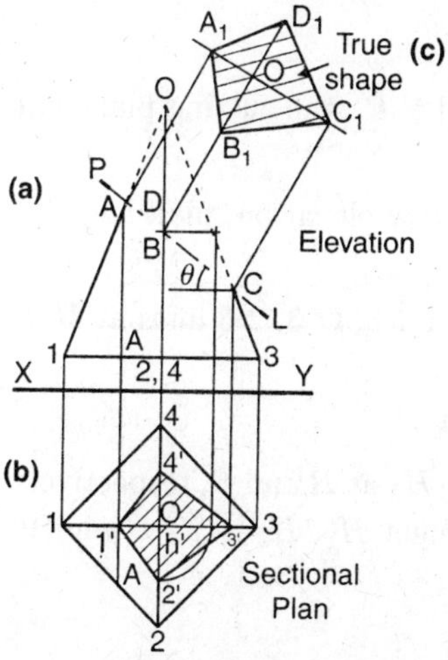

Fig. 12-5

(d) In the plan, join 1-2′, 2′-3′, 3′-4′ and 4′-1′ and draw the hatch lines in the enclosed area.

(e) Draw perpendicular lines about PL from A, (B, D) and C. Draw A, C parallel to PL at a certain distance. We go the points A_1 and C_1.

(f) From plan, take $O4'$ or $O2'$ and mark in the true shape so that $O2' = OB_1 = OD_1$. Also, $OD = OD_1$.

(g) To get the true shape, join A_1, B_1, C_1 and D_1 and draw hatch lines [Fig. 12-5(c)].

12.4.4 Section-Plane Perpendicular Both to H.P. and V.P.

Example

Refer to Fig. 12-6.

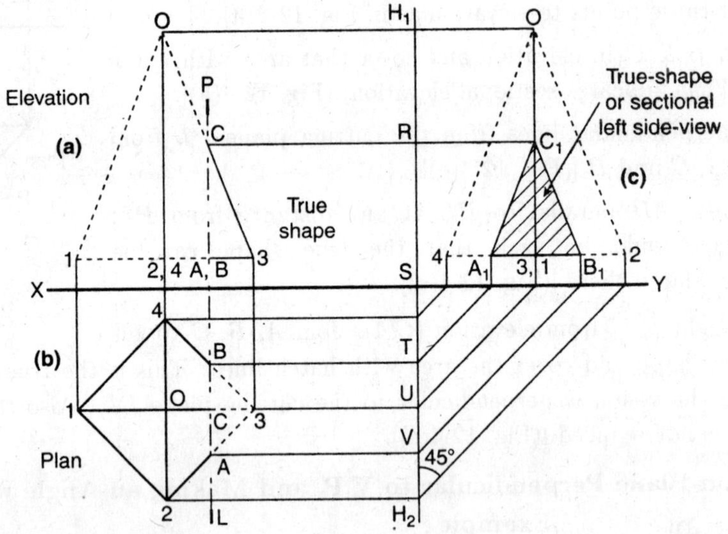

Fig. 12-6

Solution : Since the cutting plane is perpendicular both to H.P. and V.P., the cutting plane line PL is shown both in elevation and plan,

(a) In this case, we shall get the true shape in the side view. Draw elevation, plan and side view before section in faint lines. Number the points.

(b) The cutting plane line PL intersects O3 at C (elevation) and 4-3, O-3, 2-3 lines at B, C and A, respectively, (Plan).

(c) To get the true shape, draw a line H_1H_2 perpendicular to XY.

(d) From C, (A, B) in elevation, draw horizontal lines to meet H_1H_2 at R and S, respectively. Similarly, draw horizontal lines from B, C, A (in plan) to meet H_1, H_2 at T, U and V, respectively.

(e) Draw the side view of the pyramid.

(f) From T, U and V draw lines at angle 45° to meet side view at A_1, (3,1) and B_1, respectively.

(g) From C in elevation, draw **horizontal** line to meet the line O-(3,1) at C_1.

(h) Join A_1, B_1 and C_1 with the line O-(3,1) at C_1 and draw hatch lines in the enclosed area, which will be the **true shape of section** [Fig. 12-6(c)].

Examples

1. A cube, 40 mm each side, is placed on the ground with one of the sides making an angle of 30° to V.P. It is cut by a plane, perpendicular to H.P., parallel to V.P. and 25 mm away from the furthest corner of the cube from the V.P. Draw the sectional elevation, plan and true-shape.

Solution :

(i) Draw the plan first. Here, we get a square figure, one side of which makes 30° to V.P., as shown in the Fig. 12-7(a).

(ii) Number the corners as (1,5), (2,6), (3,7) and (4,8) respectively [Fig. 12-7(a)].

(iii) Draw the elevation [Fig. 12-7(b)].

(iv) The cutting plane line is PL, perpendicular to H.P. and parallel to V.P. [Fig. 12-7(a)].

(v) From the points (A, D), (B, C), project the lines to get the sectional elevation. The section is $ABCD$, which is the true shape [Fig. 12-7(b)].

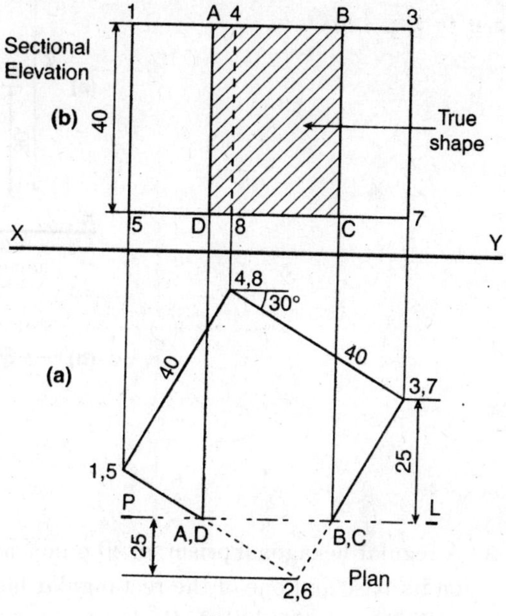

Fig. 12-7

(vi) Draw the hatch lines on enclosed area A-B-C-D.

2. An equilateral triangular pyramid, of each side 40 mm and height 65 mm, is placed on the ground on its base and one side of the base is making 30° with V.P. A cutting plane, perpendicular to V.P. and parallel to H.P., is passing through a distance of 30 mm from the base.

Draw the sectional plane and elevation.

Solution : Refer to Fig. 12-8.

(i) Draw the plane and elevation of the pyramid according to given conditions.

(ii) Number the corners and apex as 1, 2, 3 and O respectively.

(iii) We will find the cutting line PL in elevation and get the true shape of section in plan, which is also vertical trace (V.T.).

(iv) Cutting plane line PL intersects the lines O1, O2 and O3 at A, B and C, respectively in elevation.

(v) From the elevation, project the points A, B and C to plan to meet the lines O1, O2 and O3 at A_1, B_1 and C_1, respectively.

(vi) Join A_1, B_1, and C_1 of plan and cover that enclosed space with hatch-lines. This is the true-shape of section, as the cutting plane is parallel to H.P.

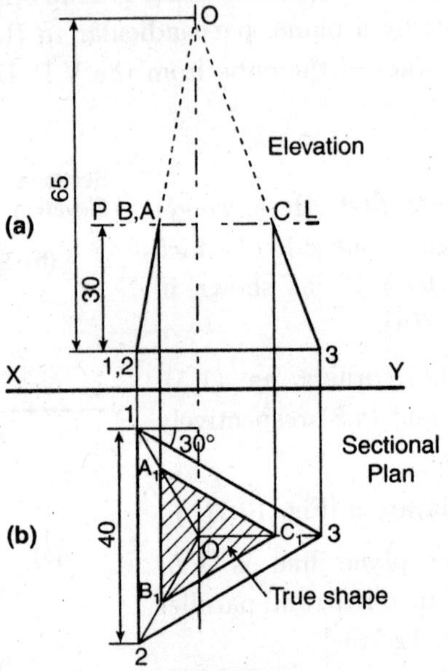

Fig. 12-8

3. A regular hexagonal prism, of 30 mm side of end and 40 mm height, is placed on the ground on its base and one of the rectangular faces parallel to V.P. A cutting plane, perpendicular to H.P. and 45° with V.P., is passing at a distance of 10 mm from the outer-most right corner. Draw plan, sectional-elevation and true-shape of section.

Solution : Refer to Fig. 12-9.

(i) Draw the prism with the given conditions (plan first, then elevation). Number the corners of top end (1, 2, 3, 4, 5, 6) and bottom end (7, 8, 9, 10, 11, 12).

(ii) We will get the true-shape of section in a view other than plan and elevation, as the cutting plane is perpendicular to H.P. and inclined to V.P.

(iii) The cutting plane line P-L (in plan) is perpendicular to H.P., 45° with V.P. and passing through a point 10 mm away from the point (4,10).

(iv) The sectional elevation A-B-C-D is the apparent shape, which is to be hatched [Fig. 12-9(b)].

(v) To get the true shape, draw perpendicular lines from the points (A, B) and (C, D) in plan [Fig. 12-9(a)].

(vi) Draw a line parallel to PL at any distance and complete the area as $ABCD$ (true shape). The distance AB is equal in both elevation and true-shape. Only AD will be enlarged in true-shape, as compared to elevation. Complete the area with hatch lines. The cut-out portions in elevation and plan should be shown faint lines or dotted lines.

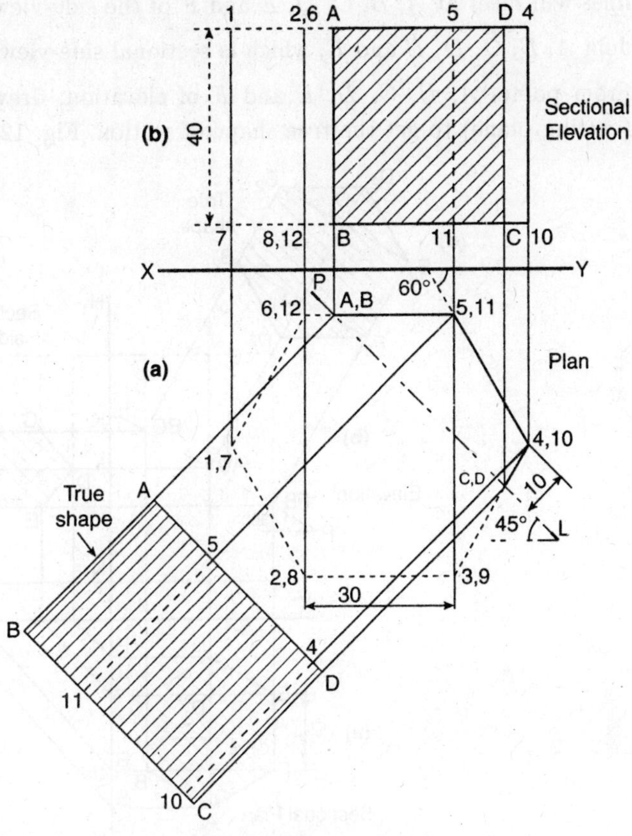

Fig. 12-9 (First-Angle)

4. A regular hexagonal prism, each side of end is 30 mm and 90 mm long, rests with its base on H.P. so that, one of the side of base is making 30° with V.P. It is cut by a plane perpendicular to V.P., inclined at 30° to H.P. and passing through the mid-height. Draw front view, sectional top-view, sectional end-view and true-shape of section.

Solution : Refer Fig. 12-10.

(i) With faint lines, draw the elevation, plan and left side-view of the hexagonal prism, according to the given conditions [Fig. 12-10(a), (b) and (c)].

(ii) In the elevation, cutting plane line PL is perpendicular to V.P. and making 30° with H.P. [Fig. 12-10(b)].

(iii) Draw hatch lines on total area of the plan, i.e., $ABCDEFA$ [Fig. 12-10(a)].

(iv) To get the left side-view, draw a vertical line H_1, H_2.

(v) Draw horizontal lines from A, B, C, D, E and F in both elevation and plan to meet the line H_1H_2 from the top to bottom at K, L, M, N, P, Q, R and S, respectively.

(vi) Draw lines from P, Q, R and S subtending 45° with XY-line from those points, draw vertical lines. Draw horizontal lines from K, L, M and N. These two sets of lines will meet at A, B, C, D, E and F of the side-view [Fig. 12-10(c)].

(vii) Join A, B, C, D, E and F, which is sectional side-view [Fig. 12-10(c)].

(viii) From points A, B, C, D, E and F of elevation, draw lines perpendicular to PL (cutting plane) to get the true shape of section [Fig. 12-10(d)].

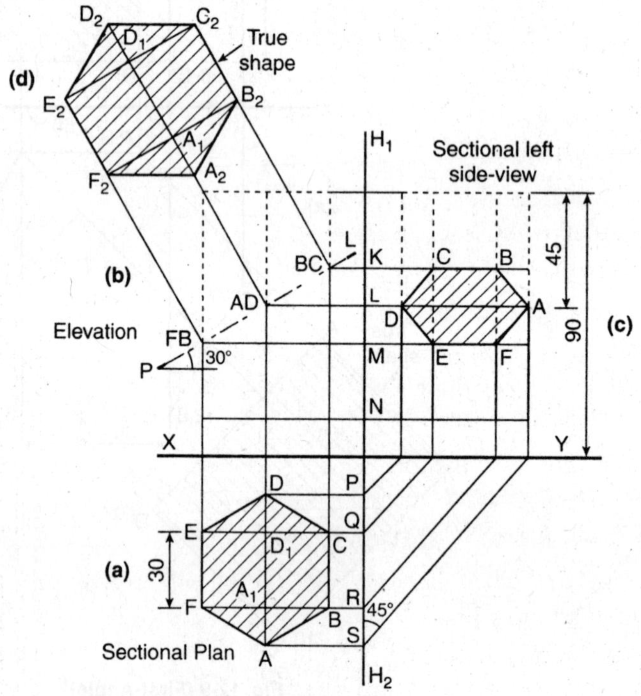

Fig. 12-10 (Problem No. 4)

(ix) Cut the line extended upon A_2D_2, equal to AD of plan at any distance from PL.

(x) From plan to true shape, cut distances $A_2A_1 = AA_1$, and $D_2D_1 = DD_1$. From A_1 and D_1, draw perpendicular lines to meet other two extended lines at B_2, F_2 and C_2, E_2, respectively.

(xi) Join A_2, B_2, C_2, D_2, E_2 & F_2, which is the true-shape of section [Fig. 12-10(d)]. Draw hatching-lines in the covered area.

5. A regular pentagonal pyramid, each side of base 30 mm and axis 70 mm long, is placed on the ground on its base and one of the sides of base is parallel to V.P. A section-plane perpendicular to V.P. and inclined at an angle of 30° with H.P. is passing through the mid-point of the axis. Draw plan, elevation and true shape of the section.

Solution : Refer to Fig. 12-11(a) and (b). The drawing steps are as under :

(i) Draw plan and elevation of a pentagonal pyramid, placed on the ground and one of its sides of base parallel to V.P. [Side 2-3 in Fig.(a)]

(ii) Number the corners as 1, 2, 3, 4 and 5 and apex O, and join O with the corner-points 1, 2, 3, 4, 5. Project 1, 2, 3, 4, 5 and O to elevation.

(iii) Draw the section-plane line PL passing through the mid-point of O5 at E, 30° to H.P. Number the cutting lines cut by PL line on O1, O2, O3, O4 and O5 as A, B, C, D and E, respectively on elevation [Fig. 12-11(b)].

(iv) Project A, B, C and D points to plan [Fig. 12-11(a)].

(v) To get E on plan, draw O1 line on plan, parallel to XY or V.P. as OK by a compass, radius O1 and O as centre.

(vi) Project the point K as 5 on the elevation. Join OS, which is the true inclined length of the edge O5 (This is also the true length of other four edges).

(vii) Draw EF line parallel to XY to meet the SO line at F. Measure the length EF and draw OE (Plan) = EF.

(viii) Join A, B, C, D and E of plan by dark lines and draw hatch line in the enclosed area. Draw 1A, 2B, 3C, 4D and 5E lines in plane with dark lines [Fig. 12-11(a)].

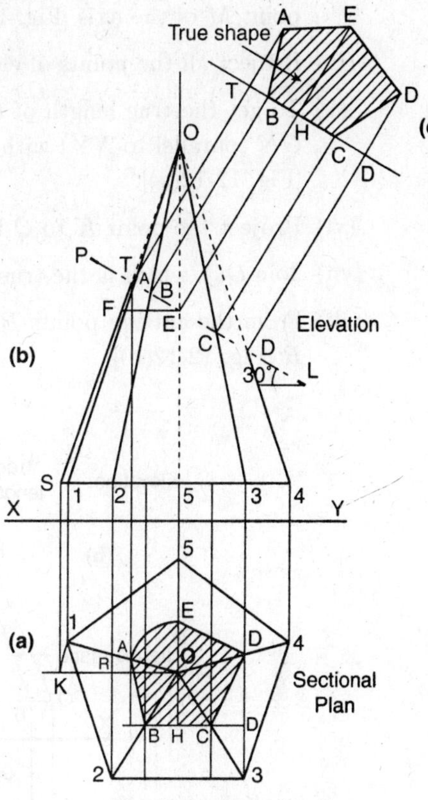

Fig. 12-11 (Prob. No. 5)

(ix) Draw 1A, 2B, 3C and 4D by dark lines below PL line in elevation. E5 line will be dotted, as it is not visible [Fig. 12-11(b)].

(x) To get the true shape of section, draw TU line parallel to PL above elevation.

(xi) From the points A, B, C, D and E on PL, draw lines perpendicular to PL to meet TU line at T, B, C, U and H, respectively. Cut the lengths TA, HE and UD equal to TA, HE and UD of plan. Join A, B, C, D and E of true shape by dark lines and draw hatch lines in the enclosed area [Fig. 12-11(c)].

6. A right hexagonal pyramid, of base-side 30 mm each and height 60 mm, rests on the ground with two of its parallel base-sides perpendicular to V.P. It is cut by an inclined plane, parallel to its slant-surface, perpendicular to V.P., and cuts the axis at its mid-height. Draw its sectional plan and sectional left-side view. **(State Council, 1996)**

Solution : Refer Fig. 12-12.

(i) With faint lines, draw the plan first, then elevation and left side-view before section [Fig. 12-12(a), (b) and (c)].

(ii) Number the corners and apex with 1, 2, 3, 4, 5, 6 and O respectively [Fig. 12-12(a)].

(iii) Cutting plane PL is parallel to triangular face 1-O-6, and passing through the mid-point M of the axis [Fig. 12-12(b)].

(iv) Project all the points of elevation to plan except the points on O5 and O2 lines.

(v) To get the true length of the edges O5 and O2 points, rotate the line O5 or O6 to OK (parallel to XY) with O as centre and O6 or O1 as radius ($\because$ O5 = O6 = O1) [Fig. 12-12(a)].

(vi) Project the point K to Q in elevation [Fig. 12-12(a), (b) and (c)].

(vii) Join OQ, which is the true length of any one of longer edges (O1, O2, etc.).

(viii) From the cutting points $B, E, (M)$ of elevation, draw a line parallel to XY meet at R [Fig. 12-12(b)].

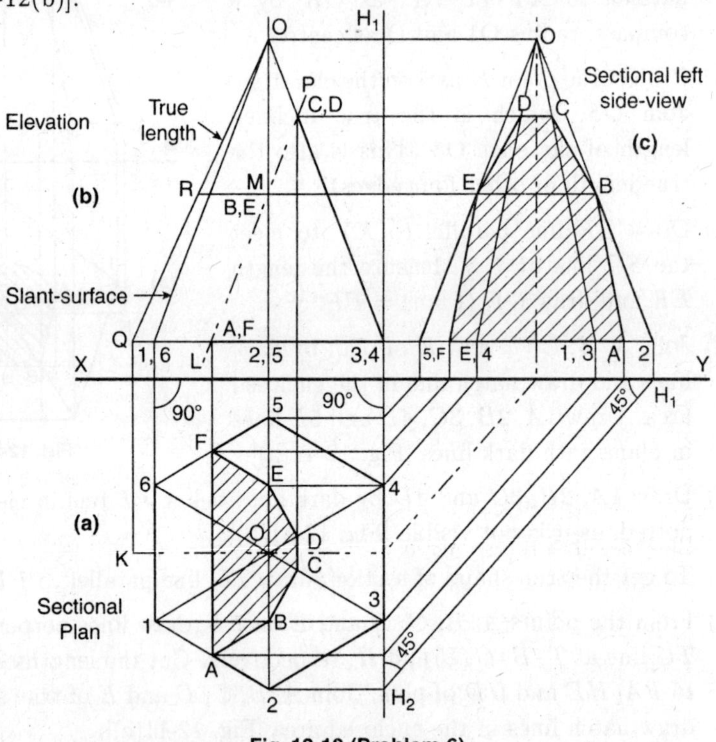

Fig. 12-12 (Problem 6)

(ix) Measure RM and draw $RM = OE = OB$ in plan.

(x) Join the cutting points on plane A, B, C, D, E and F, and hatch the covered area [Fig. 12-12(a)].

(xi) Cut-out portion will be shown by faint or dotted lines and other portions by dark lines [Fig. 12-12(a)].

(xii) Draw the line H_1H_2, at an angle 45° to XY, at the right-side of elevation and plan.

(xiii) Draw the left-side view without section [Fig. 12-12(c)].

(xiv) Now, project from the cutting points of plan and elevation [Fig. 12-12(a) and(b)] in order to get the sectional left-side view [Fig. 12-12(c)].

(xv) Join the points A, B, C, D, E and F in the side-view and draw hatch-lines on the enclosed area. See Fig. 12-12(c).

7. A right pentagonal pyramid, of base-edges 40 mm, height 70 mm, rests on its base on the ground, with a base-side parallel to V.P. It is cut by a plane inclined at 60° to V.P., perpendicular to H.P. and at a distance of 6 mm away from the vertex. Draw its plan, sectional elevation and true-shape of the section.

Solution : Refer to Fig. 12-13. The drawing steps are as follows :

(i) Draw the line 5-4 = 40 mm and parallel to X-Y. Draw the regular pentagon 1-2-3-4-5, whose centre is at O. Join O-1, O-2, O-3, O-4 and O-5. Also join 1-2, 2-3, 3-4, 4-5, 5-1. This is the plan. See Fig. 12-13(a).

(ii) From the plan, draw the elevation, taking the height of the pentagon as 70 mm. See Fig. 12-13(b).

(iii) The cutting-plane is inclined at an angle of 60° to V.P. and 6 mm away from the vertex. With the centre at O and radius = 6 mm, draw an arc at the left side of O. Draw the cutting-plane line P-L, making an angle of 60° to X-Y and tangential to the arc drawn. See Fig. 12-13(a).

(iv) The cutting plane P-L cuts the line 2-3 (in plan) at A, the line O-2 at B, O-1 at C, and 1-5 at D. See Fig. 12-13(a). From the points A and D, draw projection lines to meet the base of pyramid in elevation at A_1 and $5'$ respectively. See Fig. 12-13(a) and (b).

(v) The line O-2 is perpendicular to X-Y. With centre at O and radius = OB, draw an arc to meet the horizontal line OH at B_1. With centre at O and radius = O-1, draw an arc to meet the horizontal line O-H at H. Draw projection from H to meet the base-line in elevation at $1'$. O-$1'$ is the true length of the slant-height O-1 in elevation. See Fig. 12-13(b).

(vi) From the point C on O-1 in plan, draw projection to meet the slant-height O-1 at C_1. From the point B_1 on OH (in plan), draw projection to meet the true slant-height O-$1'$ at B_2. Draw a horizontal line B_2-B_3 to meet the axis O-2 (in elevation) at B_3.

(vii) Join $5'$-C_1, C_1-B_3, B_3-A_1 in order. Draw 45° hatch lines in the enclosed area, which is the sectional elevation. See Fig. 12-13(b).

(viii) To get the true shape, draw perpendiculars from the points A, B, C, D of the cutting plane line P-L. See Fig. 12-13(a) and (c). Draw a line A'-D' parallel to A-D in plan. See Fig. 12-13(c).

(ix) Take the heights B'-B'' and C'-C'' [Fig. 12-13(c)] equal to the vertical distances B_3-2 and C_1-C_2 in the elevation [Fig. 12-13(b)].

(x) Now, join A'-B', B'-C', C'-D' in order. Hatch the enclosed area by 45° lines. A'-B'-C'-D' is the true-shape of section. See Fig. 12-13(c).

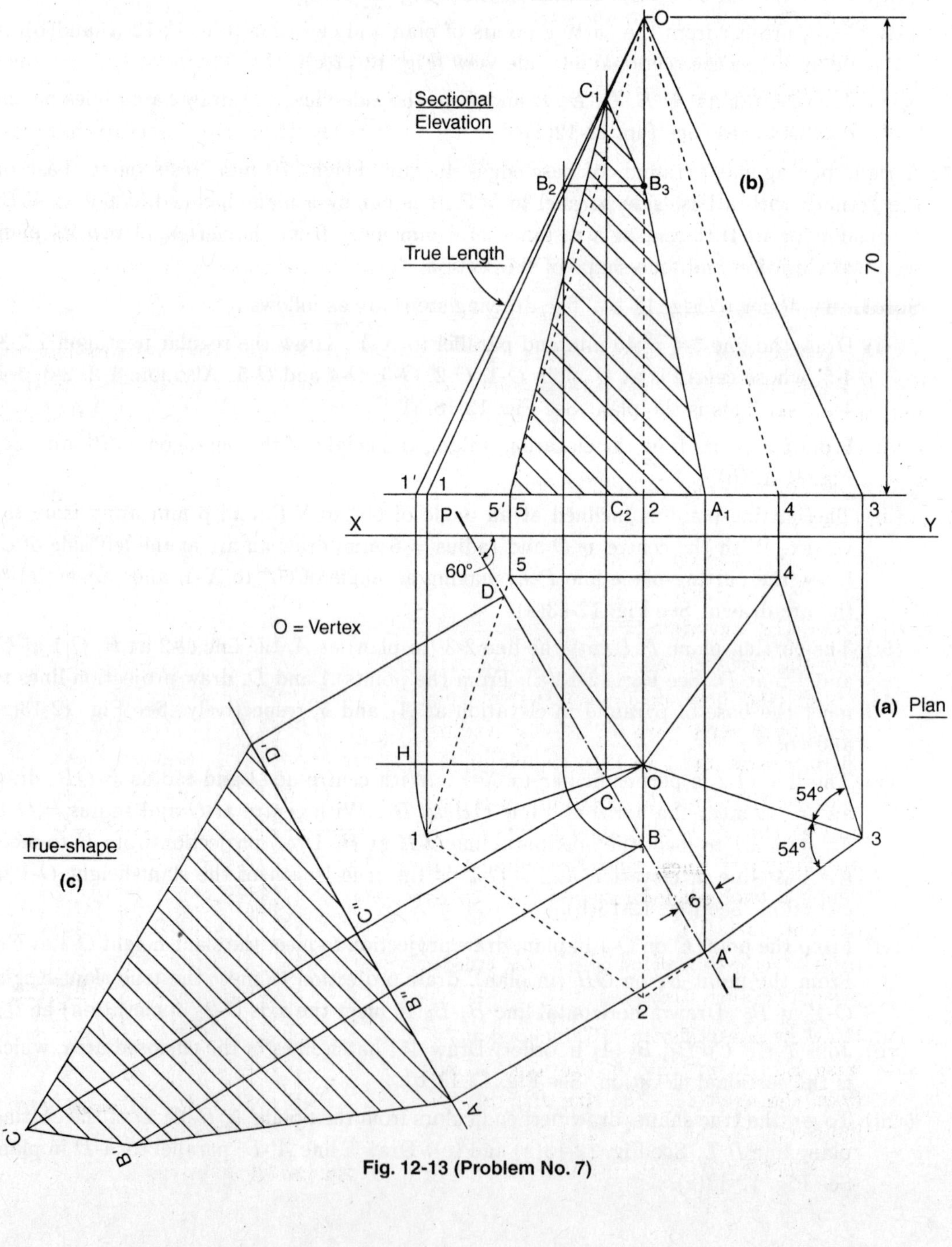

Fig. 12-13 (Problem No. 7)

Different sections, obtained by cutting a right circular cone, are as follows :

(a) When the cutting-plane is **parallel to the base** (or at right angles to the axis), the **section** is a **circle**.

(b) When the cutting-plane is inclined at an angle **less than the base-angle** of the cone, the **section** is an **ellipse**.

(c) When the cutting-plane is inclined at an angle **equal to the base-angle** of the cone, the **section** is a **parabola**.

(d) When the cutting-plane is inclined at an angle **more than the base-angle** of the cone but not passing through the vertex, the **section** is a **hyperbola**.

(e) When the cutting-plane is parallel to the axis or perpendicular to the base of cone, and passes through the **vertex**, the section may be an **isosceles** or **equilateral triangle**, depending upon the **height** and **diameter of base** of the cone.

In all the above cases a to e, the **cutting-plane** is perpendicular to V.P. also.

8. A right circular cone, diameter of base 70 mm and height 90 mm, rests with its base on H.P. It is cut by a section-plane perpendicular to V.P., inclined at an angle of 30° to H.P. and cutting the axis at a point 40 mm from the apex (i.e., vertex). Draw elevation and sectional plan.

Solution : Refer to Fig. 12-14. The drawing steps are as follows :

(i) Place the cone on its base on the ground. See Fig. 12-14(ii).

(ii) Draw the top view and elevation faintly. See Fig. 12-14(i) and (ii).

(iii) Divide the base-circle into twelve equal parts with set-square (30°/60°). Number the divisions into 1 to 12 in the plane.

(iv) Project all points to elevation base line. Join all the points with apex O. These lines are imaginary lines, which are supposed to define a twelve-sided pyramid (i.e., cone). See Fig. 12-14(ii).

[**Hint :** For every problem on cone and cylinder, perform the same operations.]

(v) Draw section-plane line PL inclined at 30° with H.P. at a point on the axis, 40 mm from the apex O. The line PL cuts the lines O1, O2, O3, ..., O12 at 1′, 2′, 3′, ..., 12′, respectively. Fig. 12-14(ii).

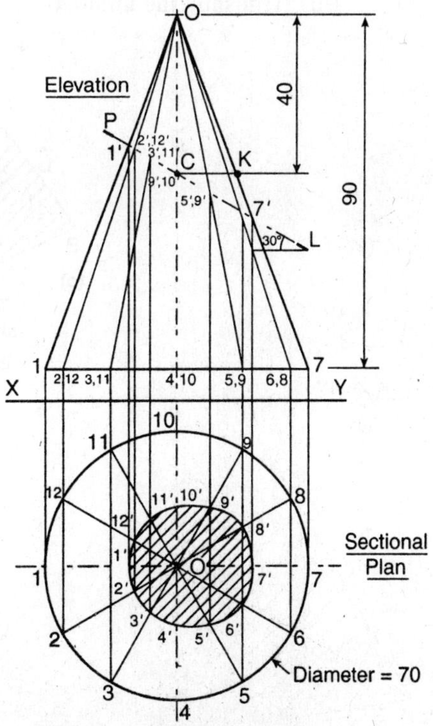

Fig. 12-14 (Problem No. 8)

(vi) Project all those points 1′, 2′, 3′, ..., 12′ to plan to their respective points. We get the points 1′, 2′, 3′, ..., 12′ to plane to their respective points. We get the points 1′, 2′, 3′, ..., 12′ in the plan. Fig. 12-14(i).

(vii) To get the length of O-10′ and O-4′ lines in the plan, draw parallel to XY and measure that length, then CK = O4' = O-10′ (on plan). Fig. 12-14(i) and (ii).

(viii) Join the points 1′, 2′, 3′, ..., 12′ on plan by dark curved line and draw hatch lines in the enclosed area. This is the sectional plan. Fig. 12-14(i).

9. A right circular cone, diameter 70 mm and height 90 mm, rests with its base on H.P. It is cut by a section-plane, perpendicular to V.P., parallel to one of its end-generators (i.e., parallel to one of the slant-heights in elevation), and passing through a point on the axis 30 mm from the vertex. Draw the front view, sectional plan, and true-shape of section.

Solution : Refer to Fig. 12-15. The drawing steps are as follows :

(i) Draw the plan and elevation of the cone, placed on the ground on its base. Divide base-circle into twelve equal parts, which are numbered as 1, 2, 3, ..., 12 [Fig. 12-15(a)].

(ii) Draw the section-plane line PL parallel to O1, which is the slant-side or end-generator (see elevation). This line is cutting the imaginary lines O-3, O-4, O-5, O-6, O-7, O-8, O-9, O-11 at C, D, E, F, G, H, J, K, L and the base circle at A and B [Fig. 12-15(a) and (b)].

(iii) Transmit the numbers A, C, E, F, G, H, J, L, and B to the plan [Fig. 12-15(a)].

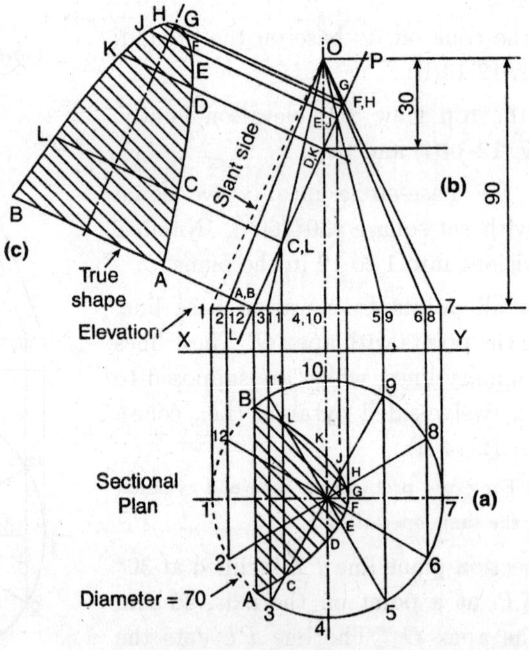

Fig. 12-15 (Problem No. 9)

(iv) To get the D and K-points in the plan, measure the length DR or KR in the elevation. Then $DR = KR = OK = OD$ (in the plan) [Fig. 12-15(a)].

(v) Join $A, C, \ldots B$ by dark line and draw the hatch-lines on enclosed area.

(vi) Right-portion of the PL line, both in plan and elevation, will be shown as dark lines and the left-portion as dotted lines.

(vii) To get true-shape of the section, follow the previous procedure. The true-shape of the curve is a **parabola**. Get the length AB, CL, etc. of true shape from those lengths in plan [Fig. 12-15(c)]. The outer curved line of the true-shape is a **parabola**.

10. A right circular cone, diameter of base 70 mm and height 90 mm, rests with its base on H.P. It is cut by a section-plane perpendicular to both the H.P. and the V.P. and 10 mm away from the axis. Draw the front view, top view, sectional end-elevation and true-shape of section.

Solution : Refer to Fig. 12-16. The drawing steps are as follows :

(i) Draw the plan, elevation and right-hand side view faintly, placing the cone on the ground on its base. Since the cutting-plane is perpendicular **both to H.P. and V.P.**, the **section on the left side-view** will be the true-shape. The outer curve of the true-shape is a **rectangular or equiangular hyperbola**. See Fig. 12-16(c).

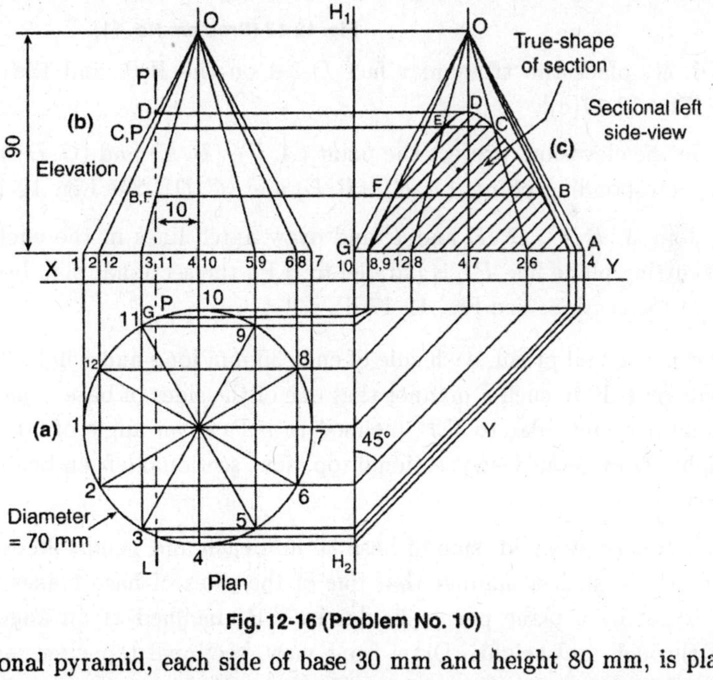

Fig. 12-16 (Problem No. 10)

11. A regular hexagonal pyramid, each side of base 30 mm and height 80 mm, is placed with one of its triangular faces on the H.P. and its axis is parallel to the V.P. It is cut by a horizontal plane bisecting the length of the axis. Draw the elevation, sectional plan and true-shape.

Solution : Refer to Fig. 12-17. The drawing steps are as follows:

(i) Draw the plan and elevation of the pyramid, keeping base on the ground and one of the sides of the base perpendicular to V.P. [Fig. 12-17(a) and (b)].

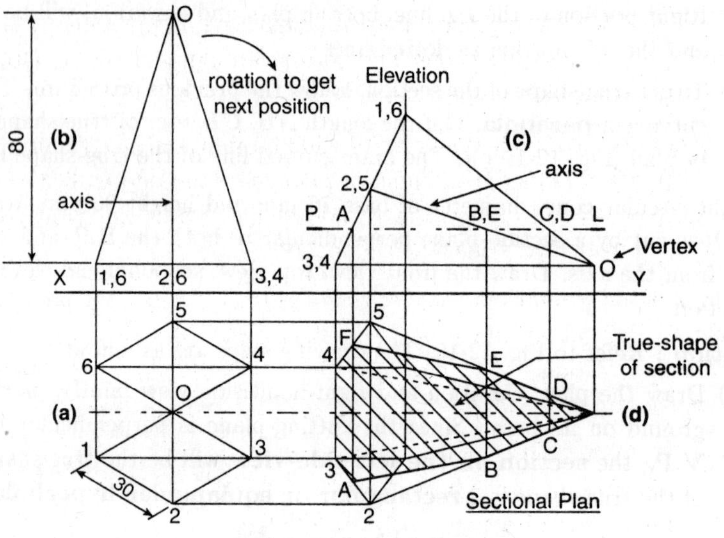

Fig. 12-17 (Problem No. 11)

(ii) Now, place the triangular face O-3-4 on the H.P. and the axis parallel to V.P. [Fig. 12-17(c)].

(iii) In the elevation, project the point (A, F), (B, E) and (C, D) to the plan to get the corresponding points (A, F), (B, E) and (C, D). See Fig. 12-17(d).

(iv) Join A, B, C, D, E and F and draw hatch lines in the enclosed area. Since the cutting plane line PL is parallel to H.P., the sectional plan becomes the true-shape of the section. See Fig. 12-17(d).

12. A regular hexagonal prism, each side of ends 30 mm long and height 90 mm, rests with one of its ends on H.P. in such a manner that one of the sides of base is parallel to V.P. It is cut by a plane perpendicular to V.P. inclined to H.P. at an angle of 30° and passing through mid-height. Draw front view, sectional top view, sectional left end-elevation and true-shape of section.

13. A regular square pyramid, side of base 50 mm each and height 90 mm, is resting with its base on H.P. in such a manner that one of the sides of base makes an angle of 45° with V.P. It is cut by a plane perpendicular to V.P., inclined at an angle of 30° to H.P. and passing through mid-height. Draw front view, sectional top view, sectional left end-view and true-shape of section (use third-angle projection).

14. A regular hexagonal pyramid, side of base 30 mm each and height 80 mm, is resting with its base on H.P. in such a manner that one of the sides of base is parallel to V.P. The pyramid is cut by a plane perpendicular to V.P., inclined to H.P. at an angle of 30° and passing through the mid-height. Draw elevation, sectional plan, sectional side view and true-shape of section.

15. A right circular cone, diameter of base 70 mm and height 90 mm, rests with its base on H.P. It is cut by a section-plane perpendicular to V.P. and cutting the base at a distance of 18 mm from the axis of the cone, and passes through the vertex. Draw front view, sectional end-elevation and true-shape of section.

16. A right circular cylinder, diameter of ends 60 mm and height 80 mm, is resting with one of its ends on the H.P. The cylinder is cut by a plane perpendicular to V.P., inclined at an angle 45° to H.P. and intersecting at mid-height. Draw the front-view, sectional top-view, sectional side-view and true-shape of section.

17. A solid sphere, 70 mm dia., is resting on H.P. It is cut by a plane perpendicular to V.P., inclined to H.P. at an angle of 45° and passing through a point on the vertical axis in elevation at a distance of 15 mm from the centre of sphere. Draw elevation, sectional plan and true-shape of section.

Solution : [**Note :** (a) If we place a sphere on the ground and look from any angle or look towards any plane, we will get a **circle**. This circle represents the outer shape of a sphere.

(b) If we cut the sphere by a plane at any angle, the true-shape of section is a circle.]

Refer to Fig. 12-18. The drawing steps are as follows :

(i) Draw plan and elevation of two circles of 70 mm dia. on both sides of XY [Fig. 12-18(a) and (b)].

(ii) Draw centre lines.

(iii) Draw section plane line PL in elevation according to given conditions [Fig. 12-18(a)].

(iv) Draw a half-circle above PL line, and diameter is equal to the length of line 1-7 (within PL).

(v) Divide the half circle into six equal parts with the help of set-square (30°/60°).

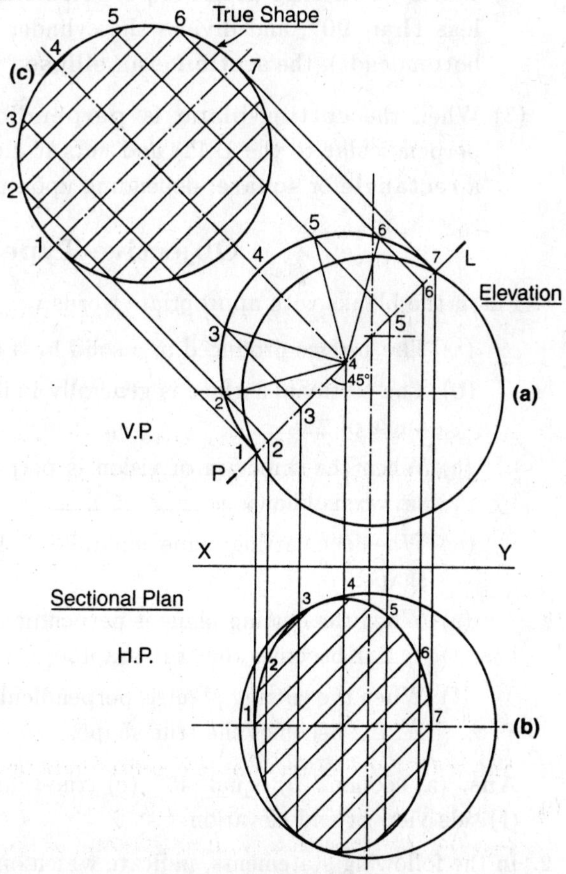

Fig. 12-18 (Problem No. 17)

(vi) Number the divisions as 1, 2, 3, 4, 5, 6 and 7. If we take the full circle, then the numbers would be 8, 9, ..., 12. This is avoided, as the figure is symmetrical about *PL* line.

(vii) Draw true-shape above *PL* line, which is a circle of diameter equal to the diameter of the half-circle [Fig. 12-18(a) and (c)].

(viii) Project the numbers to *PL* line and from this line project those numbers to plan. We will get the length of points equal to twice the lengths of points on half circle to *PL* line [Fig. 12-18(b)].

(ix) Draw hatch-lines on enclosed area of plan and true shape. See [Fig. 12-18(b) and (c)].

Different sections obtained by cutting a right circular cylinder, standing on its one end on H.P. (or ground).

(1) When the **cutting-plane is parallel to the H.P.** (or perpendicular to the axis) and cuts the cylinder at any position between the two ends, the **section** is a **circle.**

(2) When the **cutting-plane is perpendicular to V.P., inclined at an angle to H.P., less than 90°**, and divides the cylinder into two pieces (without cutting the top or bottom ends), the **section** is an **ellipse.**

(3) When the **cutting-plane is perpendicular to the ends of the cylinder** (i.e. perpendicular to the H.P.) and cuts it along the axis into two pieces, the **section** is a **rectangle** or **square**, depending upon the **height** and **diameter** of the cylinder.

Objective-Type Questions

1. Fill in the blanks with appropriate words :

(a) The surface produced in a solid by a cutting plane is called ______.

(b) The sectional-surface is generally indicated by ______ spaced lines, inclined at an angle of ______.

(c) When the direction of vision is perpendicular to the cutting plane, the sectional surface is known as ______ ______.

(d) When the cutting plane is parallel to H.P., the sectional ______ becomes the ______ shape.

(e) When the cutting plane is perpendicular to H.P. and parallel to V.P., the sectional ______ becomes the true-shape.

(f) When the cutting plane is perpendicular both to H.P. and V.P., the sectional ______ ______ becomes the true-shape.

Ans. (a) section, (b) equal, 45°, (c) true-shape or true-form, (d) plan, true, (e) elevation, (f) side-view or end-elevation.

2. In the following statements, indicate which one is true or false by writing T or F :

(a) Sectional elevation is obtained, when a truncated solid is looked from the front.

(b) When a right circular cone is cut by a plane parallel to the base of cone, the section is an ellipse.

(c) When a right circular cone is cut by a plane perpendicular to the base of cone and passing through the vertex, the section is a triangle.

(d) In a right circular cone, when the cutting-plane is inclined to the base at an angle equal to the base-angle of the cone, the section is a parabola.

(e) When the cutting-plane is inclined to the axis of the cone and parallel to one of the generators, the section obtained is a parabola.

(f) In a right circular cone, when the cutting-plane is inclined to the base at an angle less than the base-angle of the cone, the section is an ellipse.

(g) The section at any place of a solid sphere is an ellipse.

(h) In a right circular cone, when the cutting-plane is inclined to the base at an angle greater than the base-angle of the cone but not passing through vertex, the section is a hyperbola.

(i) When a right circular cone is cut by a plane, perpendicular to the base or end, and divides the cone into two pieces, after passing along the axis, the true-shape of section is an isosceles or an equilateral triangle.

Ans. (a) T, (b) F, (c) T, (d) T, (e) T, (f) T, (g) F and (h) T.

State Council Model Questions (West Bengal)

1. A cube of 50 m long edges is resting on the H.P. on one of the faces with the vertical faces equally inclined to the V.P. It is cut by a section plane which is inclined at 55° and passing through the mid-axis. Draw (a) sectional plan (b) True shape of the section.

2. A right square prism, base-side 40 mm each and height 80 mm, is resting with one of its rectangular faces on H.P., so that one of the edges of end is parallel to V.P. The prism is cut by a plane inclined at 60° to H.P., perpendicular to V.P. and passing through the mid-axis.

Draw : (a) Elevation, (b) Sectional plan, (c) True-shape of the section.

3. A right square pyramid, base-side 40 mm each and height 70 mm, is resting on H.P. on its base with all base-edges equally inclined to V.P. It is cut by a vertical section-plane parallel to V.P. and 9 mm away from the axis towards the observer. Draw top view and sectional front view and the true-shape.

4. A right circular cone, base-diameter 50 mm and axis 65 mm long, rests with its base on H.P. It is cut by a section-plane perpendicular to V.P., inclined at an angle of 45° to H.P. and passing through a point on the axis 35 mm above the base.

Draw : (a) Front view, (b) Sectional top view, (c) True-shape of the section.

5. A right circular cone, diameter of base 50 mm and axis 70 mm, is resting on its base on H.P. It is cut by a section plane perpendicular to the V.P., inclined at 75° to the H.P. and passing through the apex (or vertex).

Draw : (a) Front view, (b) Sectional top view, (c) True-shape of the section.

6. A right pentagonal pyramid, of base-side 40 mm and height 70 mm is resting on H.P. with one side of base parallel to V.P. The pyramid is cut by a section-plane, which is vertical and parallel to V.P. and at a distance of 10 mm from the vertex towards the observer.

Draw : (a) Sectional elevation, (b) Plan and (c) True-shape of section.

7. A right circular cylinder, diameter of end 50 mm and height 70 mm, is resting with one end on H.P. It is cut by a plane, which is inclined at 30° to H.P., perpendicular to V.P. and passing through the mid-height.

Draw : (a) Elevation, (b) Sectional plan and (c) True-shape of the section.

8. A sphere of 60 mm diameter is resting on H.P. It is cut by a plane inclined at 30° to H.P., perpendicular to V.P. and passing through the vertical axis at a distance of 15 mm from the centre of the sphere.

Draw : (a) Elevation, (b) Sectional plan and (c) true shape of the section.

State Council Model Questions (West Bengal)

Fill in the blanks with appropriate words :

1. When a sphere is cut by a section-plane, the true-shape of the section is ______.

2. The true-shape of the section produced is ______, when a right cone is cut by a plane parallel to the axis and passing through the apex (or vertex).

3. Section-plane, cutting all the generators of a right cylinder and making an angle with axis other than 90°, produces ______.

4. When a right circular cone is cut by a section-plane parallel to its base, the true-shape of the section is a ______.

5. A right circular cone, with an apex-angle α, is cut by an inclined plane at an angle β to the horizontal. $(\mathbf{90^\circ} - \frac{\alpha}{2})$ = Base-angle of the cone.

(a) When β is greater than $[90^\circ - (\alpha/2)]$, the intersection curve is ______.

(b) When β is equal to $[90^\circ - (\alpha/2)]$, the intersection curve is ______.

(c) When β is less than $[90^\circ - (\alpha/2)]$, the intersection curve is ______.

(d) When β is 90° and passing through the apex, the intersection curve is a ______.

6. When a right circular cylinder is cut by a section-plane, which passes through the mid of axis and makes an angle 45° to the H.P., the true-shape of section will be ______.

7. The projection of section on a plane parallel to the section-plane will show the ______ of the section.

8. The true-shape of the section is revealed in the top view, when the section-plane, cutting a solid, is parallel to _______________. **(AMIE—Winter, 2001)**

Ans. (1) a circle, (2) an isosceles or equilateral triangle, (3) an ellipse, (4) circle, (5) (a) a hyperbola, (b) a parabola, (c) an ellipse, (d) an isosceles or equilateral triangle, (6) an ellipse, (7) true-shape, (8) base or end or H.P.

9. Choose the **correct** answer from the alternatives :

(i) When the cutting plane is inclined to the axis of the right cone and parallel to one of the generators, the section obtained is (a) an ellipse (b) a parabola (c) hyperbola.
Ans. (b) **(AMIE—Summer, 2000)**

(ii) If the frustum of a pentagonal right pyramid stands on its base in H.P., with one of the sides of base parallel to V.P., its front view will be a (a) triangle (b) rectangle (c) trapezium (d) pentagon.
Ans. (c) **(AMIE—Winter, 2000)**

(iii) If a sphere is cut by a plane inclined to H.P. and perpendicular to V.P., then the true shape of the section is (a) circular (b) elliptical (c) semi-circular (d) semi-elliptical.
Ans. (a) **(AMIE—Winter, 2002)**

10. A regular pentagonal pyramid, sides of base 40 mm each and height 100 mm, lies on one of its triangular faces on H.P. with its axis parallel to V.P. A plane perpendicular to H.P. and inclined at 30° to V.P. cuts the pyramid and passes through a point 47 mm from the base along the axis. Draw the sectional elevation and true-shape of the section.
(AMIE—Winter, 2000)

11. A sphere, of 60 mm diameter, rests on H.P. It is cut by a section-plane perpendicular to V.P. and inclined at 60° to H.P. The section-plane passes through a point on the vertical axis of the sphere at a height of 20 mm from the bottom end of the vertical axis. Draw the sectional plan, elevation and true-shape of the section. **(AMIE—Summer, 2002)**

12. A pentagonal prism, base 25 mm side, axis 50 mm long, is resting on its base on horizontal plane, having an edge of the base parallel to V.P. It is cut by a plane inclined at 45° to H.P. and perpendicular to V.P. and passing through the centre of the axis. Draw the true shape of the section and also the development of the lateral surface of the truncated prism.
(Winter, 2006)

13. A cylinder, of 50 mm diameter and 70 mm long, is lying with one of its generators on H.P. such that its axis is parallel to both H.P. and V.P. A section-plane, perpendicular to V.P. and inclined at 30° to H.P., passes through a point on the axis 30 mm from one of its ends. Draw the sectional top-view and true-shape of section. **(Winter, 2005)**

Chapter 13

Development of Surfaces

13.1 Introduction

The shape of a geometrical solid (e.g., prism, pyramid, cylinder, cone, etc.) can be prepared with a metal-sheet, plastics sheet or a piece of thick paper. When the **total surface** of the solid **is projected on a single plane**, it is called the **development of the surface**.

In chapter 11, the students were advised to prepare the paper-models of simple solids. As such, it may be hoped that, the students have already acquired some idea about surface-development.

In order to draw the development of the geometrical solids, some methods are to be followed. In this chapter, we have tried to explain those methods for simple solids only. For understanding the development of surfaces, the knowledge of projection and section of solids are required.

13.2 Applications of Surface-Development

The surface-development may be used for the manufacture of different types of simple **containers or vessels** (e.g., buckets, boilers, steel chimneys, acid and water-tanks, petrol tanks) made of thin or thick metal-sheet, and complicated car-bodies, aeroplanes, ships, space capsules, etc.

13.3 Classification of Surfaces

In general, the different **surfaces of solids** may be classified into **two groups** as under :

(i) **Surfaces**, which may be developed from the **accurate measurements of the solids**, such as **prism**, **pyramid**, **cylinder**, **cone**, etc.

(ii) **Surfaces**, which may be developed from the approximate measurements of the solids, such as sphere, paraboloid, etc. This portion is not included in the syllabus of the engineering Diploma Course in West Bengal, and therefore, is not discussed here.

The surfaces in serial no. (i) above can also be sub-divided into two classes :

(a) **Parallel line development method**. This is used for the development of cube, prism, cylinder, etc.

(b) **Radial line development method**. This is used for the development of pyramid, cone, and their sections, It is mainly based on **true length** of slant sides or generators.

Thus, **true lengths** must be calculated before drawing the development of such surfaces.

It should be pointed out that, when only the side-surfaces of a solid are to be developed, then it is known as **lateral surface-development** (or **lateral-development**). Similarly, when total surfaces of a solid (lateral and bottom, or lateral, top and bottom) are to be developed, it is known as **complete or total surface-development**. For example, a cube has four side-surfaces (i.e., four squares), with the top and bottom-surfaces (i.e., two squares), i.e., six squares as the total development.

13.4 Surfaces of Simple Solids and their Cut-Out (Truncated) Portions

We shall explain the development of the following simple solids :

(a) The Cube

(b) Right prisms of the following types :

- **(i) Triangular,**
- **(ii) Square,**
- **(iii) Pentagonal and**
- **(iv) Hexagonal**

(c) Right pyramids of the following types :

- **(i) Triangular,**
- **(ii) Square,**
- **(iii) Pentagonal and**
- **(iv) Hexagonal**

(d) Right cylinder

(e) Right cone.

N.B. Please refer Article 11.3 (Preparation of the models of solids).

13.4 (a) Cube

It is a special form of square prism, where length = breadth = thickness or height.

The students should try to draw the surface-development of truncated cube with the help of parallel line method.

Problem 1 : A cube, each side 50 mm, is placed on the ground with one of its face. One of the square surfaces of the cube is parallel to (or perpendicular to) the V.P. A section-plane perpendicular to V.P. and inclined at an angle of 45° to H.P. passes through the mid-point of any one of the vertical faces. Draw the development of the remaining solid after cutting by the section-plane (marked by 'A' in Fig. 13-1)

Solution : Refer to Fig. 13-1. The drawing steps are as follows :

(i) According to the given conditions, draw the elevation and plan [Fig. 13-1(a) and (b)].

(ii) Draw the cutting-plane line PL (on the elevation), which cuts the vertical line 2-6 at the point Q and the top-surface lines 1-2 and 3-4 at T [Fig. 13-1(b)].

(iii) Draw four squares (each side 50 mm) at the right side of elevation. Also, draw two more squares (one above and another below the extreme left square). Before cutting the cube, this is the total development. See Fig. 13-1(d). The isometric view is shown in Fig. 13-1(c).

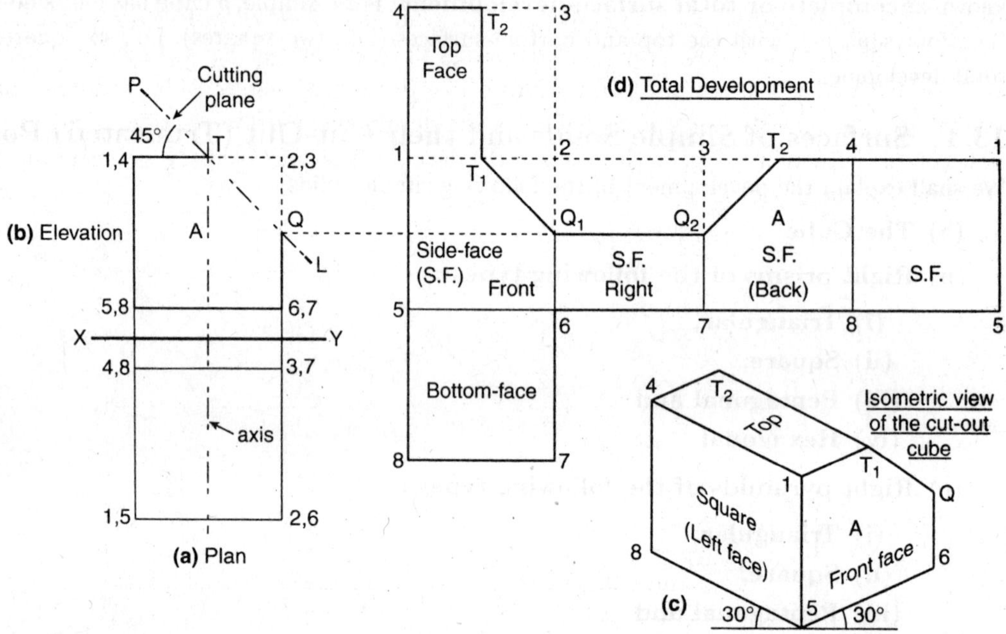

Fig. 13-1

(iv) From the junction of the cutting plane PL and the vertical line 2-6 in elevation i.e. from Q [Fig. 13-1(a)], draw horizontal line to meet the lines 2-6 and 3-7 in the development at Q_1 and Q_2 respectively.

(v) Draw a vertical line to meet 4-3 and 1-2 on the top face of the cube.

(vi) Similarly, obtain the mid-point of the line 3-4 [Fig. 13-1(d)] in the development (point T_2).

(vii) Join the above points in order by straight lines. Thus, we get the development of the truncated cube [Fig. 13-1(d)].

(viii) After the removal of the right corner of the cube beyond the line PL, cover the remaining area of the development by dark lines [Fig. 13-1(d)].

N.B. This procedure of drawing the development is called **parallel line development** method.

13.4(b)(i) Triangular Prism

In this case, **the top** and **bottom-surfaces** are **two equilateral triangles**, and lateral surface consists of **three equal rectangles**. To draw the development, parallel line method is to be followed in this case.

Problem 2 : An equilateral triangular prism is placed on the ground on one of its ends in such a manner that one of its rectangular faces is perpendicular to V.P. A cutting plane perpendicular to V.P. and inclined at an angle of 45° to H.P. passes through the top right corner in elevation. Develop the remaining surface of the solid after section. The side of end measures 30 mm and height 70 mm.

Solution : Refer to Fig. 13-2(a). We draw the plan and elevation with given conditions. The cutting plane line PL is now drawn, which passes through the top right-corner in elevation, and cuts the left longer edge at one point. Draw three rectangles, side by side, with one equilateral triangle at the extreme bottom left. This is the development before section. Project from the point on the left cutting edge in elevation to the developed surface. The development after section is covered by dark lines [Fig. 13-2(b)].

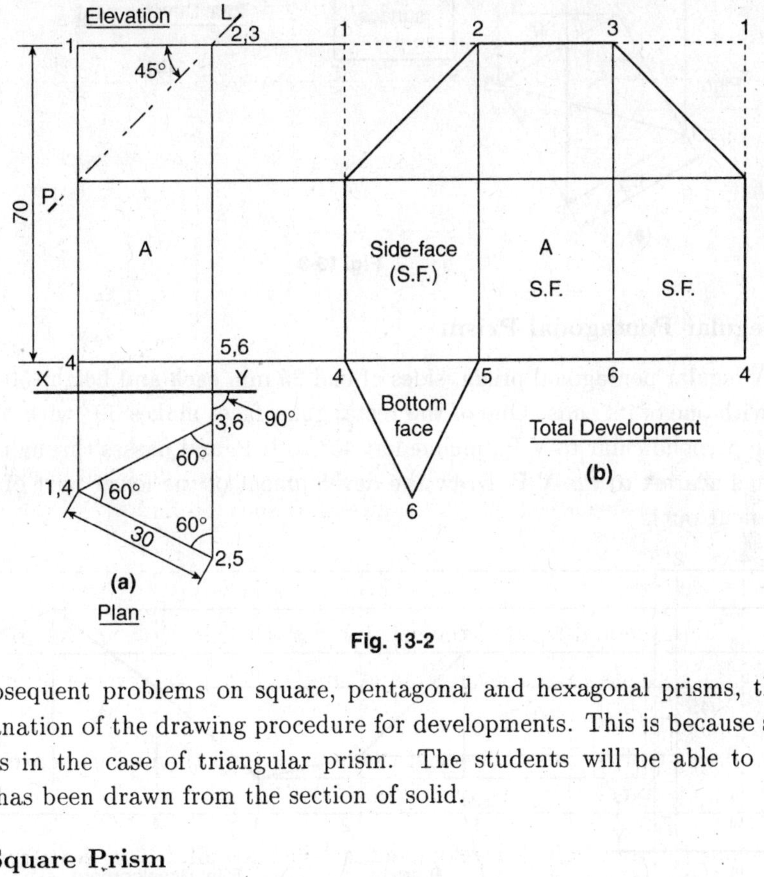

Fig. 13-2

N.B. For subsequent problems on square, pentagonal and hexagonal prisms, there will be no detailed explanation of the drawing procedure for developments. This is because same procedure is followed, as in the case of triangular prism. The students will be able to follow how the development has been drawn from the section of solid.

13.4(b)(ii) Square Prism

Problem 3 : A square prism is placed with its one end on the ground in such a way that, one of the sides of end makes 30° with V.P. Each side of end is 30 mm and height 60 mm. A cutting plane is perpendicular to V.P., inclined at an angle of 30° to H.P. and passes through the mid-point of the axis. Develop the lower part of the truncated prism.

Solution : Refer to Fig. 13-3(a). The developed surface is shown by dark lines in Fig. (b). The cut-out portions are shown by dotted lines.

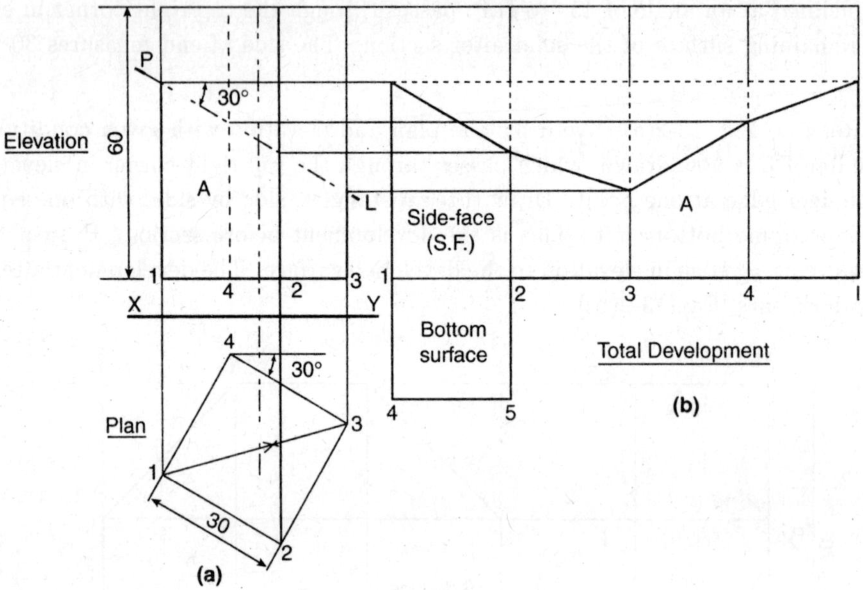

Fig. 13-3

13.4(b)(iii) Regular Pentagonal Prism

Problem 4 : A regular pentagonal prism, sides of end 25 mm each and height 50 mm, is placed on the ground with one of its ends. One of the rectangular faces makes 30° with V.P. The prism is cut by a plane perpendicular to V.P., inclined at 45° to H.P. and passes through the mid-point of the longer edge nearest to the V.P. Draw the development of the lower part of the truncated prism (i.e., after cut-out).

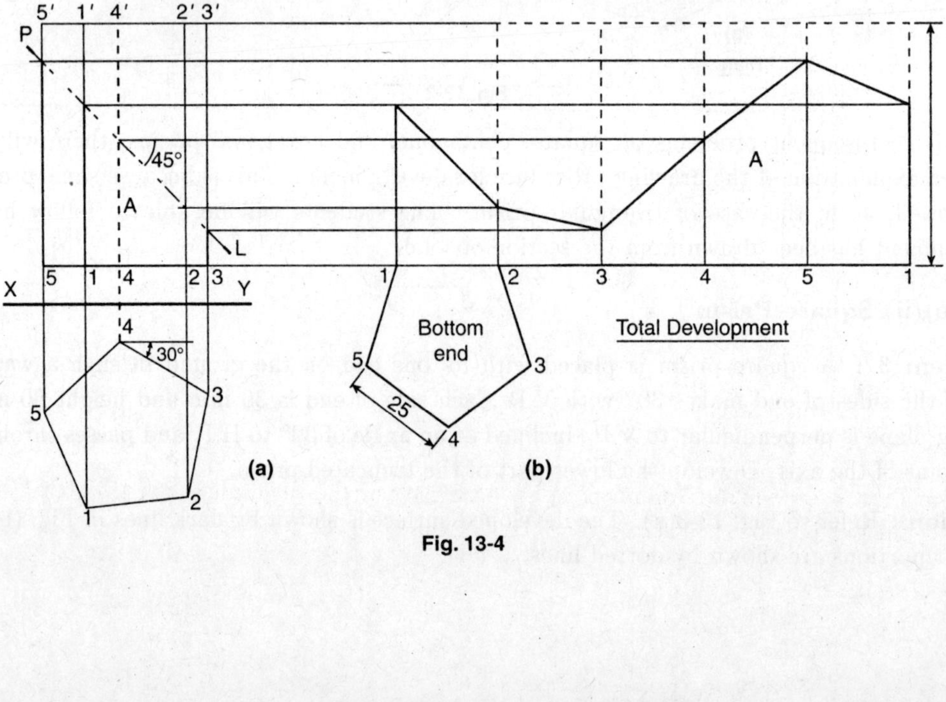

Fig. 13-4

Solution : The longer edge nearest to V.P. (4-4′) is shown dotted [refer Fig. 13-4(a)] in the elevation. The development is shown by dark lines in the right side of elevation [Fig. 13-4(b)]. The cut-out portions are shown by dotted lines. As the top-end has been cut-out, it is not shown in the figure.

13.4(b)(iv) Regular Hexagonal Prism

Problem 5 : A regular hexagonal prism is placed on the ground with one of its ends. One of the sides of end is parallel to V.P. Length of the sides of end is 25 mm each and height 60 mm. The prism is cut by a plane perpendicular to V.P., inclined at 60° to H.P. and passes through a point on the axis at a height of 10 mm from the top end. Draw the development of the total surface of the left-portion of the cutting plane (The portion at the right side of the cutting plane is assumed to be removed).

Solution : Refer to Fig. 13-5(a). The following steps are as follows :

(a) Draw the elevation and plan. Draw the section-plane *P-L* according to the condition of the problem. In the elevation, all the vertical edges (1-1′, 2-2′,. . ., 6-6′) are true-lengths, as these are parallel to V.P.

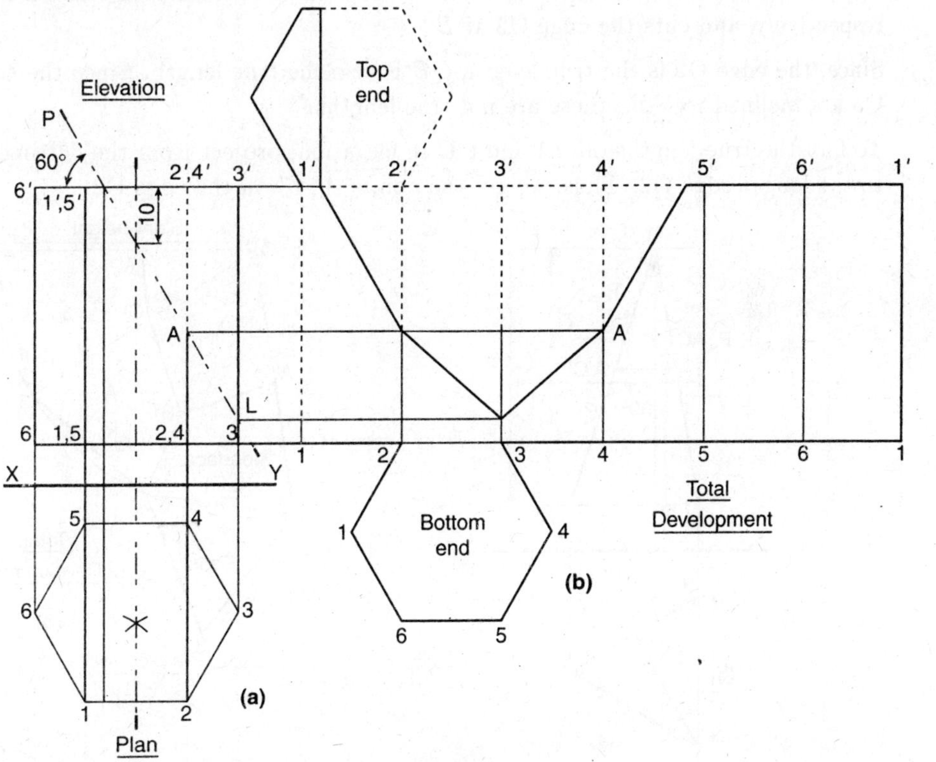

Fig. 13-5

(b) Then draw the development. Draw dark lines at the remaining part [Fig. 13-5(b)]. The cut-out portions are shown by dotted lines.

13.4(c) Right Pyramids

The **lateral-surface of a regular pyramid consists of a number of equal isosceles triangles**, placed side by side. The bottom surface may be an equilateral triangle, a square, a regular pentagon or a hexagon, according to the number of base-sides of the pyramid.

For drawing the development, the **true-length of the longer edge** should be found out geometrically. This will be explained in problem 6.

13.4(c)(i) Right Triangular Pyramid

Problem 6 : An equilateral triangular pyramid, sides of base 25 mm each and height 40 mm, is placed with its base on the ground. One of the side of base is perpendicular to V.P. The pyramid is cut by a plane perpendicular to V.P., inclined at an angle of 30° to H.P. and bisecting the axis. Draw the development of the lower part of the truncated pyramid.

Solution : Refer to Fig. 13-6(a). The drawing steps are as follows :

(i) Draw plan and elevation with given conditions. **Since the longer edge O-2 in plane is parallel to V.P., the length O-2 in elevation is the true length (TL)** of the longer edge of the pyramid. The cutting plane PL cuts the longer edges O1 and O3 at A and C respectively, and cuts the edge O2 at B.

Since, the edge O2 is the true length, OB is also the true length. Since the edges O1 and O3 are inclined to V.P., these are not true lengths.

To find the true lengths of OA and OC in elevation, project from the cutting points A, C to the longer edge O2. Thus, we get the points A', C' on the edge O2.

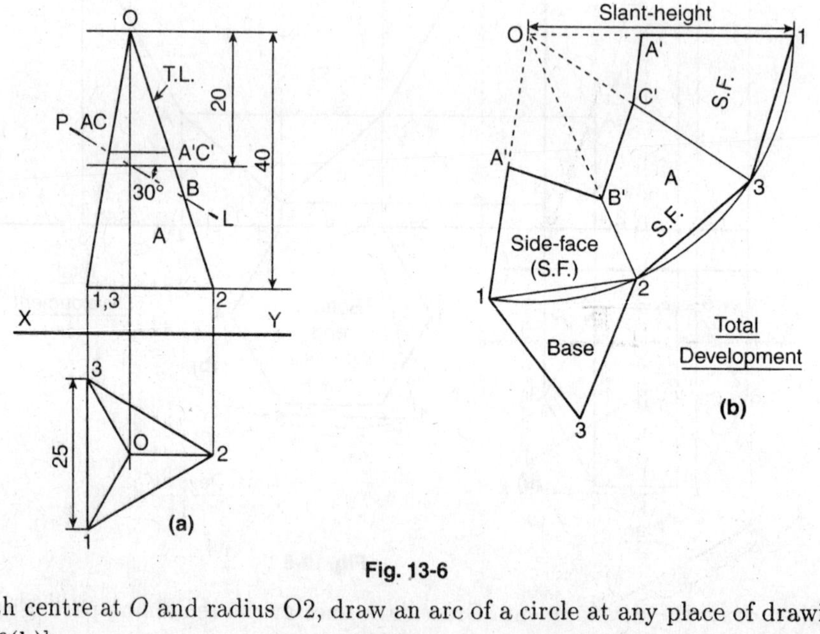

Fig. 13-6

(ii) Now, with centre at O and radius O2, draw an arc of a circle at any place of drawing sheet [Fig. 13-6(b)].

(iii) With radius 25 mm (side of base of pyramid) cut out three equal portions of the arc at 2, 3 and 1.

(iv) Join 1-2,2-3,3-1 by straight lines. Join also O-1,O-2,O-3,O-1, which are longer edges of the pyramid (true length).

(v) On the line 1-2, draw an equilateral triangle 1-3-2, which is the base.

(vi) With radius OA', cut lengths OA' on lines O1 at A' and A'.

(vii) Similarly, with radius OB, cut length OB'on the edge O2.

(viii) Join $A'B'$, $B'C'$, $C'A'$ by straight lines. Keep the lines OA', OB', OC' and OA' by faint lines.

(ix) The development of the lower part of the pyramid is shown by dark lines [Fig. 13-6(b)]. The cut-out portions are shown as dotted lines.

13.4(c)(ii) Square Pyramid

Problem 7 : A square pyramid, each side of base 25 mm and height 50 mm, is placed with its base on the ground. One of the sides base makes an angle of 60° to V.P. The pyramid is cut by a plane, perpendicular to V.P., inclined at 45° to H.P. and bisecting the axis. Draw development of the lower part of the truncated pyramid.

Solution : Refer to Fig. 13-7(a). Here, all the longer edges are inclined to V.P., and therefore, they are not true lengths. The drawing steps are as follows :

(i) From the apex O in plan, draw a line OK, parallel to XY.

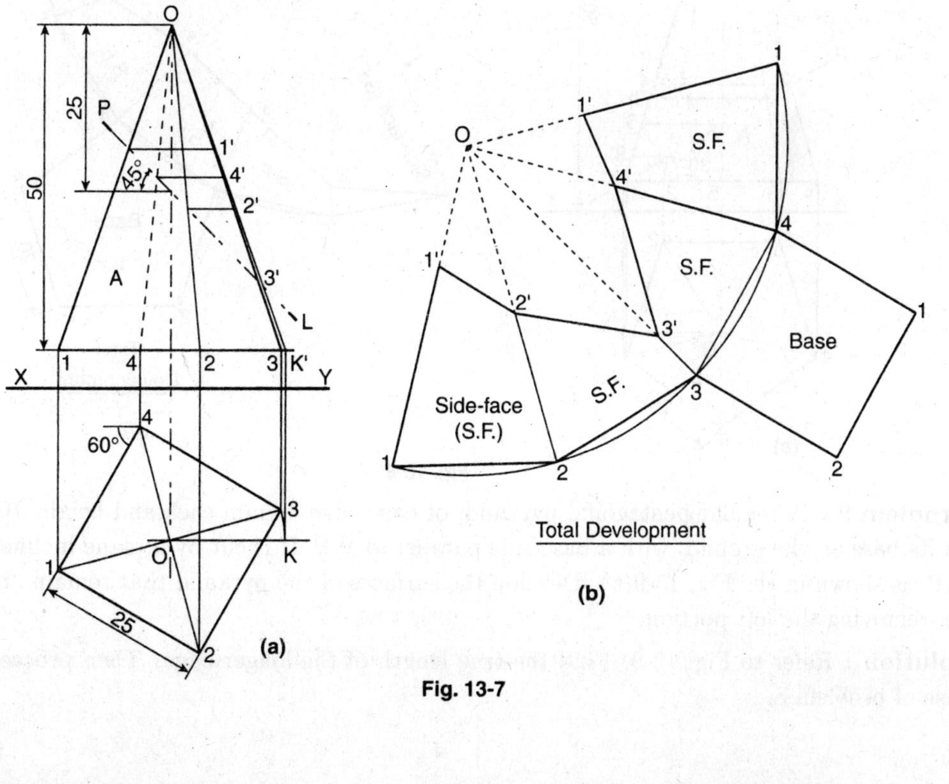

Fig. 13-7

(ii) Project K to elevation, which is K'.

(iii) Join OK' in elevation, which is the true length of the longer edge.

(iv) Project from the junction points of cutting plane and longer edges in elevation to the true length OK'. These points are $1'$, $4'$, $2'$ and $3'$.

(v) Now proceed to draw the development as in the case of triangular pyramid [Fig. 13-7(b)].

13.4(c)(iii) Regular Pentagonal Pyramid

Problem 8 : A regular pentagonal pyramid, each side of base 20 mm and height 50 mm, is placed with its base on ground. One side of base is parallel to V.P. and two triangular faces point towards the observer. It is cut by a plane perpendicular to V.P., inclined at 30° to H.P. and passing through the right bottom corner in elevation. Draw the development of the pyramid to the left of the cutting plane PL.

Solution : Draw the plan and elevation. Refer to Fig. 13-8(a). Proceed to draw the plan, elevation and development, as in the case of a square pyramid. Find the true length. The development of the lower part is shown by dark lines [Fig. 13-8(b)]. The upper truncated part is shown by dotted lines.

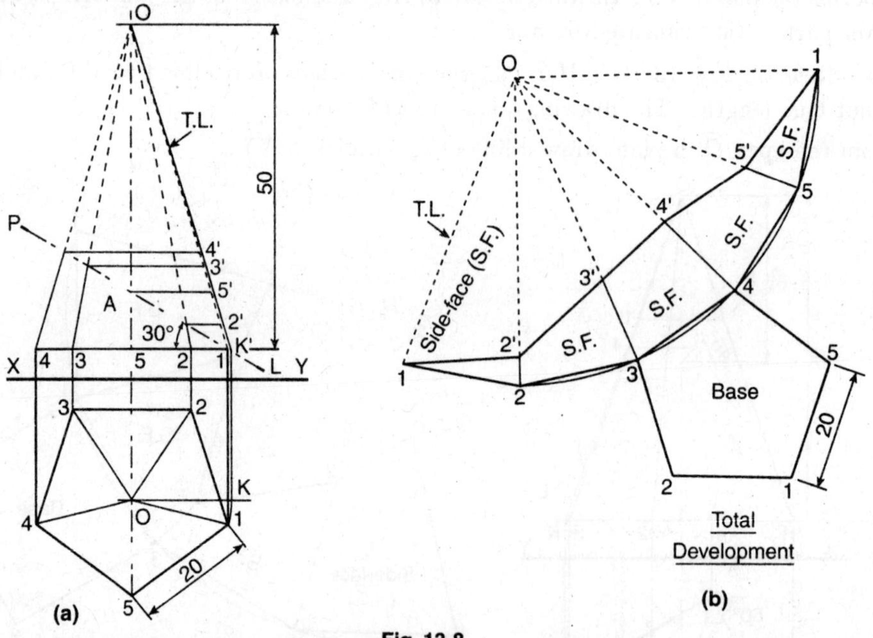

Fig. 13-8

Problem 9 : A regular pentagonal pyramid, of base-edge 30 mm each and height 70 mm, rests on its base on the ground, with a base-side parallel to V.P. It is cut by a plane inclined at 60° to V.P. as shown in the Fig. 13-9(a). Develop the surfaces of the pyramid that remain after section, i.e. removing the left portion.

Solution : Refer to Fig. 13-9. Find the true length of the longer edge. Then proceed as in the case of problem 8.

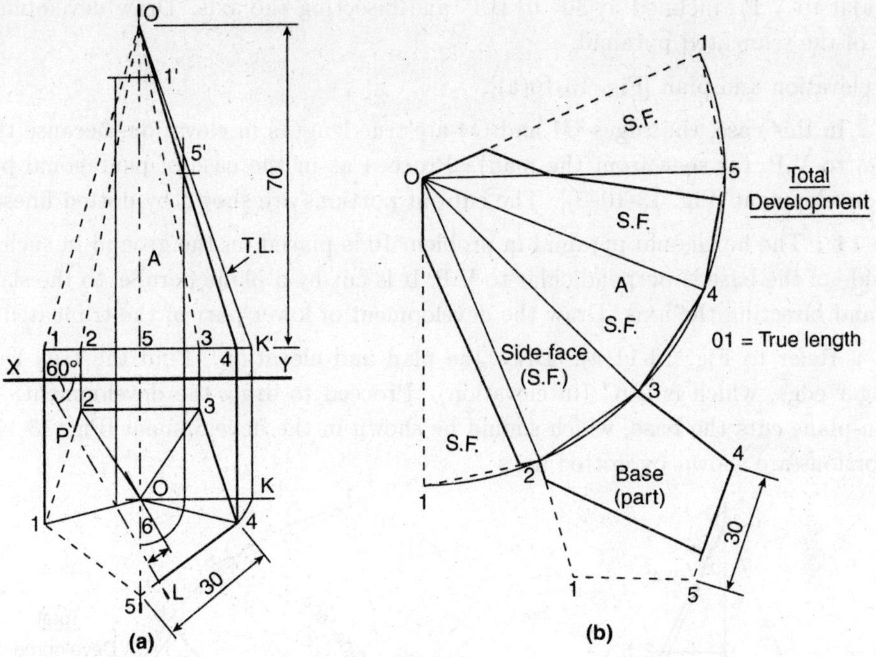

Fig. 13-9

13.4(c)(iv) Regular Hexagonal Pyramid

Problem 10 : A regular hexagonal pyramid, each side of base 25 mm and height 50 mm, is placed with its base on ground, so that one side of the base is parallel to V.P. It is cut by a plane

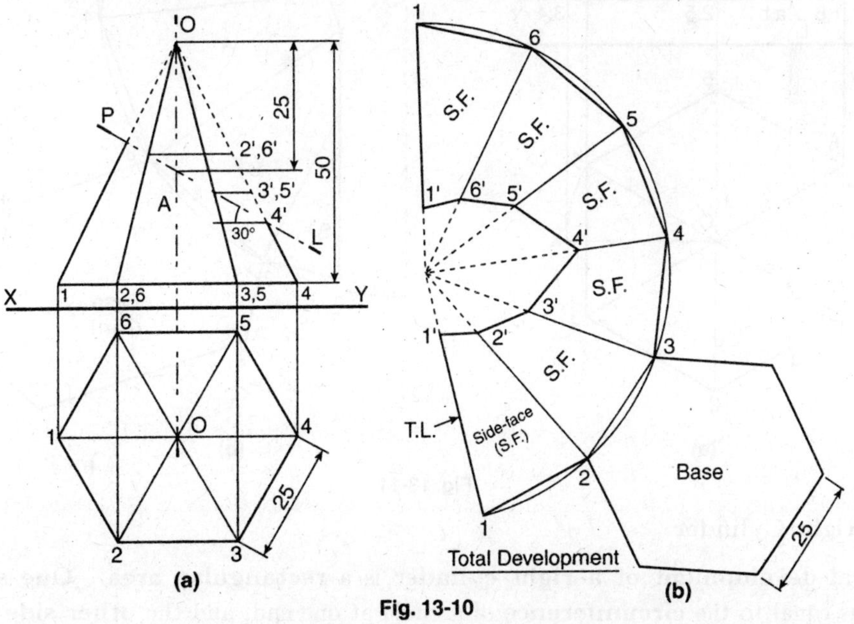

Fig. 13-10

perpendicular to V.P., inclined at 30° to H.P. and bisecting the axis. Draw development of the lower half of the truncated pyramid.

Draw the elevation and plan [Fig. 13-10(a)].

Solution : In this case, the edges O1 and O4 are true lengths in elevation, because these edges are parallel to V.P. (as seen from the plan). Proceed as in the case of pentagonal pyramid to draw the development [Fig. 13-10(b)]. The cut-out portions are shown by dotted lines.

Problem 11 : The hexagonal pyramid in problem 10 is placed on the ground in such a manner that one side of the base is perpendicular to V.P. It is cut by a plane parallel to the slant edge in elevation and bisecting the axis. Draw the development of lower part of the truncated pyramid.

Solution : Refer to Fig. 13-11(a). Draw the plan and elevation. Find the true length (TL) of the longer edge, which is OK' (in elevation). Proceed to draw the development. Note that the section-plane cuts the base, which should be shown in the development [Fig. 13-11(b)]. The cut-out portions are shown by dotted lines.

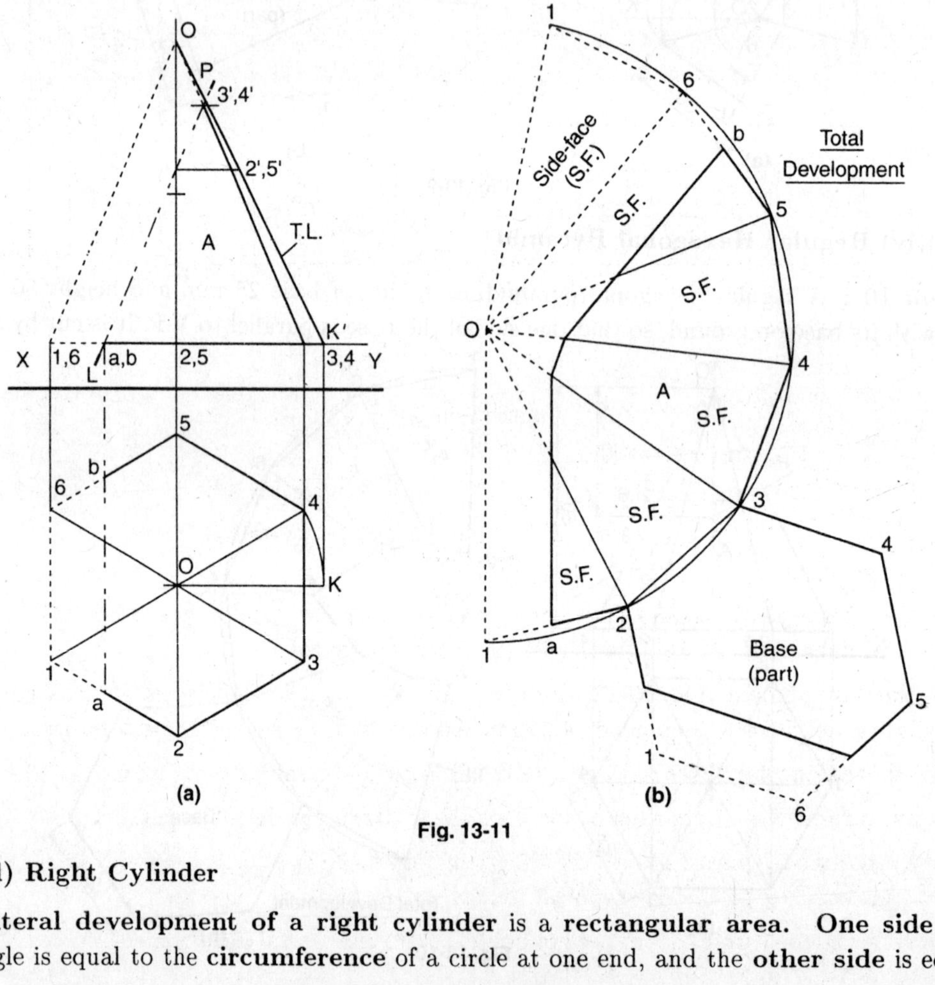

Fig. 13-11

13.4(d) Right Cylinder

The **lateral development of a right cylinder** is a **rectangular area. One side** of the rectangle is equal to the **circumference** of a circle at one end, and the **other side** is equal to the **height of the cylinder**. The top and bottom surfaces are two circles, whose diameters are equal to the diameter of the cylinder.

Problem 12 : A right cylinder (diameter of end 30 mm and height 60 mm) is placed with its one end on the ground. It is cut by a plane, perpendicular to V.P., inclined at 45° to H.P. and bisecting the axis. Draw the development of the lower half of the truncated cylinder.

Solution : Refer to Fig. 13-12(a). The drawing steps are as follows :

(i) Draw the plan and elevation with given dimensions.

(ii) Divide the circle in plan into 12 equal parts and number them from 1 to 12.

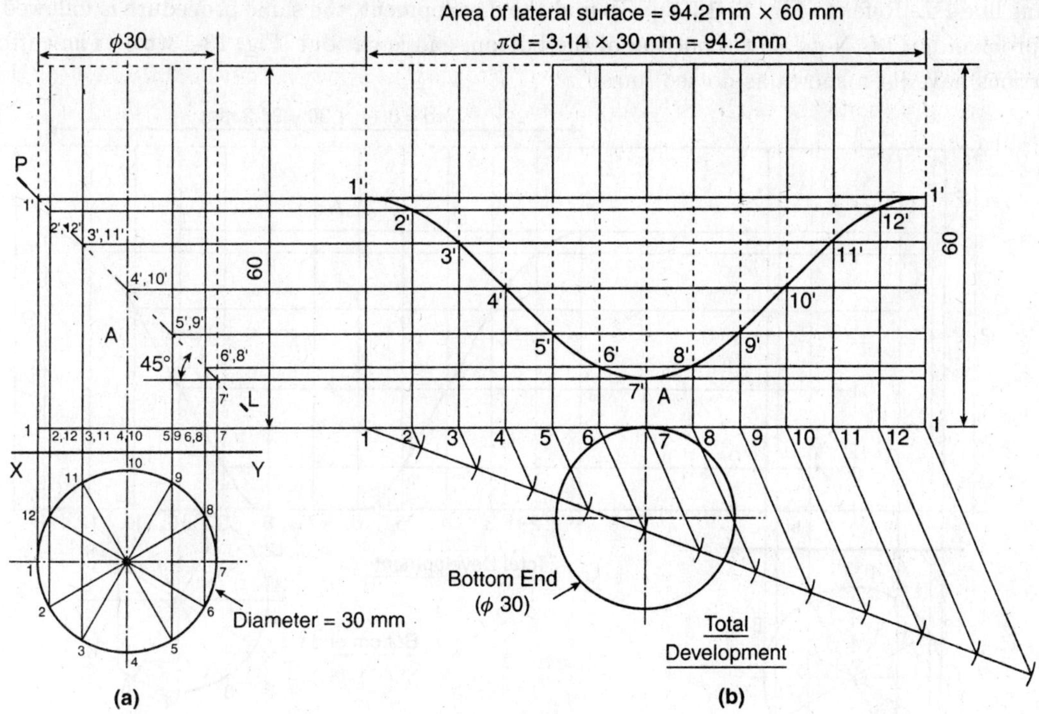

Fig. 13-12

(iii) Project from all division-points into the elevation. The section plane line PL cuts the vertical lines at $1'$, $2'$, $12'$, $3'$, $11'$, $4'$, $10'$, $5'$, $9'$, $6'$, $8'$ and $7'$.

(iv) In the development [Fig. 13-12(b)] draw a line 1-1 equal to the circumference (πd) of the cylinder, i.e., 3.14 × 30 mm or 94.2 mm. Another side of the rectangle is 60 mm.

(v) Divide the line 1-1 (94.2 mm) into 12 equal parts and number them from 1 to 12.

(vi) Draw vertical lines from each of the division-points in the development.

(vii) Draw horizontal lines from all the junction points ($1', 2', 12', \ldots, 7'$ in elevation) to meet the vertical lines in the development at $1', 2', 3', \ldots, 1'$).

(viii) Draw a smooth curve through these points to get the lateral development.

(ix) A circle (dia. = 30 mm) is drawn at bottom, touching the line (1-1). This represents the bottom end of cylinder.

As the upper part is removed, the development of the lower half is shown by dark lines and the upper part by dotted lines [Fig. 13-12(b)]. The top-end (a circle) is also not shown.

Problem 13 : The cylinder in problem 12 is cut by a section-plane (inclined at 60° to H.P.) in such a manner that, the base is also cut. Draw the development of the lower part of the truncated cylinder.

Solution : Draw elevation and plan of the cylinder with the given dimensions. Draw the sectional plane line *PL* [Refer to Fig. 13-13(a)]. To find the development, the same procedure is followed as in problem no. 12. Note that a portion of the bottom-end is cut-out [Fig. 13-13(b)]. The cut-out portions have been shown as dotted lines.

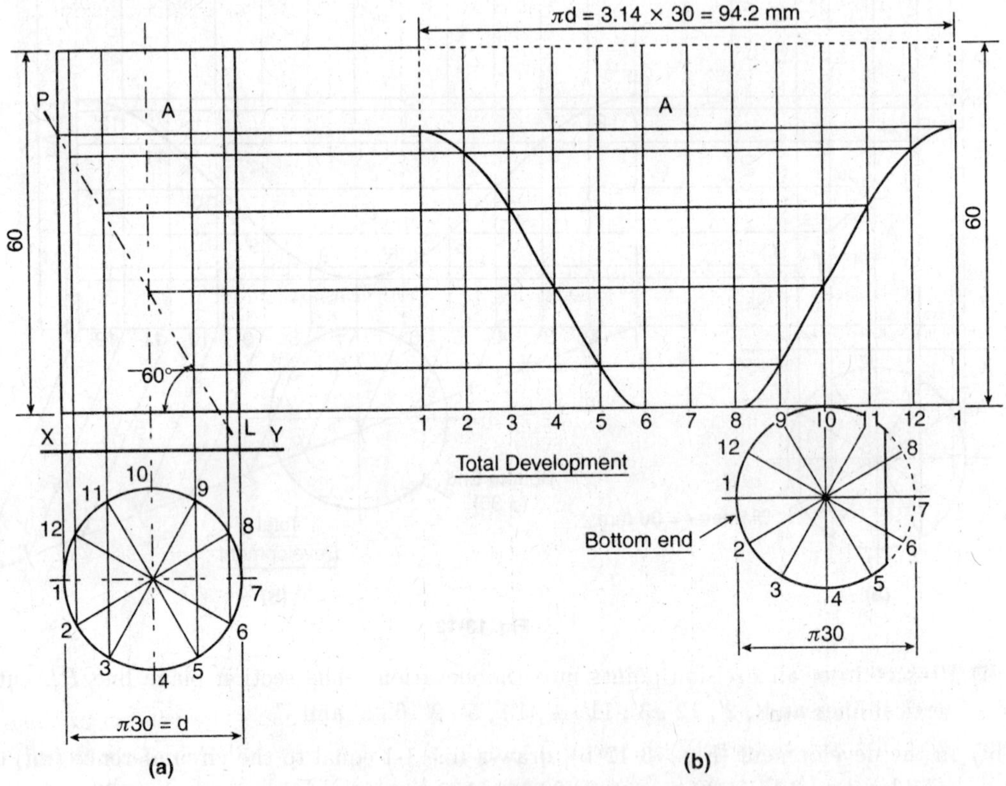

Fig. 13-13

Problem 14 : The cylinder, of same dimensions as in problem 12, is cut by **three sectional planes** as follows :

(a) The cutting plane P_1L_1 is inclined at an angle of 60° to H.P. and passes through the centre of the top-end in elevation, as shown at the right side.

(b) The cutting plane P_2L_2 is inclined at an angle of 30° to H.P. and passes through the centre of the top-end in elevation, as shown at the left-side.

(c) The cutting plane P_3L_3 is inclined at an angle of 30° to H.P. and passes through the bottom right corner in elevation.

Draw the development of the lateral surface of the cylinder after cutting by the three planes mentioned above.

Solution : Refer to [Fig. 13-14(a)]. Draw the plan and elevation. Draw also the three cutting planes P_1L_1, P_2L_2 and P_3L_3, according to the questions. Draw the lateral development for each cutting plane separately, as done in problem 12. Draw dark lines in the development for the remaining part of the truncated cylinder [Fig. 13-14(b)].

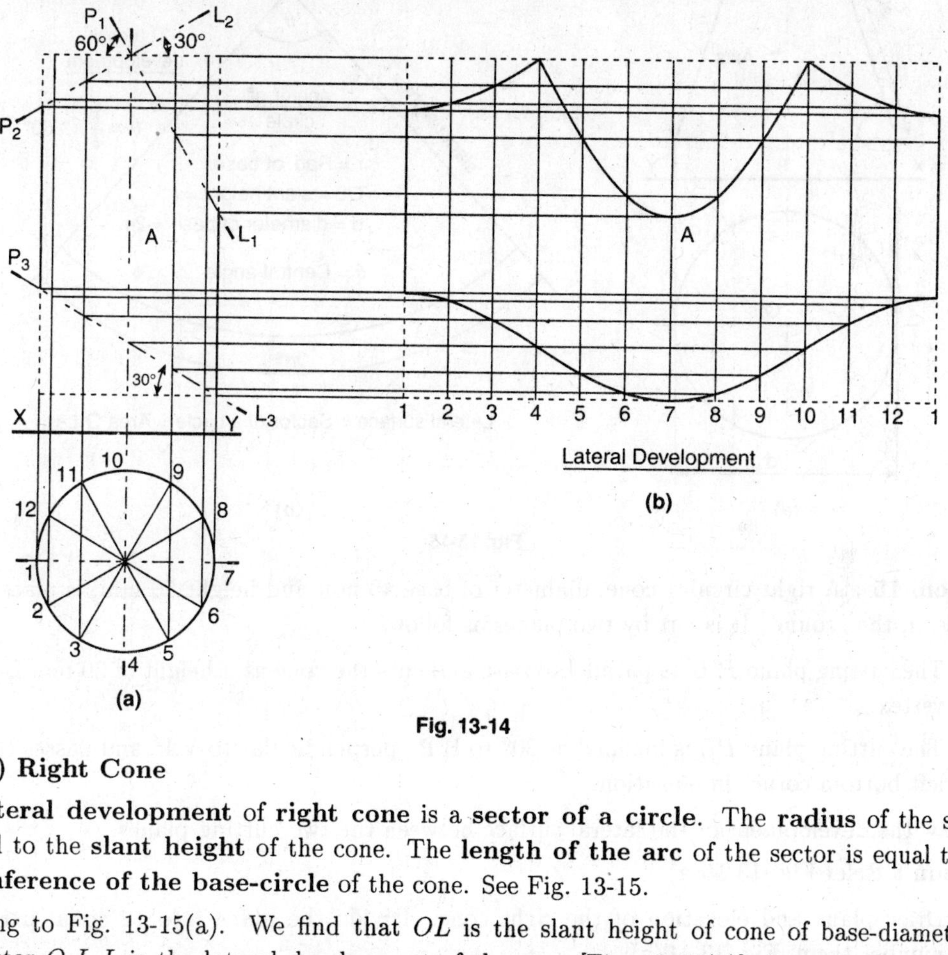

Fig. 13-14

13.4(e) Right Cone

The **lateral development** of **right cone** is a **sector of a circle.** The **radius** of the sector is equal to the **slant height** of the cone. The **length of the arc** of the sector is equal to the **circumference of the base-circle** of the cone. See Fig. 13-15.

Referring to Fig. 13-15(a). We find that OL is the slant height of cone of base-diameter d. The sector O-L-L is the lateral development of the cone [Fig. 13-15(b)]. Length of arc L-L = circumference of base circle, i.e., πd.

To find out the central angle θ, we write the formula as follows :

$$\theta = \frac{r}{l} \times 360°$$

where r = radius of base of the cone, l = slant height of the cone = OL [Fig. 13-15(b)].

Thus, by calculating the central angle θ, the lateral development of the cone can be drawn accurately.

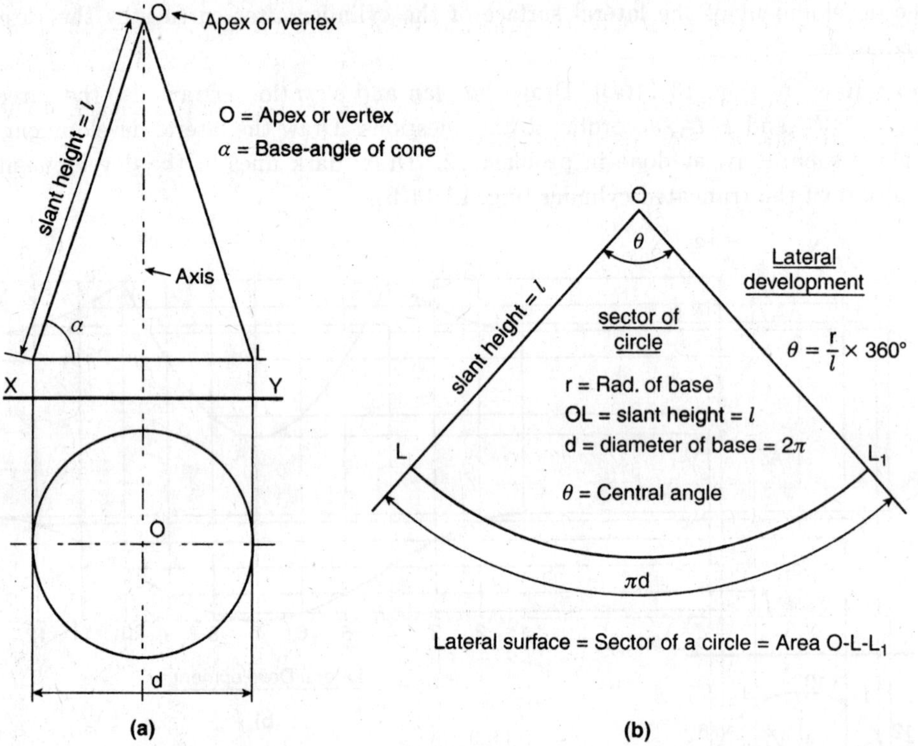

Fig. 13-15

Problem 15 : A right circular cone, diameter of base 40 mm and height 50 mm, is placed with its base on the ground. It is cut by two planes as follows :

(i) The cutting plane $P'L'$ is parallel to base and cuts the cone at a height of 20 mm from the vertex.

(ii) The cutting plane PL is inclined at 30° to H.P., perpendicular to V.P. and passes through left bottom corner in elevation.

Draw the development of the lateral surface between the two cutting planes.

Solution : Refer Fig. 13-16(a).

(i) Draw plane and elevation of the right cone. Divide the plane into 12 equal parts and number them [Fig. 13-16(a)].

(ii) Project from all the division points in plan to the base of cone in elevation and join them with the vertex.

(iii) Draw the cutting plane lines $P'L'$ and PL, according to the problem [Fig. 13-16(a)].

(iv) In the elevation, draw horizontal lines from the junction points of cutting plane and the radial lines.

Objective-Type Questions

...ks with proper words :

...eral development of a right cone is a ______ of a circle.

...he longer edge of a right pyramid is parallel to V.P., we get ______ length of ...dge.

...ateral development of a right cylinder is a ______.

...ateral development of a right pyramid consists of a number of ______.

...naking a bucket with tin-sheet, the knowledge of ______ is helpful.

...e lateral-surface of a right prism consists of a number of ______.

...e total-surface of a tetrahedron consists of ______.

...hen the longer edge of a right pyramid is inclined to V.P., we get the ______ length ...f that edge.

...n a circle, the arcs are ______ to ______ angles.

In drawing the development of a surface, ______ length must be used.

(i) sector, (ii) true, (iii) rectangle, (iv) isosceles triangles, (v) surface-development, ...rectangles, (vii) four equilateral triangles, (viii) apparent or shorter, (ix) proportional, ...ral, (x) true.

...heet-metal funnel is to be prepared as per the following details :

Larger diameter of conical portion : 80 mm

Height of the conical portion : 50 mm

Diameter of the cylindrical stem : 25 mm

Height of the cylindrical stem : 40 mm

Draw the complete development, so that the sheets may be cut accordingly.

(AMIE—Summer, 2005)

These horizontal lines meet either to the left or right slant edge of cone, (as it is the true length).

(v) The points on the slant edges are numbered as (2′,12′), (3′,11′), (4′,10′), (5′,9′) and (6′,8′).

(vi) With slant edge as radius, draw an arc of a circle, and take the central angle θ by calculation. This is the lateral development as shown in Fig. 13-16(b).

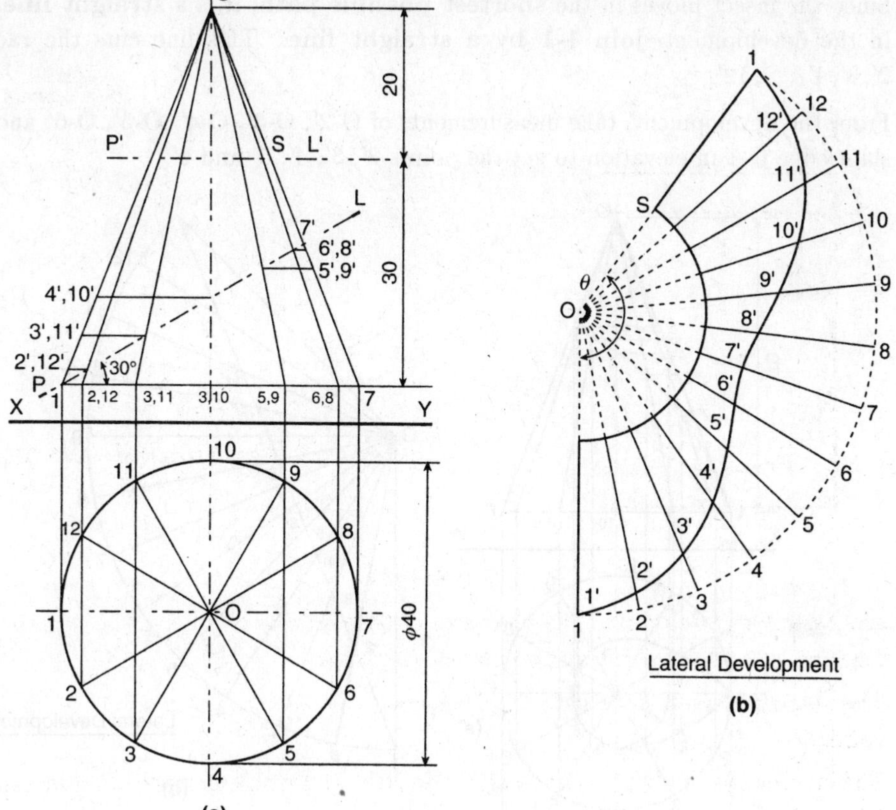

(a) (b)

Fig. 13-16

(vii) With centre at O and radius $= OS$, draw an arc over the sector already drawn.

(viii) From the elevation, take measurements O-2′,O-3′,O-4′,O-5′,O-6′,O-7′ by a compass, and cut the radial lengths on the development. Similarly, mark O-8′,O-9′,O-10′,O-11′,O-12′.

(ix) Now, draw a smooth curve through the points $1, 2', 3', 4', \ldots, 12', 1$.

(x) The development between the two cutting planes $P'L'$ and PL is shown by dark lines [Fig. 13-16(b)].

Problem 16 : A right circular cone, diameter of base 60 mm and height 70 mm, is placed with its base on the ground. An insect starts moving from a point on the circumference of the base.

After passing over the curved surface of the cone in the shortest possible path, the insect returns to the starting point. Show the path of the insect on the development, and in elevation and plan of the cone.

Solution : Refer to Fig. 13-17(a). The steps are as follows :

(i) Draw the plan and elevation and also draw the development of the lateral surface [Fig. 13-17(b)].

(ii) Since the insect moves in the **shortest possible path**, it is a **straight line**. Therefore, in the development, **join 1-1 by a straight line.** This line cuts the radial lines at $2', 3', 4', \ldots, 12'$.

(iii) From the development, take measurements of O-2′, O-3′, O-4′, O-5′, O-6′ and cut on the slant edge O-1 in elevation to get the points 2′, 3′, 4′, 5′ and 6′.

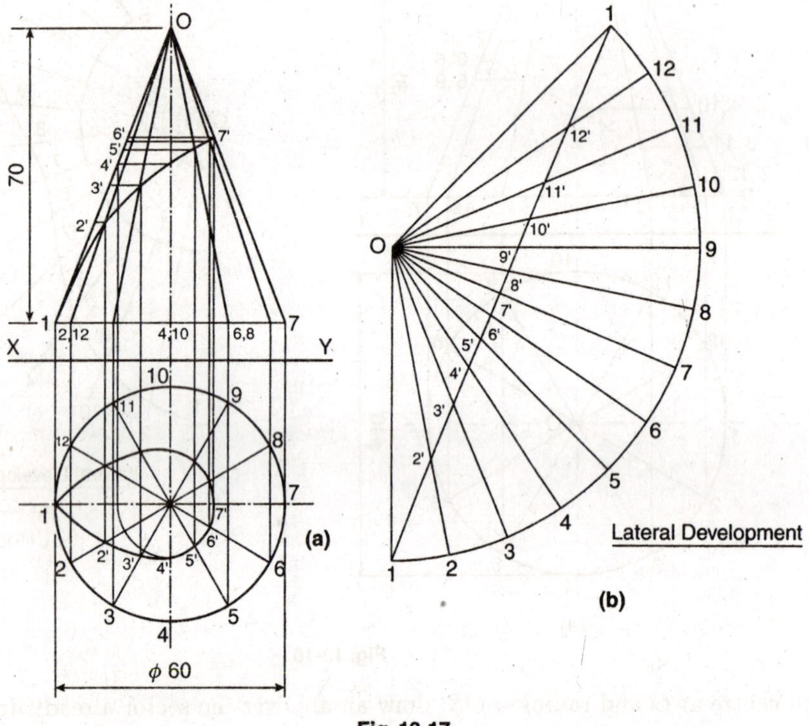

Fig. 13-17

(iv) Take measurement of O-7′ from the development and mark 7′ on the slant edge O-7 in elevation.

(v) Draw horizontal lines from the points 2′, 3′, 4′, 5′, 6′ (in elevation) to meet the corresponding radial lines in elevation.

(vi) Draw a smooth curve from 1 to 7 in the elevation, which is **the path of the insect.**

(vii) The points 8′, 9′, 10′, 11′, 12′ are not visible, because these are on the back side of the cone.

(viii) From the junction points of ho to meet the corresponding radial in the plane.

(ix) Join $1', 2', 3', \ldots, 12', 1'$ by a smoo **from top** [see plan, Fig. (a)].

Pr

1. A galvanized iron (G.I.) cylindrical drum n the total surface of the drum with the lid.

2. A regular hexagonal pyramid, each side of bas base on the ground. One side of base is perpen at 45° to H.P., perpendicular to V.P. and bisecti of the lower part of the truncated cone.

3. A right circular cone, dia. of base 30 mm and height It is cut by a plane parallel to the slant surface and development of the lower part of the truncated cone.

4. Draw the total development of a closed wooden box of s

5. A regular hexagonal prism, each side of end 30 mm and h one end on the ground, so that one side of end is perpendicu inclined at 45° to H.P., perpendicular to V.P. and bisectin development of the lower part of the truncated prism.

6. The vertical section of a right circular cone through the axis is a 50 mm and height 60 mm. The cone rests on H.P. on its base. extreme left end of the base and walks around the curved surface to the starting point. Find geometrically, the shortest path the ant the path in the front view and the top view of the cone. **(A**

7. A right circular cone, of base-diameter 75 mm and 100 mm long, axis base. An ant *P* starts from a point on its base and moves around the su and comes back to the same point through the shortest path. Find the len plan and elevation of the cone. **(AMIE–**

8. A right cone of base diameter 100 mm and axis 120 mm long stands on its The cone is cut by a horizontal plane intersecting the axis at 50 mm below th the upper portion is removed. The truncated cone is again cut by a section-pla at 30° to the base and passing through the right end of the base in elevation lower portion is removed. Draw the development of the surface of the remaining tr cone. **(AMIE—Summer,**

9. Fill in the blan
(i) The lat
(ii) When that e
(iii) The l
(iv) The
(v) In n
(vi) Th
(vii) Th
(viii) W o
(ix)
(x)

Ans.
(vi) cent

10. A s

Chapter 14

Isometric Projection

14.1 Introduction

With the help of our knowledge about projection of solids, section of solids and surface-developments, we can visualize some geometrical solids, such as cube, prisms, pyramids, cone, cylinder, etc. If we assume that, even this knowledge is not sufficient to visualize geometrical or other solid objects, we can take the help of **pictorial drawings. Through pictorial drawings, we can see a number of faces of geometrical or other solids at a time.** It becomes easier to have a clear idea about a solid object to us. **Pictorial drawings are divided into many classes, such as axonometric projection, oblique projection and perspective projection.** We shall concentrate on isometric projection, which comes under axonometric projection. (A chapter on perspective projection has been added in chapter 18.)

14.2 Isometric Projection

In this projection, all the sides of an object are shown to be shortened by equal amounts. Also, the adjacent sides of the object are drawn with an inclination of 30° to the horizontal. This is shown in Fig. 14-1. Here, the isometric projection of a cube is shown.

14.3 Isometric Process

In order to get the isometric view, let us take an example of a cube. We are to proceed as follows :

(i) Take the paper-model of a cube.

(ii) Place the cube with one square face on the ground, so that all vertical faces are equally inclined to V.P (i.e., makes 45° to V.P), as shown in Fig. 14-2(a), plan.

(iii) Number the top and bottom corners as 1, 2, 3, 4 and 5, 6, 7, 8 respectively.

(iv) In the elevation, the solid diagonal 1–7 makes an angle of 35°-16′ with the base or ground, 35°-16′ (35 degree and 16 minute). 1° (one degree) = 60′ (sixty minutes).

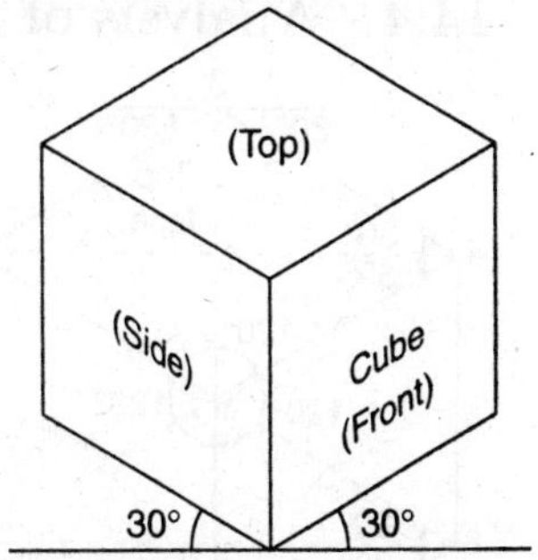

Fig. 14-1

(v) Now, tilt the cube about the corner 5, so that the solid diagonal 1–7 is parallel to H.P (see Fig. 14-2(b), elevation). Thus the bottom face of the cube is inclined at 35°-16′ with H.P. or ground.

(vi) The plan of Fig. 14-2(b) is rotated through 90°, which is shown in Fig. 14-2(c).

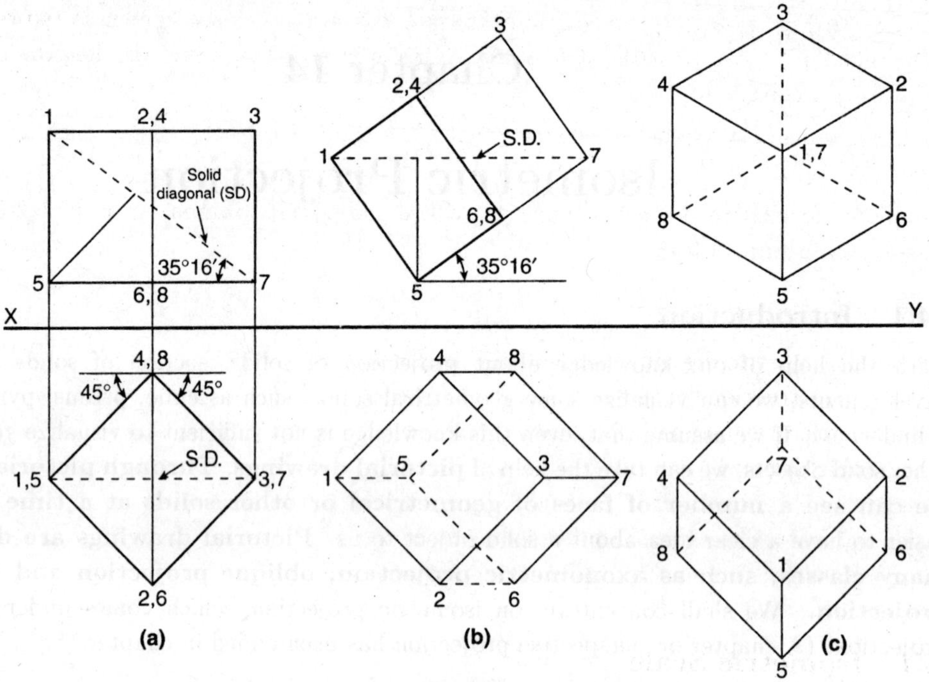

Fig. 14-2

(vii) Finally, project from plan of (c) to get the elevation. This view in Fig. 14-2(c) represents the solid diagonal 1-7, perpendicular to V.P. and parallel to H.P.(This has been clearly explained in chapter 11). This view [Fig. 14-2(c), elevation] is the isometric view of a cube.

(viii) We can now visualize the three faces 1-2-3-4, 1-2-6-5 and 1-4-8-5 of the cube. All these surfaces have been distorted, and are not the true shapes. See Fig. 14-2(c).

14.4 Analysis of Isometric View

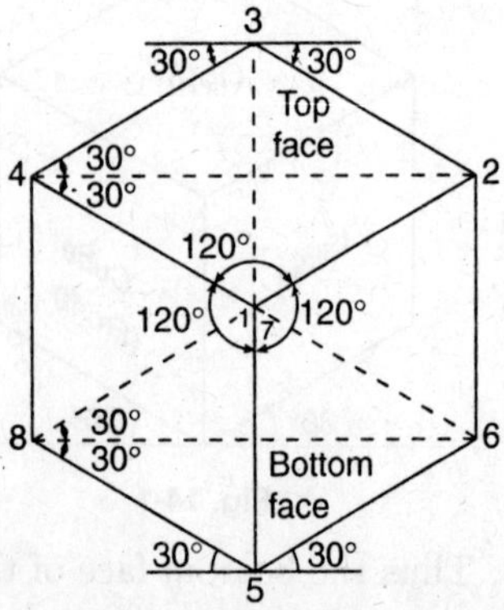

Fig. 14-3

Referring to Fig. 14-3, let us try to understand the isometric view by analysis. 1-2, 1-4, 1-5 are isometric axes, which are 120° apart.

(i) All the horizontal lines are inclined at 30° to H.P.

(ii) All the horizontal and vertical lines are equally shortened.

(iii) All the faces look similar and are equal rhombuses, and not squares.

(iv) Top and bottom longer diagonals 4-2 and 8-6 are parallel to V.P.

14.5 Isometric Axes, Lines and Planes

Referring to Fig. 14-3, the three skeleton lines 1-2, 1-4 and 1-5 are known as **isometric axes.** These make 120° angles with each other. The isometric axes represent length, breadth and thickness of a solid.

The lines, which are parallel to isometric axes, are called **isometric lines**, such as 2-3, 3-4,2-6, etc. in Fig. 14-3.

The planes formed by isometric lines are called **isometric planes**, such as 1-2-3-4, 1-2-6-5, 1-4-8-5, etc. in Fig. 14-3.

14.6 Dimensioning

In general, isometric drawings are not dimensioned. But, for construction purposes (Buildings, roads, etc), isometric drawings may be dimensioned. In an orthographic drawing, the dimensions and lettering are vertical and horizontal. These should be parallel to the corresponding axis in the isometric drawing. Fig. 14-4 shows the method of dimensioning an isometric drawing.

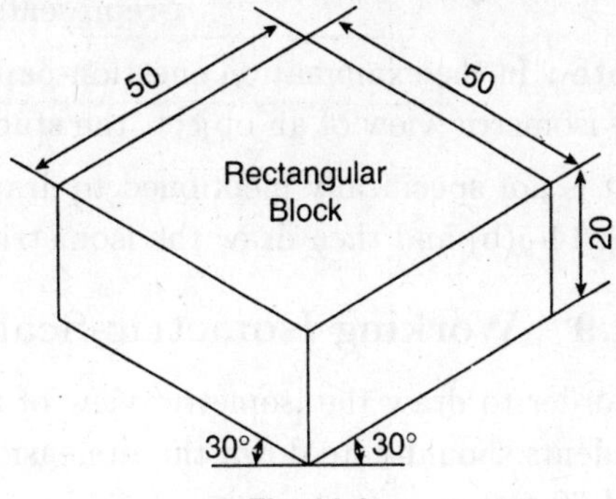

Fig. 14-4

14.7 Isometric Scale

We have already noted that the length of an object is shortened in the isometric view. Mathematically, it can be proved that this shortened ratio is 0.187. But, by using isometric scale, we can avoid this calculation. The procedure of drawing an isometric scale is explained below :

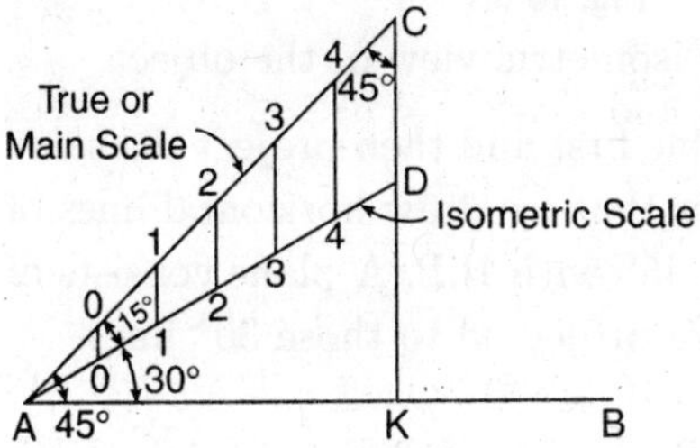

Fig. 14-5(a)

Refer to Fig. 14-5(a).

(i) We draw any line AB of any suitable length.

(ii) We draw another line AC, which is inclined at 45° with AB. Mark points directly from the standard scale (say, up to 70 mm or 80 mm). Thus, the line AC represents the true scale.

(iii) From the point C, we draw a perpendicular CK on AB.

(iv) Now, we draw a line AD, which makes an angle of 30° with AB.

(v) We draw vertical lines (i.e., parallel to CK) from the points 1, 2, 3,..., 6 to meet the line AD at 1, 2, 3,..., 6.

(vi) Now the length AD represents the isometric scale of the true length AC.

14.8 Another Method of Drawing Isometric Scale

Refer to Fig. 14-5(b). Here, the triangle ADC of Fig. 14-5(a) has been drawn separately.

The angle $CAD = 45° - 30° = 15°$. Also$\angle ACD = 45°$.

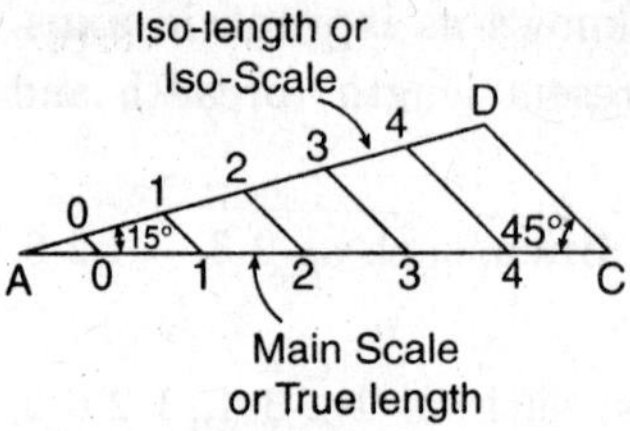

Fig. 14-5(b)

We draw the straight line AC and on it mark the true length from a standard 30 cm or 15 cm scale.

At the point A, draw a line AD, which is inclined at 15° to AC.

Also, draw another line CD, which is inclined at 45° to CA. The lines AD and CD meet at D.

From the points 0, 1, 2,..., 6, draw lines parallel to DC. These lines will meet the line AD at 0, 1, 2,..., 6, etc. The line AD represents isometric scale.

Note : In the examination question-paper, if it is asked to draw isometric scale and then to draw the isometric view of an object, the students should draw the Fig. 14-5(a).

If it is not specifically mentioned to draw the isometric scale, then the students should draw the Fig. 14-5(b) and then draw the isometric view of an object.

14.9 Working Isometric-Scale For Simplicity

In order to draw the isometric view of a given figure (i.e. elevation and plan of an object), the students should note down the dimensions first, such as 5 mm, 15 mm, 21 mm, 30 mm, 49 mm and 60 mm.

(i) Now, draw a line AC (Fig. 14-6), and mark the above distances on AC from a standard scale of length 15 cm or 30 cm.

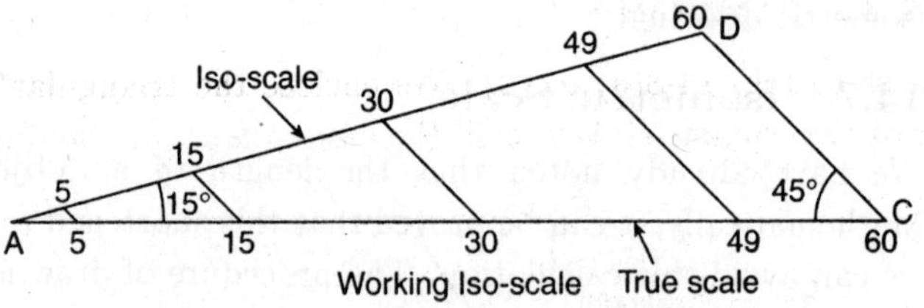

Fig. 14-6

Then draw the triangle ACD, as in Fig. 14.5(b) and get the isometric lengths for the true lengths on AC. This will help to draw the isometric view of the object.

Note : In isometric scale, we draw true measurements on 45° line first and then project the same vertically to 30° line. Physical significance of these behaviour is that we draw horizontal lines of any object as 30° with H.P. but plane consists of those lines is 45° with H.P. A plane consists of lines and lines consist of points. All the points of that plane are projected to those 30° lines.

14.10 Isometric Drawing or View of a Lamina

We are going to explain the procedure of drawing the isometric view of different lamina, according to the following order :

(i) Square (ii) Triangle (iii) Regular pentagon (iv) Regular hexagon (v) Circle.

14.11 Square Lamina

Refer to Fig. 14-7(a) and (b). The front view of a square lamina is square, which is its true shape [Fig. 14-7(a)]. We find that, in this square, there are two horizontal lines(1-2 and 4-3), and two vertical lines (1-4 and 2-3). **In the isometric view, draw always the horizontal lines as**

30°-inclined lines, but the vertical lines will remain vertical. Draw the isometric view in this manner, when the lamina is kept vertical. Thus,the lines 4-3 and 3-4 are drawn 30° inclined to horizontal [Fig. 14-7(b) and (d)]. But the lines 1-4 and 2-3 are drawn vertical. In these cases, the points 4 and 3 are called **poles**, indicated by P [Fig. 14-7(b) and (d)].

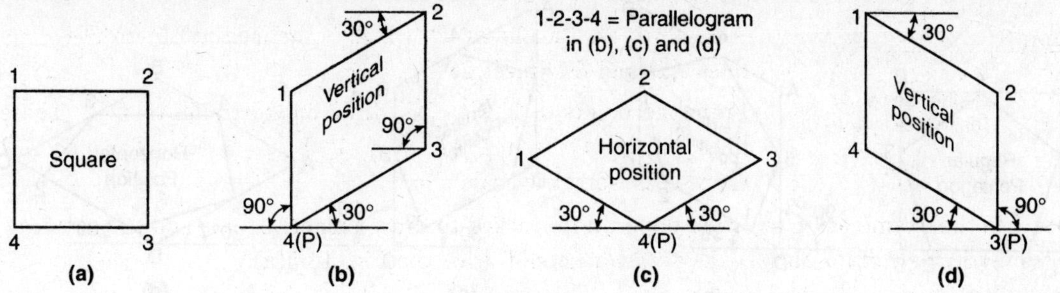

Fig. 14-7

When the lamina is kept in a horizontal position, all the lines (1-2, 4-3, 1-4, 2-3) are drawn 30° inclined to horizontal [Fig. 14-7(c)]. The same procedure will be followed in the case of other lamina (triangle, pentagon, hexagon, etc). In this case, point 4 is known as the pole indicated by P in Fig. 14-7(c).

14.12 Triangle

Refer to Fig. 14-8(a) to (d). We enclose the triangular* lamina by a rectangle, and proceed to draw the isometric view as in the case of a square lamina (See Fig. 14-7).

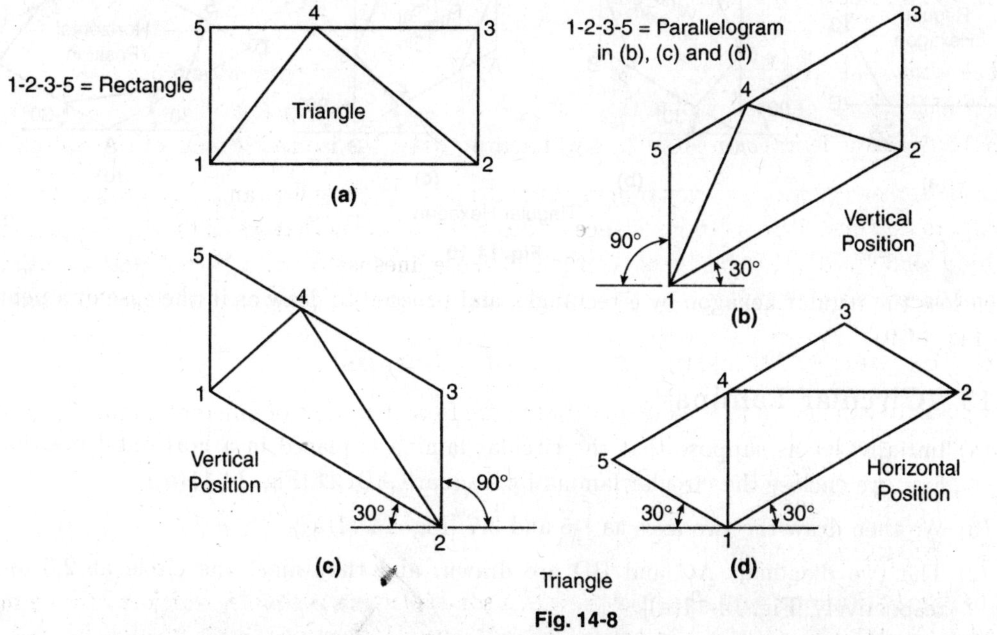

Fig. 14-8

*A 'lamina' is a plane or area of some shape, with two dimensions (length and breadth) only.

14.13 Regular Pentagonal Lamina

Refer to Fig. 14-9(a), (b), (c) & (d), we enclose the regular pentagon by a rectangle, and proceed to draw the isometric views as in the case of triangular or square lamina.

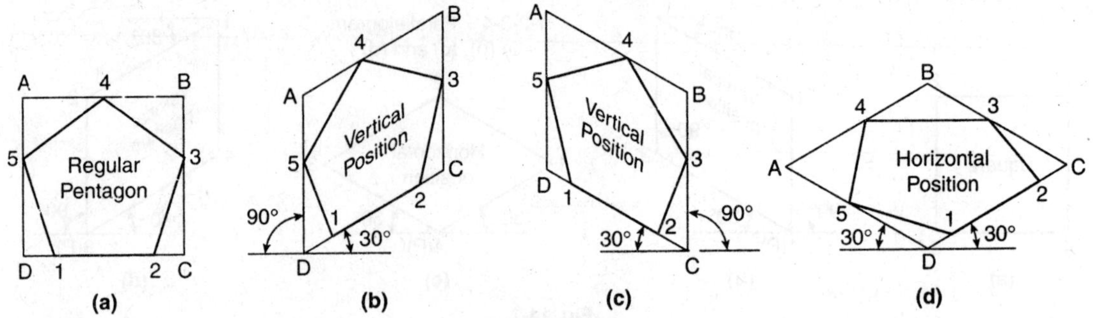

Regular Pentagon

Fig. 14-9

14.14 Regular Hexagonal Lamina

Refer to Fig. 14-10(a) to (d).

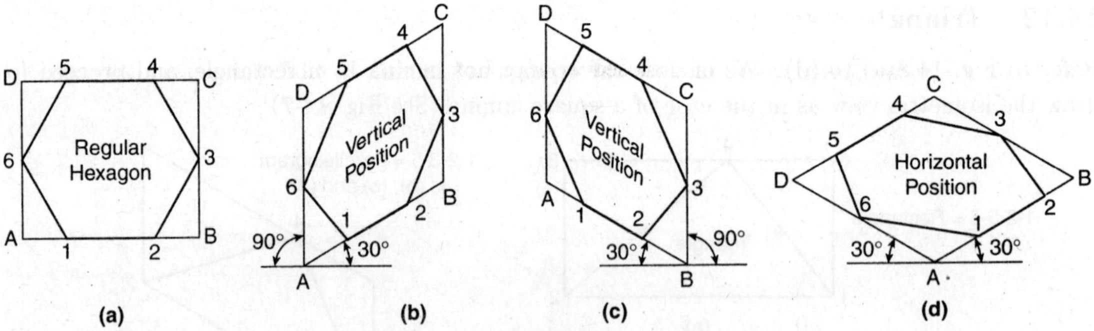

Regular Hexagon

Fig. 14-10

We enclose the regular hexagon by a rectangle, and proceed to draw as in the case of a pentagon (See Fig. 14-9).

14.15 Circular Lamina

(a) Initially, let us suppose that the circular lamina is placed in a horizontal position. At first, we enclose the circular lamina by a square ABCD [Fig. 14-11(a)].

(b) We then draw the two axes as 1-5 and 3-7 [Fig. 14-11(a)].

(c) The two diagonals AC and BD are drawn, and they meet the circle at 2,6 and 4,8 respectively [Fig. 14-11(a)].

(d) From the point 8, we have drawn horizontal and vertical lines 8-h and 8-v respectively [Fig. 14-11(a)].

(e) Now, we choose the point A as pole, and draw two lines AB and AD, making 30° with horizontal [Fig. 14-11(c)].

(f) In isometric scale, take the lengths AB and AD. Also draw parallel lines DC and BC to meet at C [Fig. 14-11(c)].

(g) From Fig. 14-11(a), take the length Dh or DV and mark Dh and Dv on the lines DA and DC respectively [Fig. 14-11(c)].

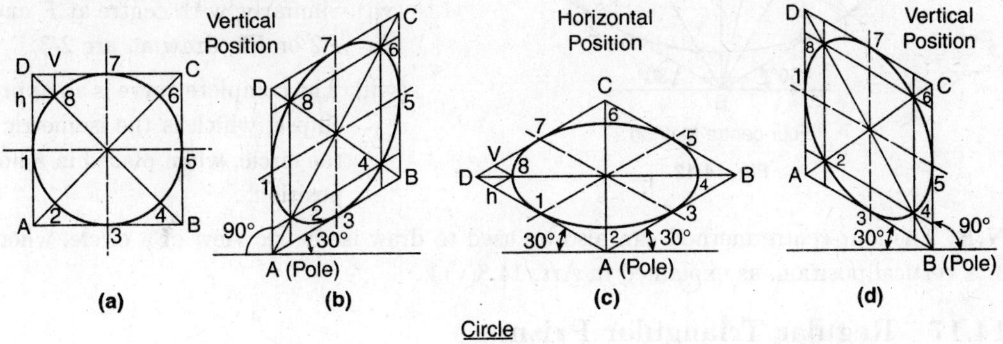

Fig. 14-11

(h) From the points h and V, draw parallel lines to meet at 8 [Fig. 14-11(c)].

(i) In the same manner, find the points 2, 4 and 6 [Fig. 14-11(c)].

(j) Through the eight points $(1, 2, \ldots, 8)$, draw a smooth curve, which is an ellipse [Fig. 14-11(c)].

(k) This is known as **eight-point method** of drawing isometric view of a circle.

When the circular lamina is in a horizontal position, this procedure of 8-point method is followed [Fig. 14-11(c)]. But, when the circular lamina is in a vertical position, the horizontal line AB is to be drawn at an angle of 30° to the horizontal. The vertical lines (DA, CB) will remain vertical [Sec Fig. 14-11(b) and (d)]. **The isometric view of the circle will be ellipse in both the cases of (b) and (d), and also in (c).**

14.16 Four-Centre Method

The isometric view of a circular lamina can be drawn easily by the four-centre method, which isexplained below.

Refer to Fig. 14-12.

(i) Draw two lines BA and BC inclined at 30° to horizontal and take measurement (in isometric scale) of BA and BC (i.e., diameter of circle). 1 and 2 are the mid-points of AB and BC respectively.

(ii) Complete the parallelogram *ABCD*. 3 and 4 are the mid-points of *DC* and *DA* respectively.

(iii) Join *D*1 and *D*2. These lines *D*1 and *D*2 meet the diagonal *AC* at *E* and *F* respectively.

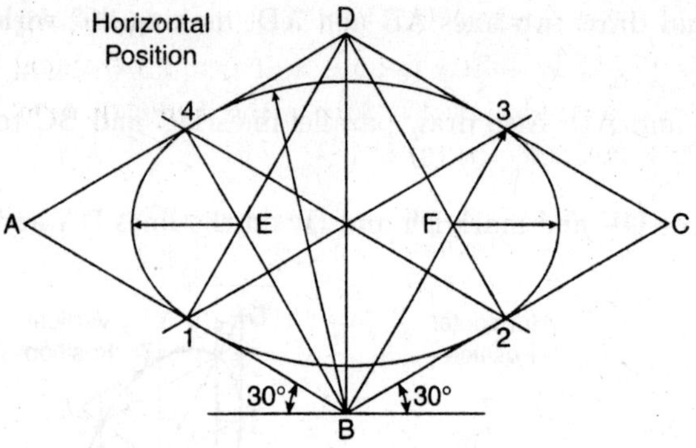

Four-centre Method

Fig. 14-12

(iv) With centre at D and radius $D1$ or $D2$, draw an arc 1-2.

(v) Similarly, with centre at B, and radius $B3$ or $B4$, draw an arc 4-3.

(vi) With centre at E and radius $E4$ or $E1$, draw an arc 1-4.

(vii) Similarly, with centre at F and radius $F2$ or $F3$, draw an arc 2-3.

(viii) The complete curve is an approximate ellipse, which is the isometric view of the circle, when placed in a horizontal position.

N.B. The four-centre method can also be used to draw isometric view of a circle, when placed in a vertical position, as explained in Art. 14.8(v).k.

14.17 Regular Triangular Prism

Refer to Fig. 14-13(a).

The elevation and plan of the triangular prism are drawn first.

(i) Convert the triangular prism into a rectangular prism as A-B-C-D-1-3-6-4 [Fig. 14-13(b)].

(ii) As in the Fig. (b), keeping AD line front, draw two lines DC and $D4$ at 30° inclined as shown. Complete the isometric view as of A-B-C-D-1-3-6-4.

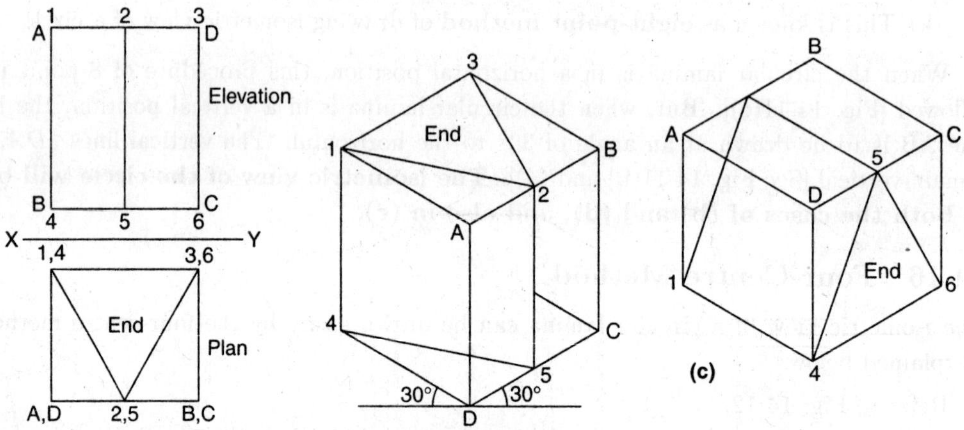

(a) **(b)** Vertical position Horizontal position (c)

Triangular Prism

Fig. 14-13

(iii) Find the coordinate 2 and 5 and then complete the isometric drawing of 1-2-3-4-5-6 (triangular prism, where the line 3–6 is not visible).

(iv) The isometric views of the prism has been drawn in horizontal position in Fig. 14-13(c).

Note : Isometric view of the same can be drawn in other method. We have already acquired the knowledge of drawing isometric view of a lamina. We know that a lamina is of two dimensional (length and width). If we add the third dimension, i.e., height, then it will be the isometric view of the solid (i.e., triangular prism) as shown in Figs. 14-14(a) (b) (c) & (d).

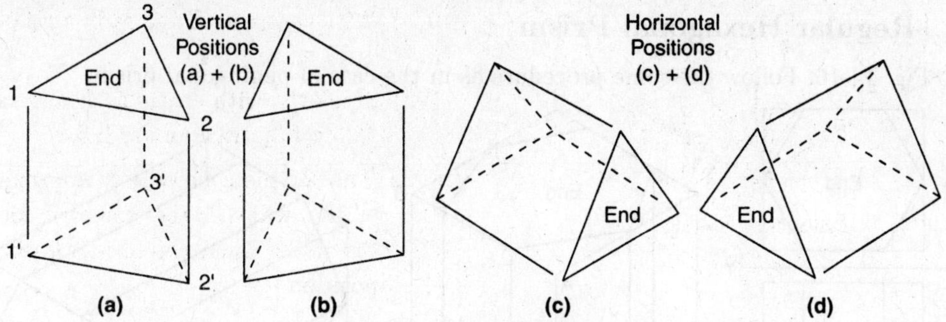

Fig. 14-14 Triangular Prism

If three vertical lines of equal height (1-1′, 2-2′, 3-3′) are drawn from the three vertices of the triangular lamina 1-2-3 and the three points (1′-2′-3′) are joined, we will get the isometric view of the triangular prism. In this figure, 1′-2′-3′ will be the other end of the solid. In order to clarify the idea, dotted lines are drawn in Fig.(a). But in the isometric view, these dotted lines should not be drawn in actual case and Fig. (b) is another example of vertical one. Figs. (c) and (d) are given for the isometric views of the solid in horizontal position. After completion of isometric views, compare them with Fig. 14-13.

14.18 Regular Pentagonal Prism

We have avoided the square prism, as all the prisms will be converted into square or rectangular prism.

Refer to Fig. 14-15.

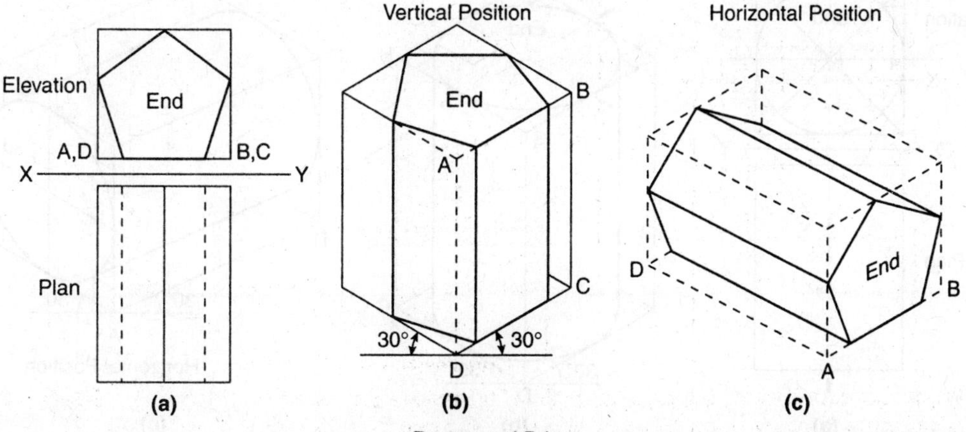

Pentagonal Prism

Fig. 14-15

(i) The elevation and plane of pentagonal prism are drawn Fig. 14-15(a).

(ii) the prism is enclosed on a rectangular box. This prism can also be drawn from the pentagonal lamina, as shown in Fig. 14-14.

14.19 Regular Hexagonal Prism

Refer to Fig. 14-16. Follow the same procedure as in the case of pentagonal prism.

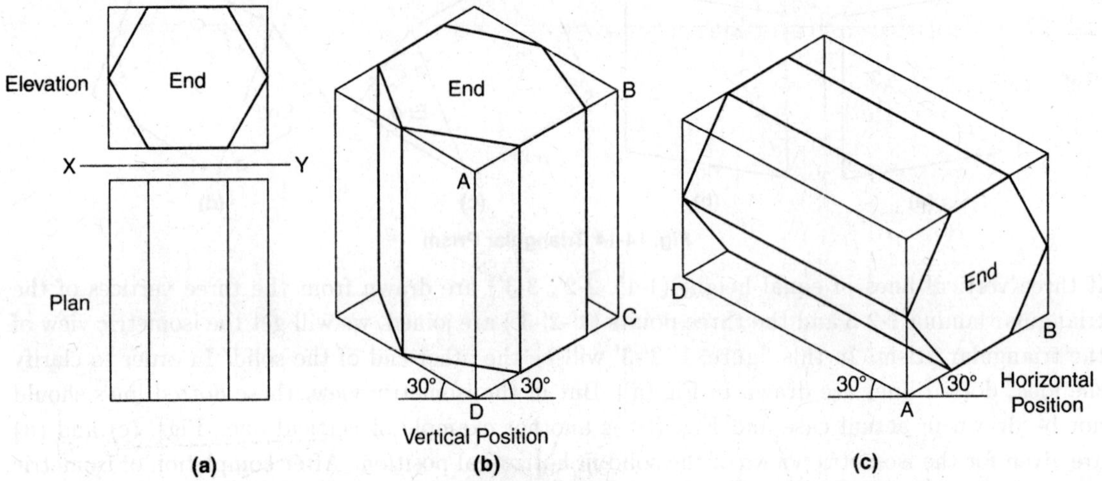

Hexagonal Prism

Fig. 14-16

14.20 Right Cylinder

Refer to Fig. 14-17.

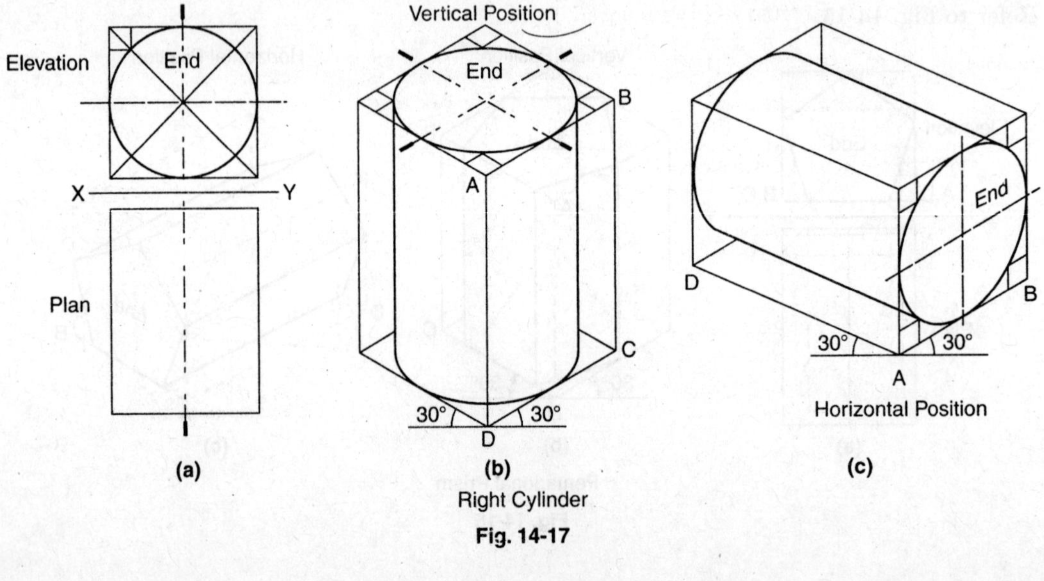

Right Cylinder

Fig. 14-17

The circular lamina may be considered as infinite number of sides, i.e., polygon of infinite number. But, in case of cylinder we divide the base circle into twelve equal parts, i.e., cylinder is assumed to be twelve-sided polygon for simplicity. Other procedures remain the same as before. Convert the cylinder into square prism. In isometric view the top and bottom of the cylinder will appear like ellipse. Therefore, draw the isometric view of the circular lamina first (which will be an ellipse) and then draw vertical lines to make it a cylinder.

14.21 Regular Triangular Pyramid

Refer to Fig. 14-18.

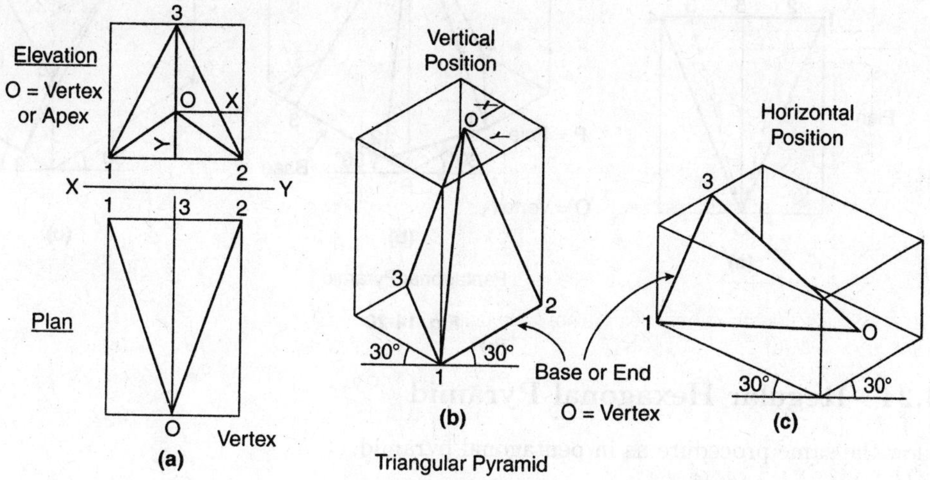

Triangular Pyramid

Fig. 14-18

14.22 Regular Square Pyramid

Refer to Fig. 14-19. All the procedures are same as in previous problems.

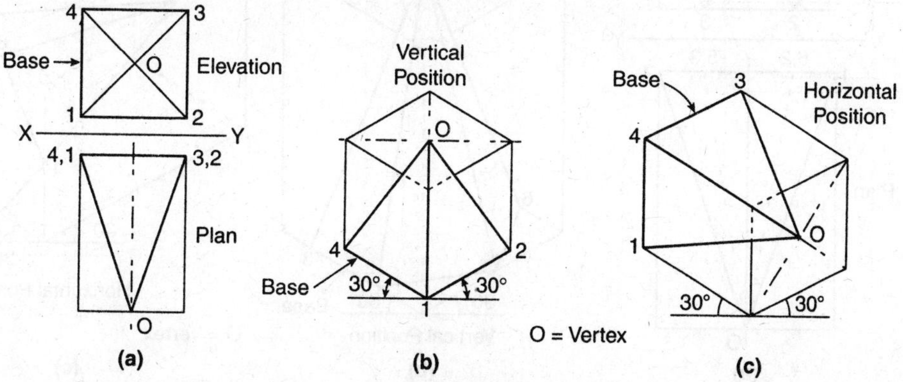

Square Pyramid

Fig. 14-19

14.23 Regular Pentagonal Pyramid

Find the apex and join the base points to get the isometric view.

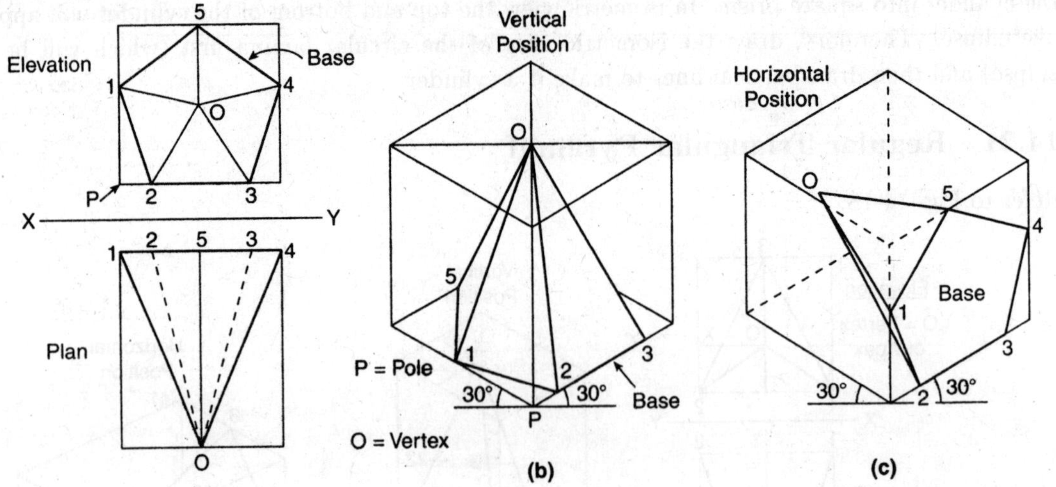

Pentagonal Pyramid

Fig. 14-20

14.24 Regular Hexagonal Pyramid

Follow the same procedure as in pentagonal pyramid.

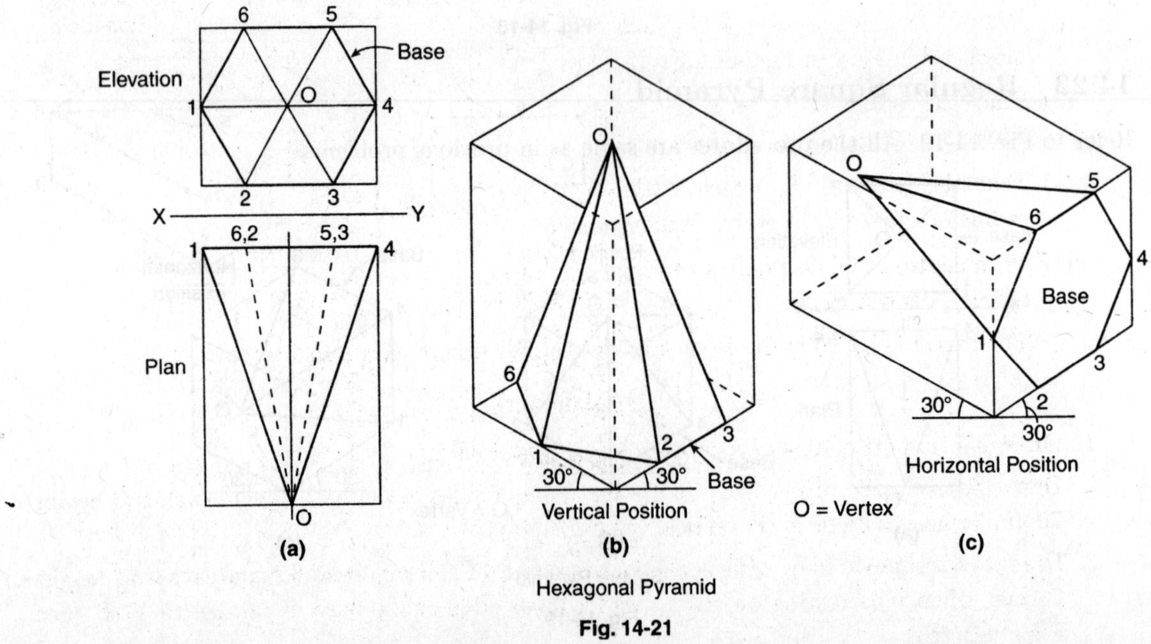

Hexagonal Pyramid

Fig. 14-21

14.25 Right Cone

We may consider the cone as pyramid of infinite number of sides of base. But we divided the base circle into twelve equal parts. So we assume the cone to be twelve-sided pyramid. But in case of cone, the base lines are curved instead of straight lines.

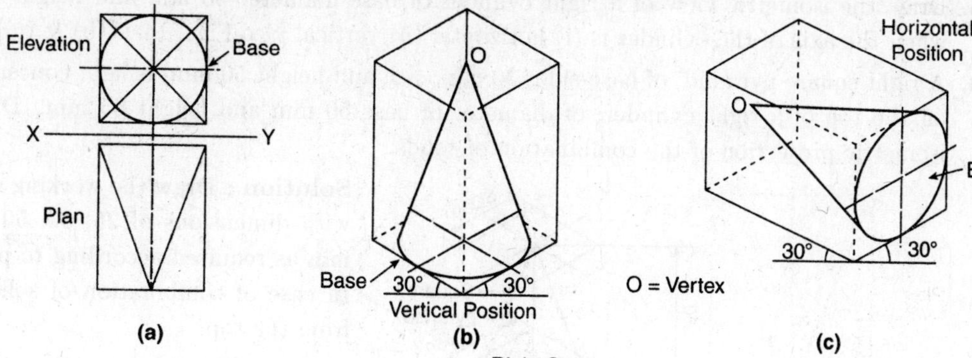

Right Cone

Fig. 14-22

14.26 Sphere

We know that, if we look from any angle, the view of a sphere is always a circle. This is also true for isometric view of a sphere. The sphere is first placed on an object or any plane. 'P' is the touching point of sphere and the plane.

To draw the isometric view of the sphere with respect to an object or body, the steps are as follows :

(i) Draw the isometric view of the solid block [a-b-c-g-e-f in Fig. 14-23(b)].

(ii) Find the point P, by joining two diagonals ac and bd (d is not seen).

(iii) Take the height PO′, which is the radius of sphere in isometric scale.

(iv) With centre at O′ and radius r (True radius), draw a circle.

(v) This circle (Fig. b) is the isometric view of the sphere shown in Fig. (a).

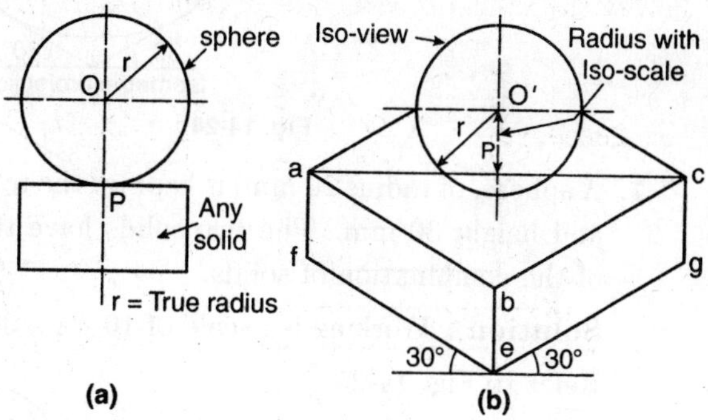

Fig. 14-23

Problems

1. Draw the isometric view of a cube of sides 40 mm each.

2. Draw the isometric projection of a square prism of sides of the base 40 mm each and height 70 mm, when the axis is (i) vertical (ii) horizontal.

3. Draw the isometric view of a hexagonal pyramid of sides of base 35 mm each and height 70 mm when it is resting on H.P., such that an edge of the base is parallel to V. P. [See Fig. 14-21(b)].

4. Draw the isometric projection of a pentagonal pyramid of base-side the 30 mm each and height 65 mm.

   Note : Position of the pyramid is not mentioned here. So use any one of the figures 14-20(b), (c).

5. Draw the isometric view of a right cylinder of base diameter 50 mm and height 70 mm, when the axis of the cylinder is (i) horizontal (ii) vertical [See Figs. 14-17(b) & (c)].

6. A right square pyramid, of base-sides 30 mm each and height 50 mm, is kept concentrically on the top of a right cylinder, of diameter of base 50 mm and height 65 mm. Draw the isometric projection of the combination of solids.

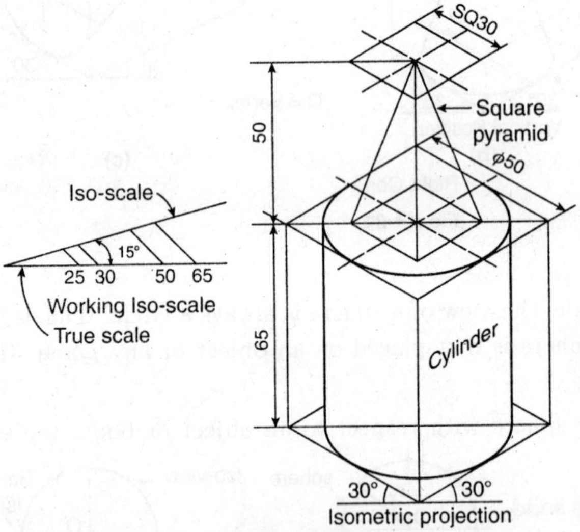

Fig. 14-24

Solution : Draw the working iso-scale with dimensions of 25, 30, 50 and 65 mm as required according to problem. In case of combination of solids start from the top.

Refer to Fig. 14-24. Here draw the isometric view of square pyramid first with dark lines placing the pyramid in a square prism. Then draw the cylinder through the centre of pyramid. **Do not erase the construction lines. In case of combination, always start from the top.** This is because we can draw the dark lines from the very beginning, and double-penciling can be avoided.

7. A sphere, of radius 20 mm, is kept on the top face of a square prism, of base sides 40 mm each and height 30 mm. The two solids have the common axis. Draw the isometric projection of the combination of solids.

   Solution : Working iso-scale of 10 mm, 30 mm and 40 mm.

   Refer to Fig. 14-25.

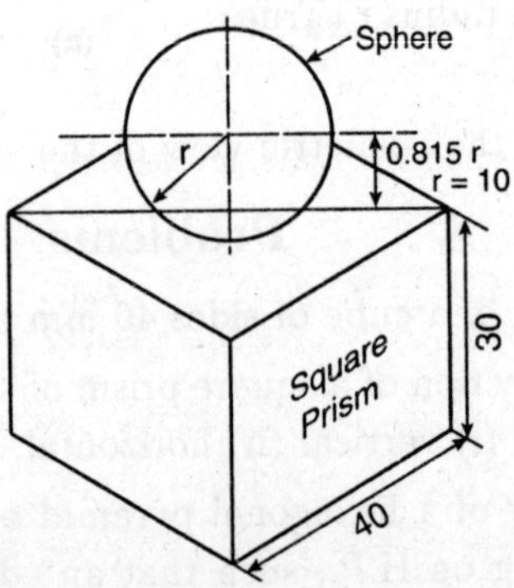

Fig. 14-25

State Council Model Questions (West Bengal)

1. Differentiate between isometric projection and isometric drawing.

Ans. In case of isometric projection, the three planes will be equally shortened and the axes equally spaced at 120° apart. An isometric drawing is similar to isometric projection, but here actual measurements are taken, without reducing their sizes, refer Fig. 14-26. Here, *O*-1, *O*-2 and *O*-3 are isometric axes.

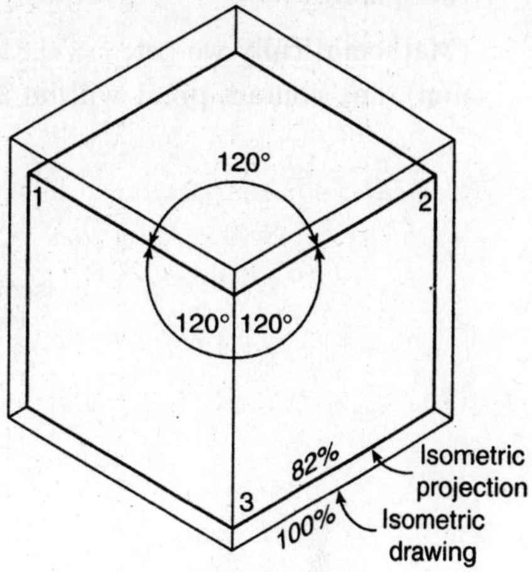

Fig. 14-26

2. Describe the four-centre method of drawing isometric projection of a circle.

Solution : Consult Art. 14-16 and Fig. 14-12.

3. Construct an isometric scale and show how it differs from true scale.

Solution : Consult Art. 14-7 and 14-8 and 14-5(a) & (b).

4. State true (T) or False (F) :

(a) In isometric projection, two dimensions of a solid are shown.

(b) The ratio of isometric length to the true length is 2 : 3.

(c) The view, drawn with the isometric scale, is called isometric projection.

(d) The view drawn with the isometric scale is called isometric drawing.

(e) The isometric projection of a circle is always a circle.

(f) The actual view of an object will be exactly the same shape of an isometric view but about 22.5 larger.

(g) Isometric projection of a sphere is a circle.

(h) The angle between the isometric axes is 120°.

(i) Isometric scale must be used while drawing isometric views of spherical object.

(j) In isometric projection, the three principal axes of the object will be equally foreshortened.

(k) True lengths are used to draw the isometric projection. **(AMIE—Summer, 2002)**

Ans. (a) F, (b) F, (c) T, (d) F, (e) F, (f) F, (g) T, (h) T, (i) T, (j) T, (k) F.

Further Problems

1. Draw an isometric view of three solids, resting one upon the other with same vertical axis, using ordinary scale. The bottom one is a cylinder, of dia. 70 mm and height 30 mm. The middle one is a cube of each side 60 mm. A sphere of dia. 50 mm is resting centrally above the cube. **(West Bengal State Council, 1995)**

Solution : Refer Fig. 14-27. Drawing of iso-scale is not essential here. But to get the contact point of the sphere and the cube, we will have to find out the isometric radius of the sphere.

(Mathematically we get, $r \times 0.815$, where $r = 25$ mm, i.e., $r \times 0.815 = 25 \times 0.815 = 20.375$ mm), i.e., contact point will be 20.375 mm (20 mm apx) below the centre of the sphere.

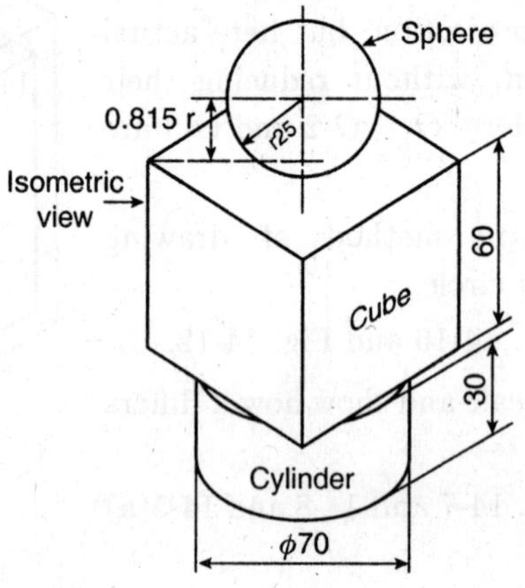

Fig. 14-27

2. Draw an isometric projection of two solids, resting one upon the other, with same vertical axis, using iso-scale, looking in the direction of arrow '*X*'. The bottom one is a hexagonal prism of base-sides 30 mm each and height 50 mm. A hemisphere of dia. 50 mm. is resting on the top surface of the prism, as shown in Fig. 14-28(a).

(West Bengal State Council, 1995)

Solution : Refer Fig. 14-28(b).

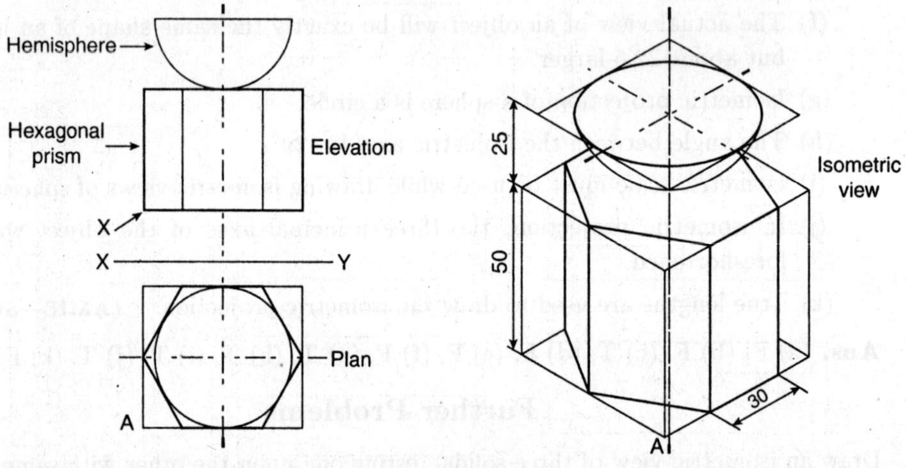

Fig. 14-28

3. Draw the isometric views, step by step, after slicing the wooden block, as shown in Fig. 14-29, (a) and (e). Fig. (e) is the isometric view and (a) is the orthographic view.

Solution : First draw the isometric solid block as in Fig. 14-29(e), and then slice-out the block. Select dimensions as you like. See Fig. 14-29(b),(c) and (d) in elevation and plan.

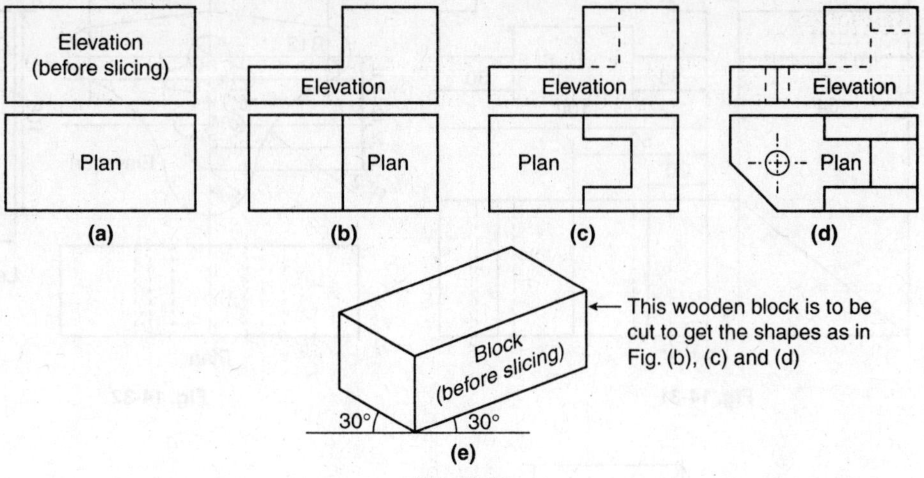

Fig. 14-29

4. Draw the isometric views of the given orthographic drawings, as shown in the Figs. 14-30, [(a) to (f)] 31, 32, 33 and 34.

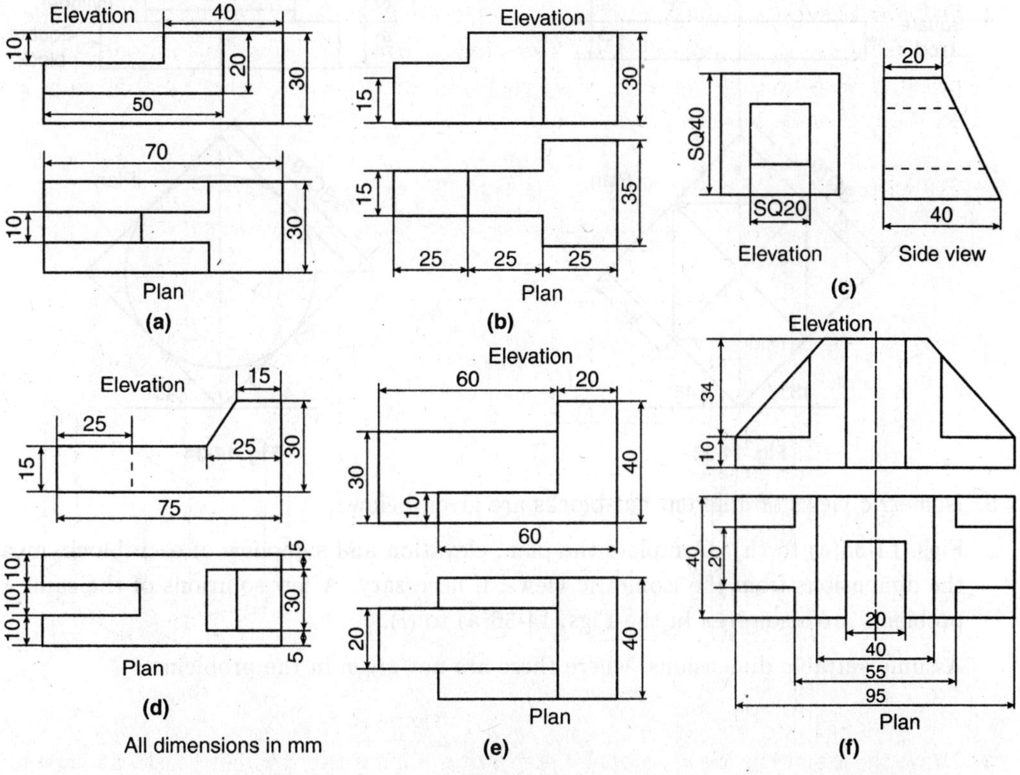

Fig. 14-30

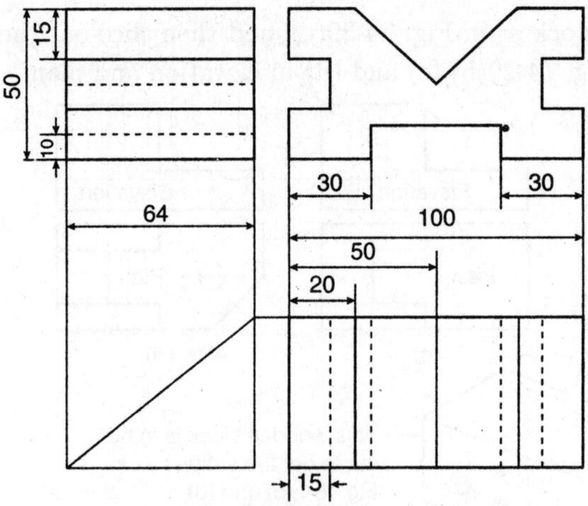

Fig. 14-31

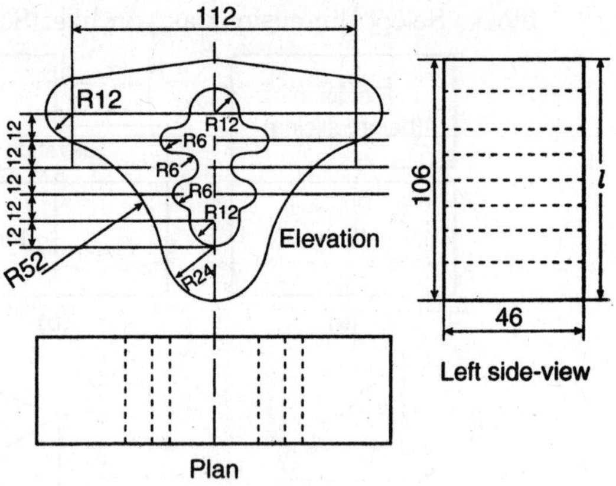

Fig. 14-32

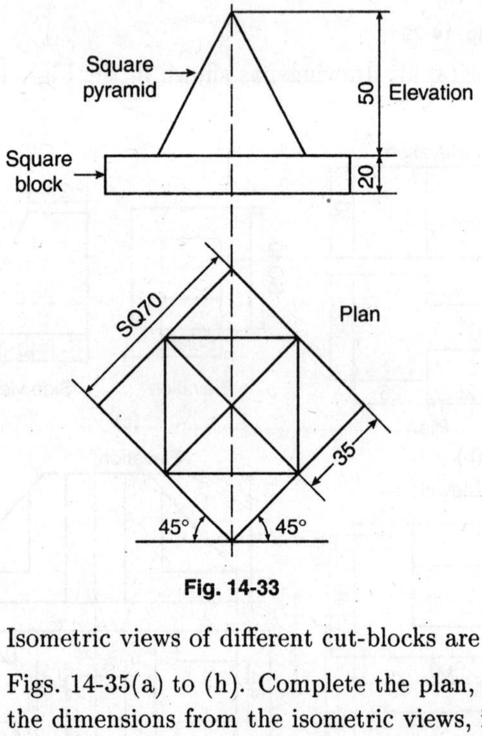

Fig. 14-33

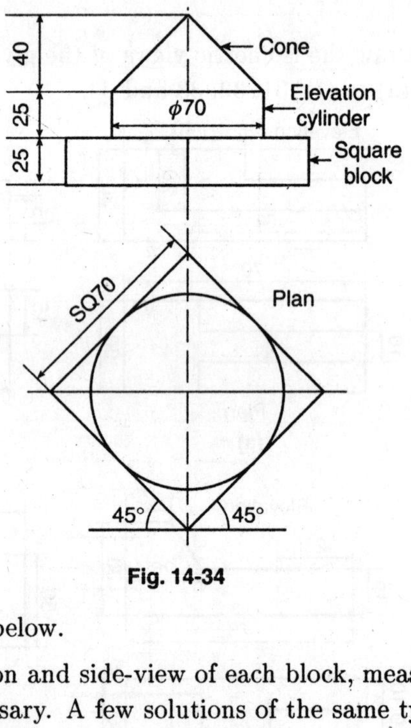

Fig. 14-34

5. Isometric views of different cut-blocks are given below.

Figs. 14-35(a) to (h). Complete the plan, elevation and side-view of each block, measuring the dimensions from the isometric views, if necessary. A few solutions of the same type of problems are also given in the Figs. 14-36(a) to (i).

Assume suitable dimensions, where these are not given in the problems.

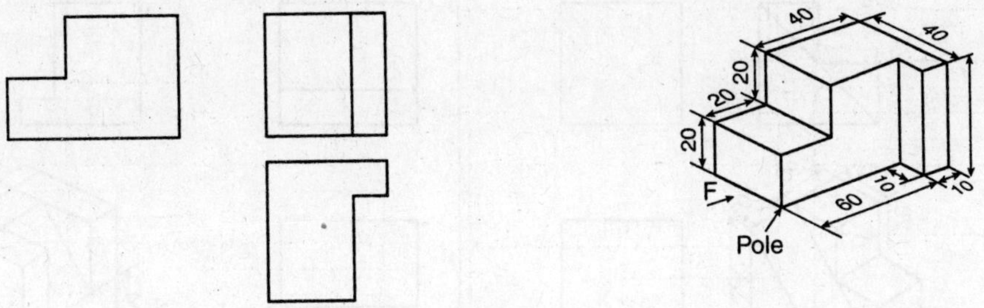

(a)

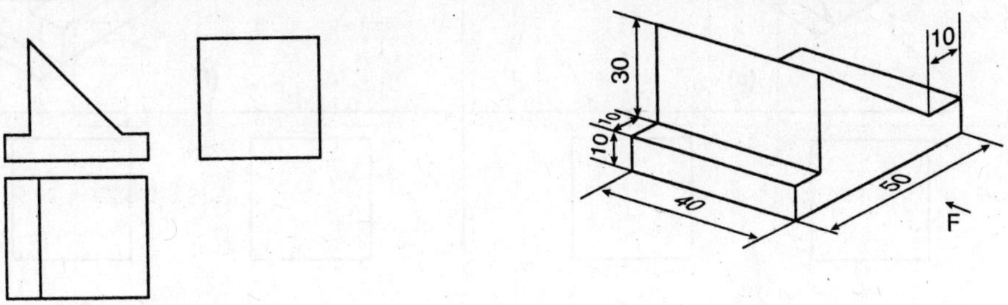

(b)

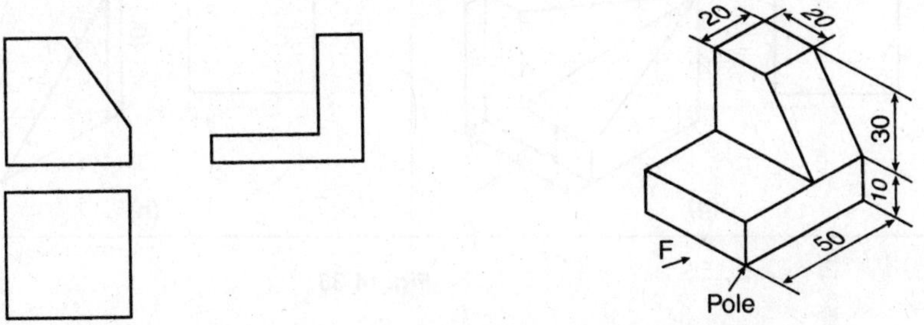

(c)

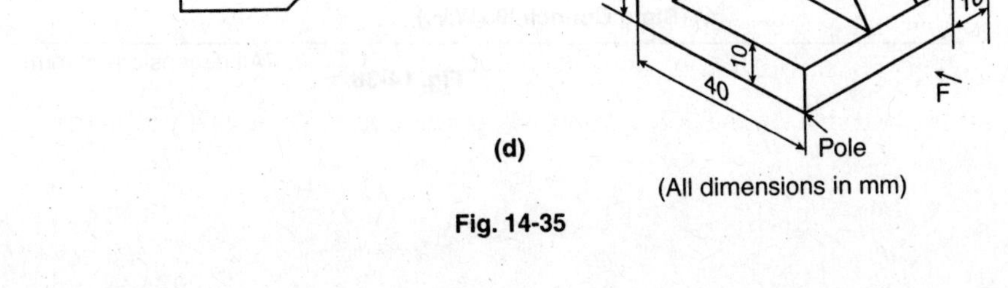

(d)

(All dimensions in mm)

Fig. 14-35

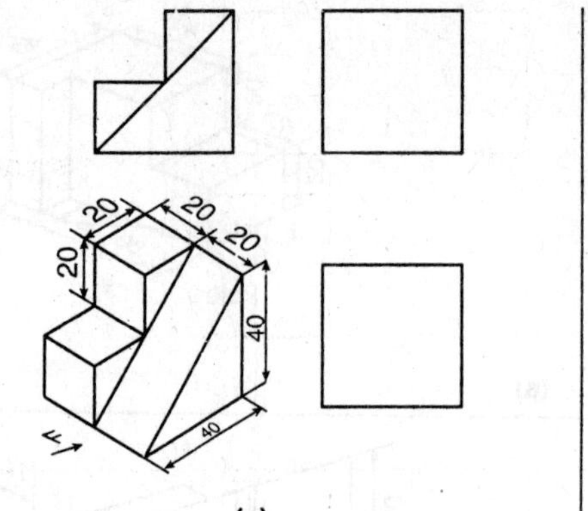

(e)

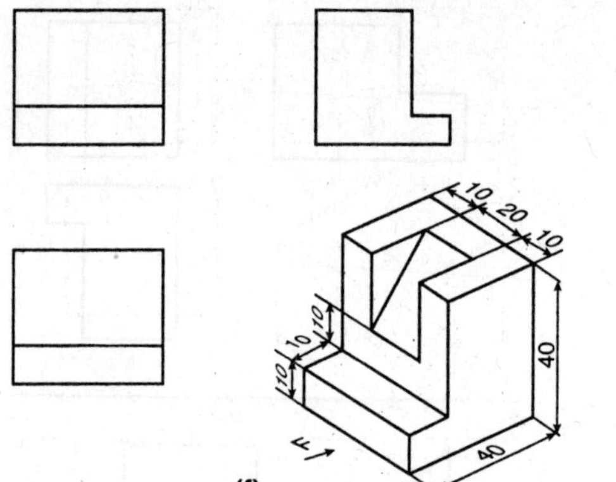

(f)

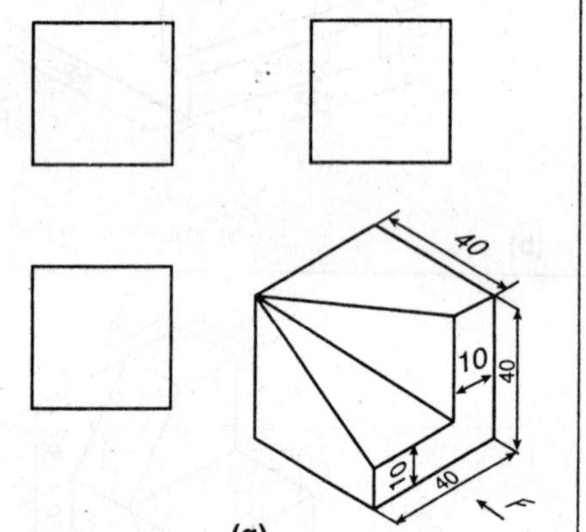

(g)

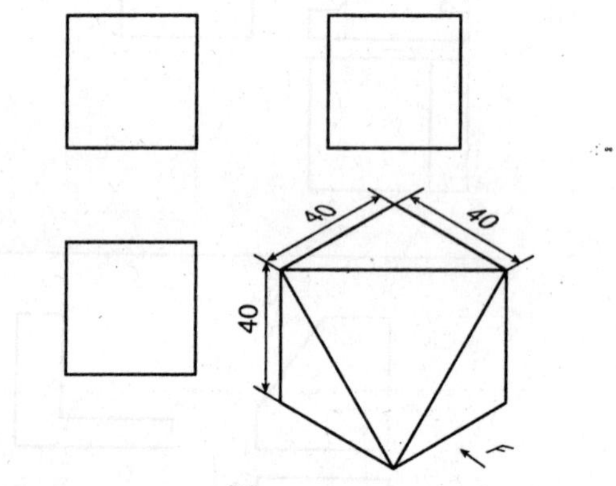

(h)

Fig. 14-35

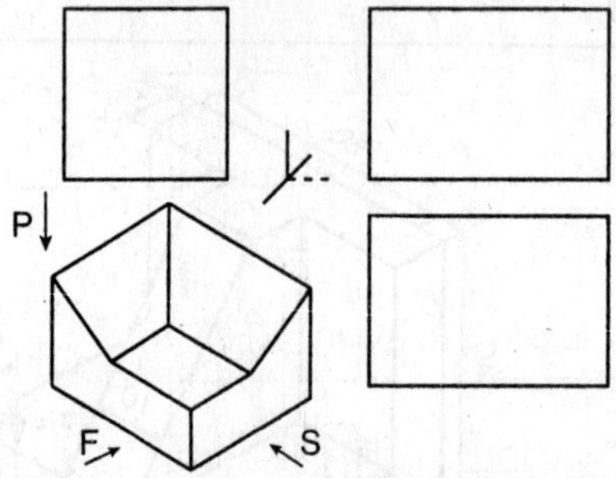

(a) (State Council-'95 W.B.)

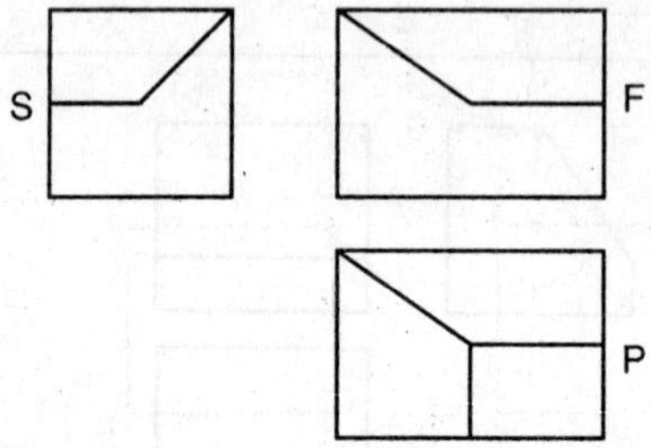

Solution (a)

(All dimensions in mm)

Fig. 14-36

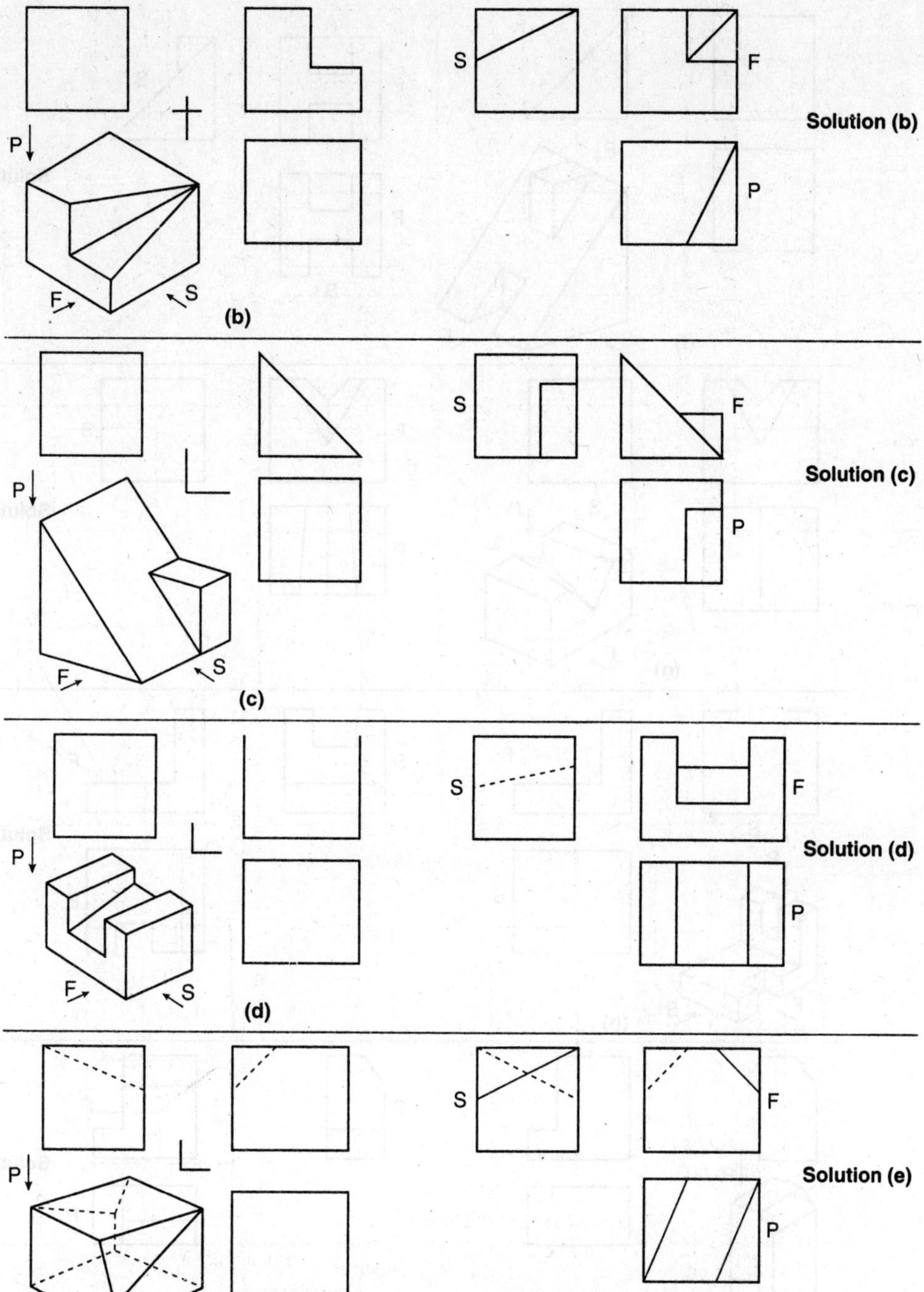

Abbreviations

F = Front elevation, P = Plan, S = Side view

Fig. 14-36

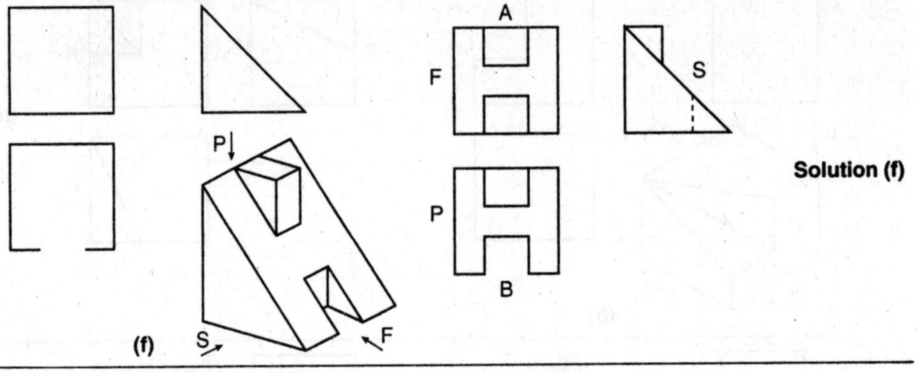

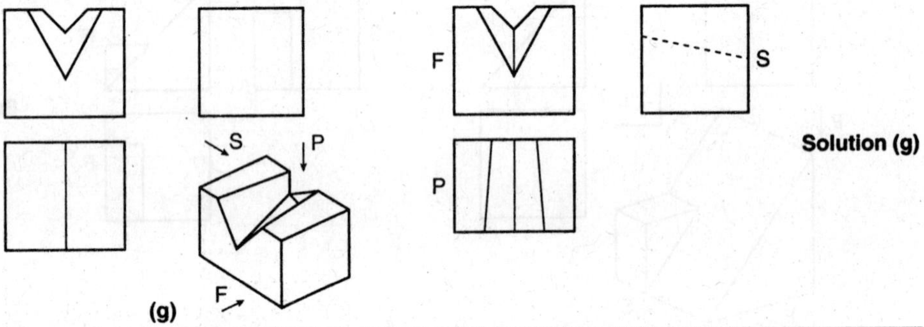

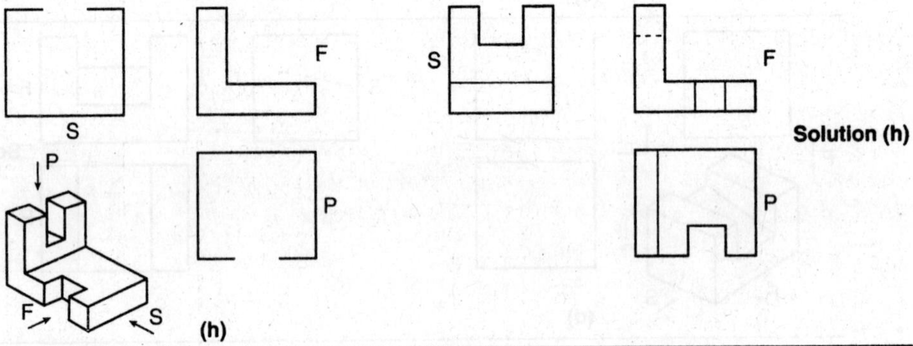

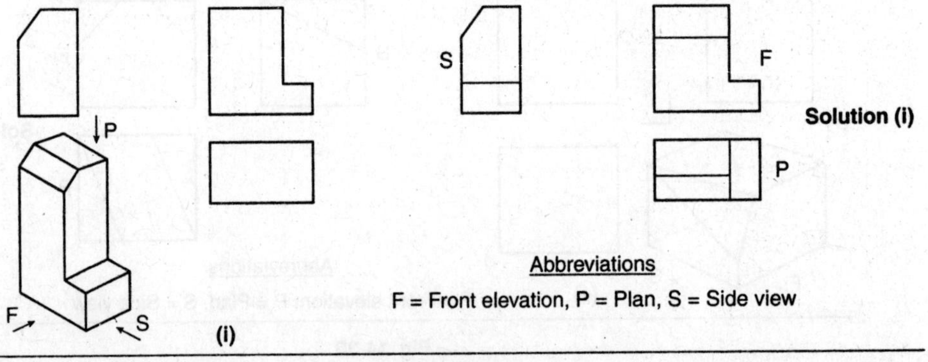

__Abbreviations__

F = Front elevation, P = Plan, S = Side view

Fig. 14-36

6. Fig. 14-37 shows three views of an object. Draw the isometric projection of the object, looking along the arrow X, showing the isometric scale used. **(AMIE—Summer, 2000)**

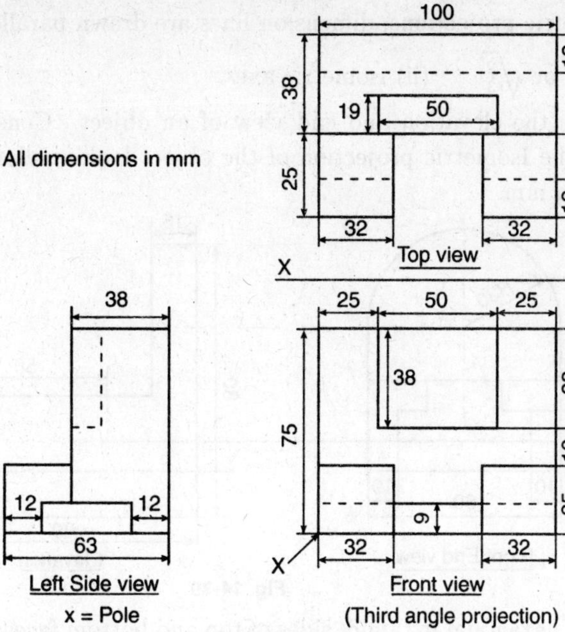

Fig. 14-37

7. The frustum of a hexagonal pyramid has base-sides 80 mm each, top sides 50 mm each and height 60 mm. It stands on its base in HP. A hemisphere is placed on the top of the frustum with flat face down and circular face touching all sides of the hexagon. Draw the isometric projection of the combination. Draw the isometric scale. **(AMIE—Winter, 2000)**

8. Fig. 14-38 shows the top and front views of a hemisphere, of 44 mm diameter, resting on the top of a frustum of a hexagonal pyramid. The sides of the top and the bottom-faces of the frustum of the hexagonal pyramid are 15 mm and 30 mm respectively and its height is 50 mm. The axes of both the solids coincide. Draw the isometric projection of the combination, showing the isometric scale. **(AMIE—Summer, 2001)**

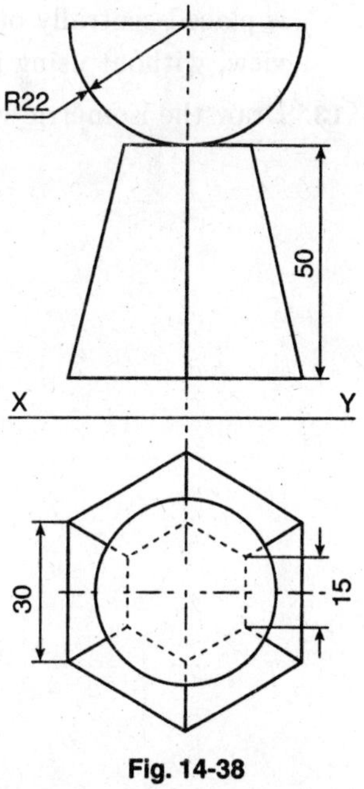

Fig. 14-38

9. Choose the correct answer :

To get the isometric length of a line, the true length has to be multiplied by

(a) $\sqrt{\frac{2}{3}}$ (b) $\sqrt{\frac{3}{2}}$ (c) $\frac{\sqrt{2}}{3}$. **(AMIE—Summer, 2001)**

Ans. (a), $\frac{\sqrt{2}}{\sqrt{3}} = 0.81$

10. Fill in the blanks :

(i) Ratio between the isometric and true length is ______.

(ii) In isometric projections, dimension lines are drawn parallel to ______ ______.

Ans. (i) 0.82 : 1 or $\sqrt{\frac{2}{3}}$ (ii) isometric axes.

11. Fig. 14-39 shows the elevation and end view of an object. Construct an Isometric scale and then draw the Isometric projection of the object looking along the arrow *X*. All the dimensions are in mm. **(AMIE—Summer, 2002)**

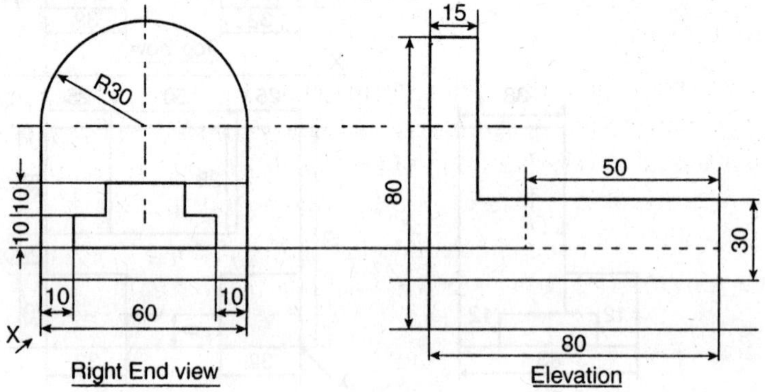

Fig. 14-39

12. The frustum of a hexagonal pyramid, sides of top and bottom faces being 15 mm and 30 mm each respectively and height 50 mm, stands vertically. A hemisphere, of 44 mm diameter, is placed centrally on the top-face with its flat-face up and horizontal. Draw the isometric view, without using isometric scale. **(AMIE—Winter, 2002)**

13. Draw the isometric view of the arrangement of the composite solid, as shown in Fig. 14-40. **(AMIE—Winter, 2006)**

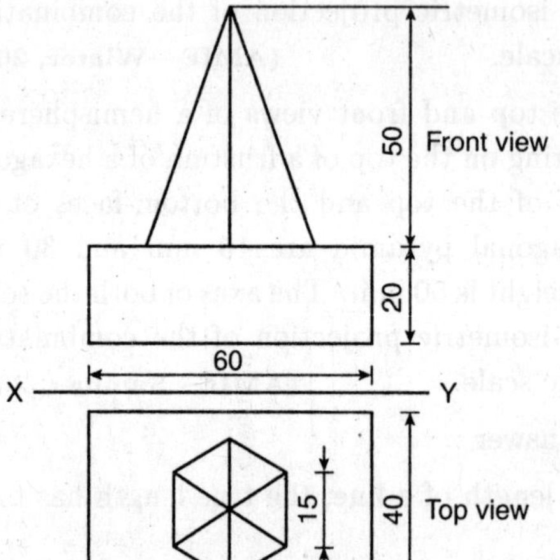

Fig. 14-40

Chapter 15

Free-Hand Sketches

15.1 Introduction

In this chapter, we like to say that the students should declare independence against drawing instruments. Dependence upon instruments makes a technical man less useful in the practical field. So, students should develop the skill of free-hand drawing. It creates an immediate impact and encourages the workers in a shop or in other engineering areas.

To achieve this skill, students should develop the confidence of drawing the **straight and curve** lines first. Also, the following **parameters are important**.

15.2 Parameters

(i) **Clear conception or ideas of the object of drawing** i.e. both the shape and size of the object.

(ii) **Sense of approximation and proportion** i.e. ratio of sizes.

(iii) **Confidence in medium** i.e. the type of soft or hard pencil, and the type of paper to be used for drawings.

(iv) **Eye-Estimation**, i.e., idea of distance by looking an object.

15.3 Lines

Straight lines may be drawn in three positions as follows :

(i) **Horizontal**, (ii) **Vertical** and (iii) **Inclined.**

The above types of lines are shown in Fig. 15-1(a), (b), (c), (d) and (e).

(a) Mark the start and finish-points with 'X'.

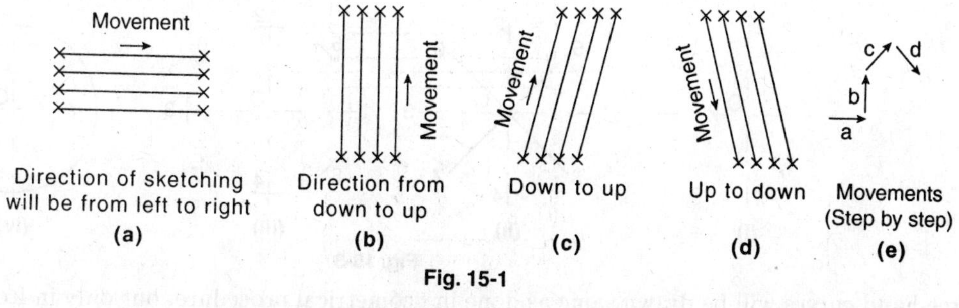

Fig. 15-1

(b) Join 'X' by faint line for double or triple or more strokes with eye-estimation. Finally, finish with dark lines.

(c) Use blunt-headed and soft pencil, i.e., HB or B, which is ideal for drawing sketches.

15.4 Square and Circle

The different steps are as under :

(a) Draw two lines, cross-wise of same length with eye estimation, as in Fig. 15-2(i).

(b) Take equal four short lengths on all sides from 0, as in Fig. 15-2(ii).

(c) Now complete the square with end-points as in Fig. 15-2(iii).

(d) Draw the circle, making lines tangent at the four points, as shown in Fig. 15-2(iv).

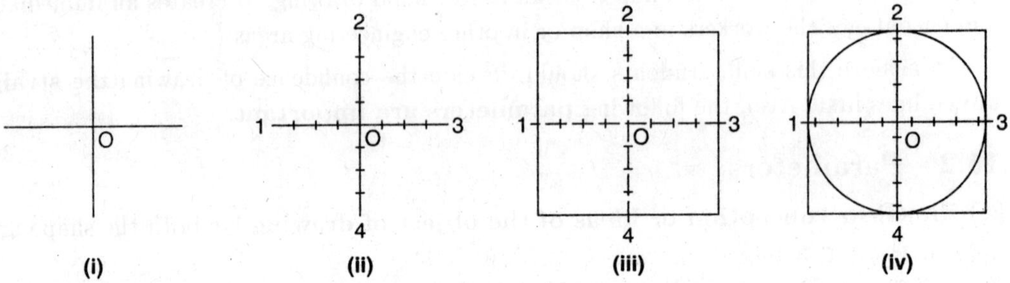

Fig. 15-2

15.5 Big-Circle

The different steps are as under :

(i) Draw the centre lines, i.e., making four quadrants, refer to Fig. 15-3(i).

(ii) Divide the quadrants into half, i.e., making 45° angles [Fig. 15-3(ii)].

(iii) Cut equal lengths from centre O. Draw short straight lines and short arcs, making short straight lines as tangents at the end-points [Fig. 15-3(iii)]. Thus, O1 = O2 = O3 = O4 = O5 = O6 = O7 = O8.

(iv) Finally, complete the circle, as in Fig. 15-3(iv).

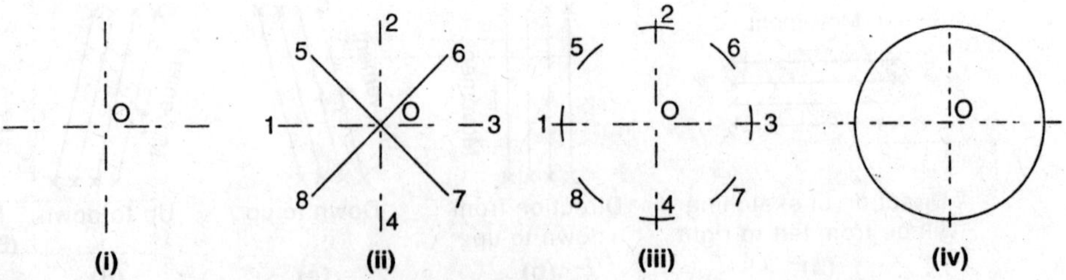

Fig. 15-3

Free-hand curves will be drawn same as done in geometrical procedure, but only in free-hand and use eye-estimation for the sizes, as far as practicable.

15.6 Rivet-Heads (For Free-Hand Practice)

Rivet-heads are used to join two metal-plates permanently. There are mainly six types of rivet-heads. Practise and memorize the proportions of rivet-heads in terms of rivet-diameter *d*.

The required proportions and angles are 0.5*d*, 0.7*d*, 1.5*d*, 1.6*d*, 15° and 60°. Draw the centre-line first, then, draw the proportions to complete the rivet-heads, as shown in Fig. 15-4(i) to (vi). Follow the same procedure for other type of rivet-heads.

(a) Snap or Cup-Head

Complete the figure step by step, as shown in the Fig. 15-4(i), (ii), (iii), (iv), (v) and (vi).

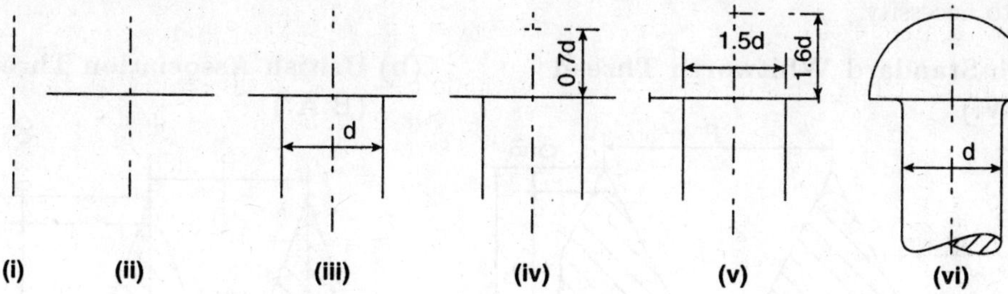

Fig. 15-4

(b) Pan-Head

Refer to Fig. 15-5, follow the same procedure as in the case of snap head and other cases also.

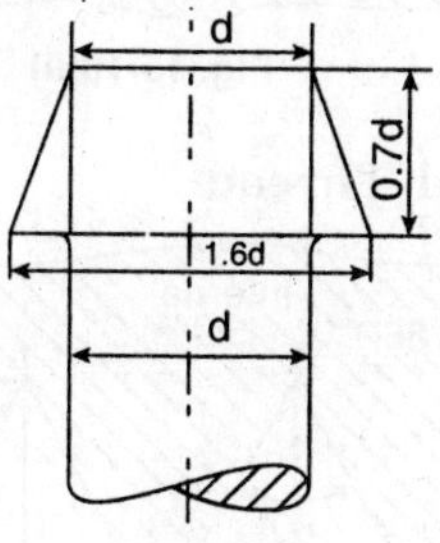

Fig. 15-5

(c) Pan-Head with Tapered Neck

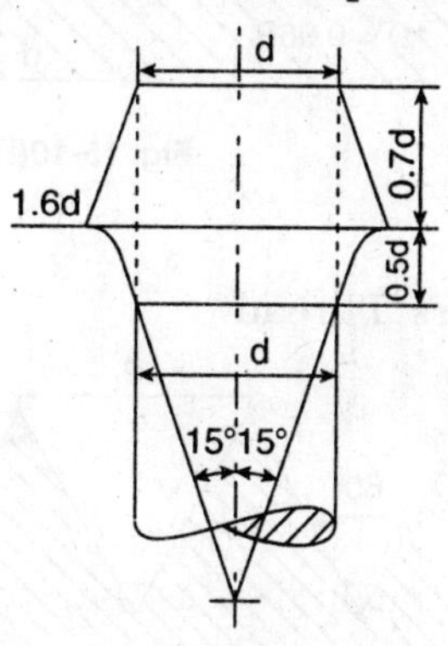

Fig. 15-6

(d) Rounded Countersunk

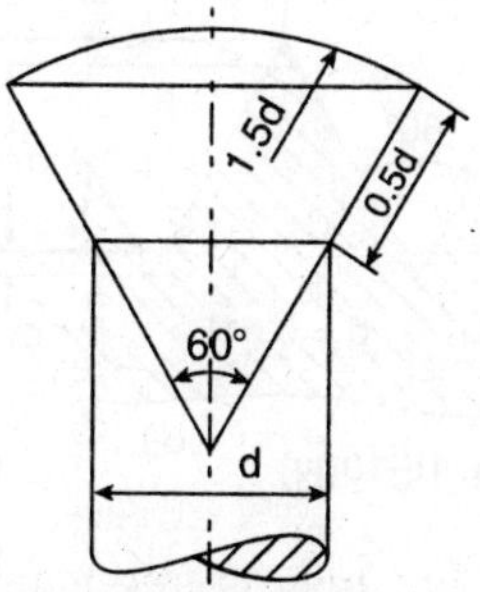

Fig. 15-7

(e) Flat Countersunk

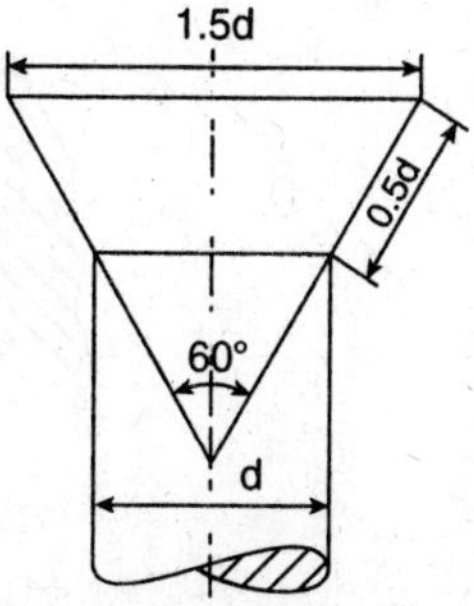

Fig. 15-8

(f) Conical

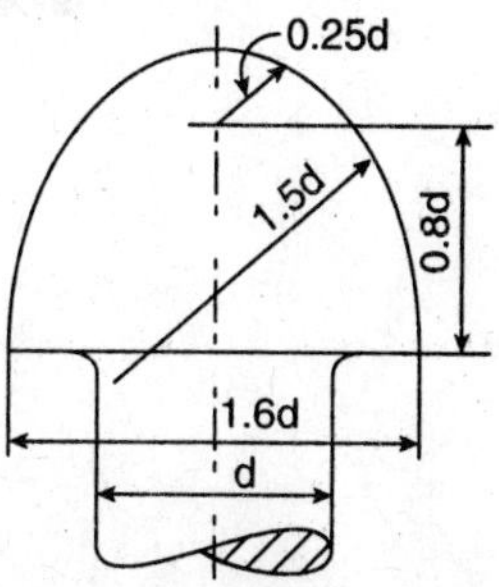

Fig. 15-9

15.7 Thread-Sections (Free Hand)

Threads are cut on engineering components (Nuts, Bolts, Screws etc.) to join metal-plates (two or more) temporarily. For all thread-sections, P = pitch of the thread. Refer to Fig. 15-10(i) to 15-10(viii).

For a single-start thread, the **axial movement** of the nut on a bolt, by one turn of the nut, is called the **pitch** of the thread.

Threads may be cut externally (Bolts, Screw pipes, etc.) or internally (Nuts, pipes). Two metal parts may be joined temporarily with the help of bolts and nuts, or screws and nuts, according to necessity.

(a) British Standard Whitworth Thread (B.S.W.)

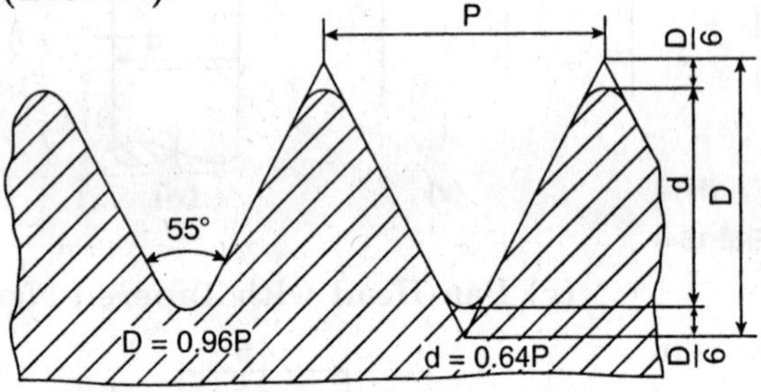

Fig. 15-10(i)

(b) British Association Thread (B.A.)

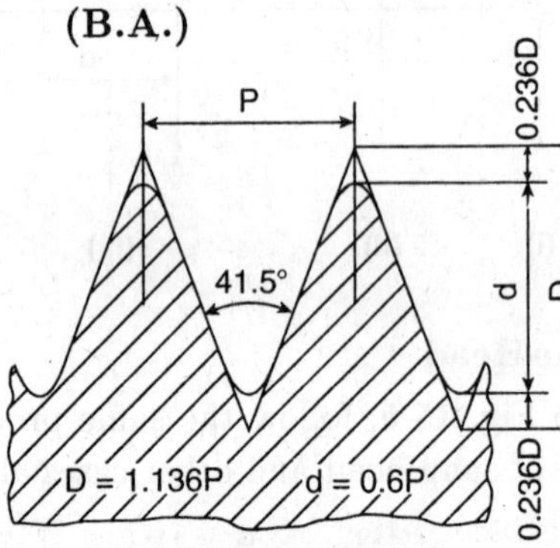

Fig. 15-10(ii)

(c) Sellers Thread

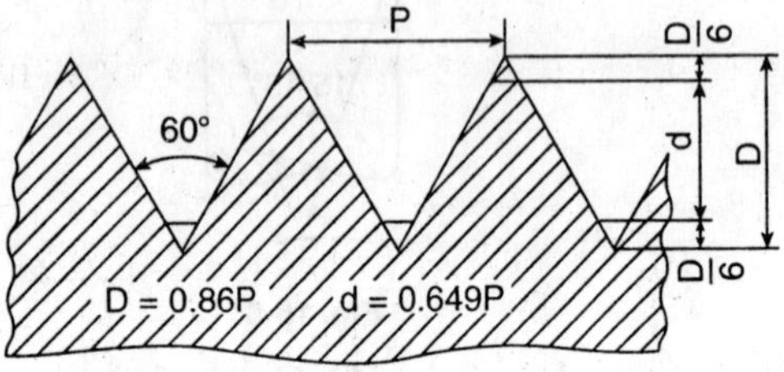

Fig. 15-10(iii)

(d) Unified Thread

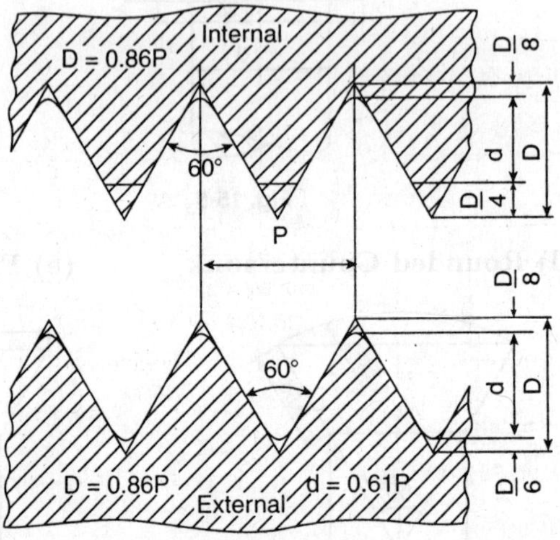

Fig. 15-10(iv)

(e) Square Thread

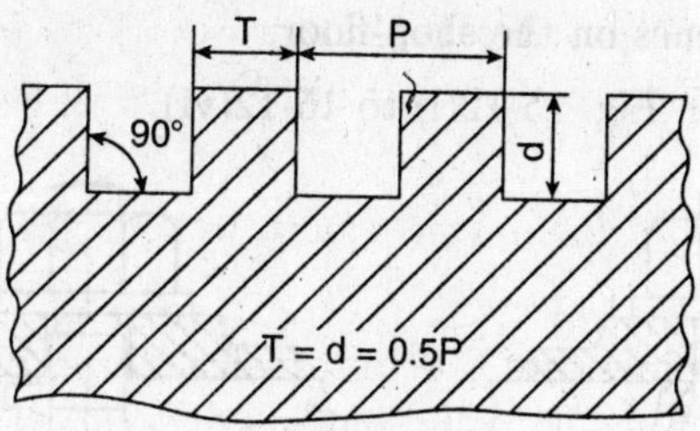

Fig. 15-10(v)

(f) Acme Thread

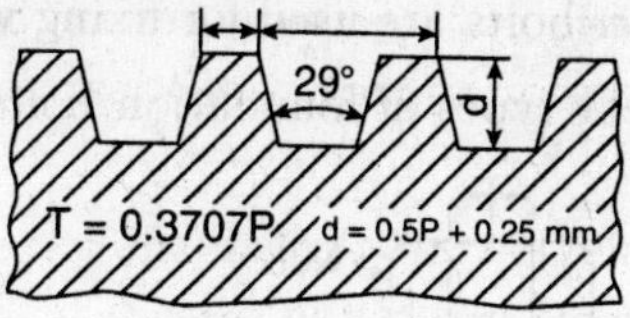

Fig. 15-10(vi)

(g) Knuckle Thread or Rounded Thread

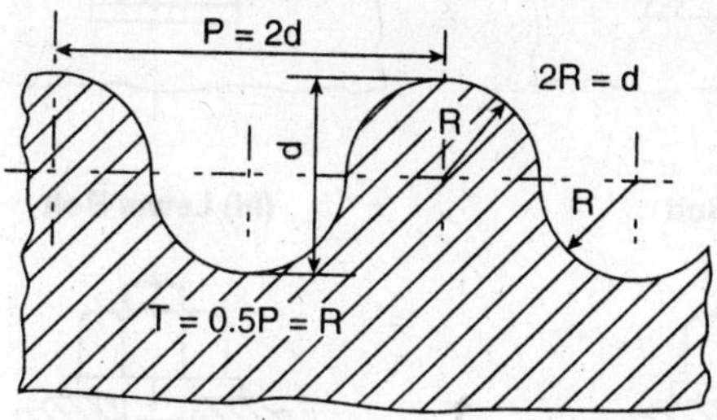

Fig. 15-10(vii)

(h) Buttress Thread

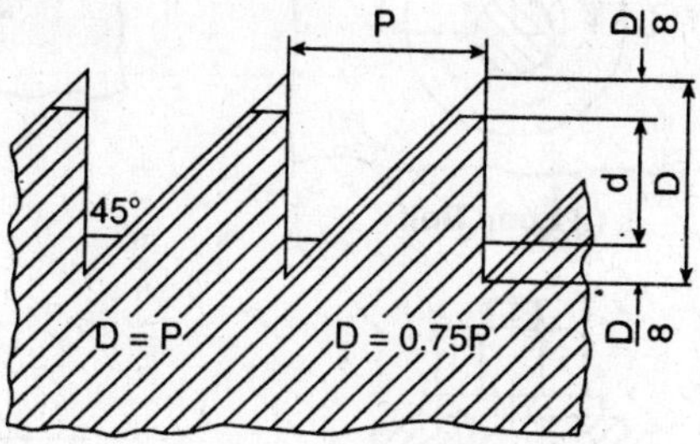

Fig. 15-10(viii)

15.8 Bolts and Nuts (Free Hand)

Refer to Fig. 15-11(i to vi).

d = diameter of bolt.

A 'stud' is an engineering component, threaded at both ends, with a cylindrical part (unthreaded) at the middle [Fig. 15-11(iv)].

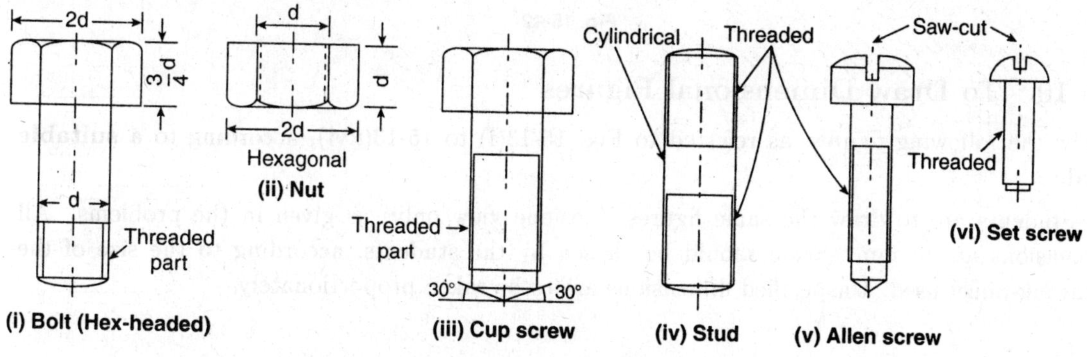

Fig. 15-11

Saw-cuts are provided on screws for turning them with the help of a screw-driver.

15.9 Foundation-Bolts (Free Hand)

Foundation-bolts are used for fixing various machines on the shop-floor.

Different types of foundation bolts are shown in Fig. 15-12(i) to 15-12(vi).

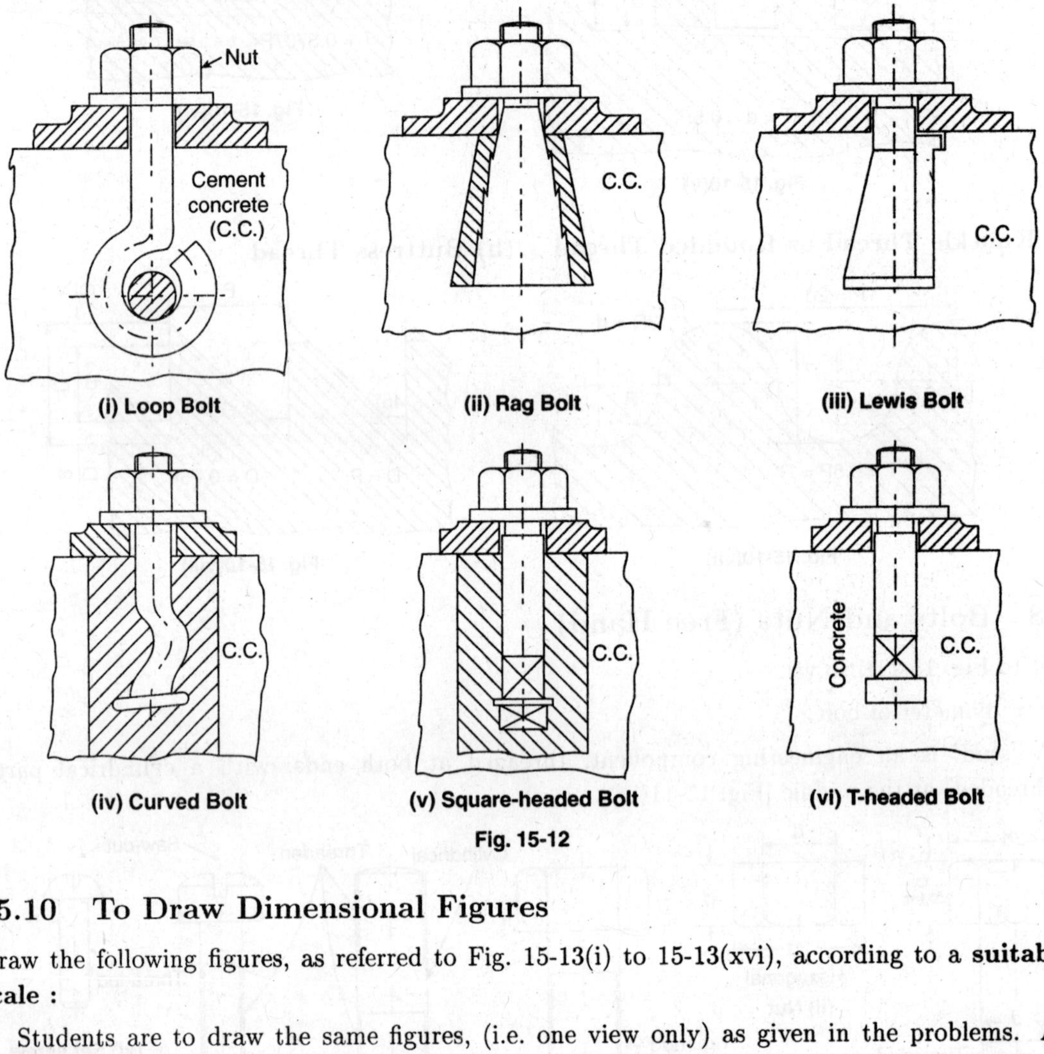

Fig. 15-12

15.10 To Draw Dimensional Figures

Draw the following figures, as referred to Fig. 15-13(i) to 15-13(xvi), according to a **suitable scale :**

Students are to draw the same figures, (i.e. one view only) as given in the problems. All dimensions are in mm. Scale should be chosen by the students, according to the size of the Drawing-Sheet used. Unspecified dimensions are to be taken proportionately.

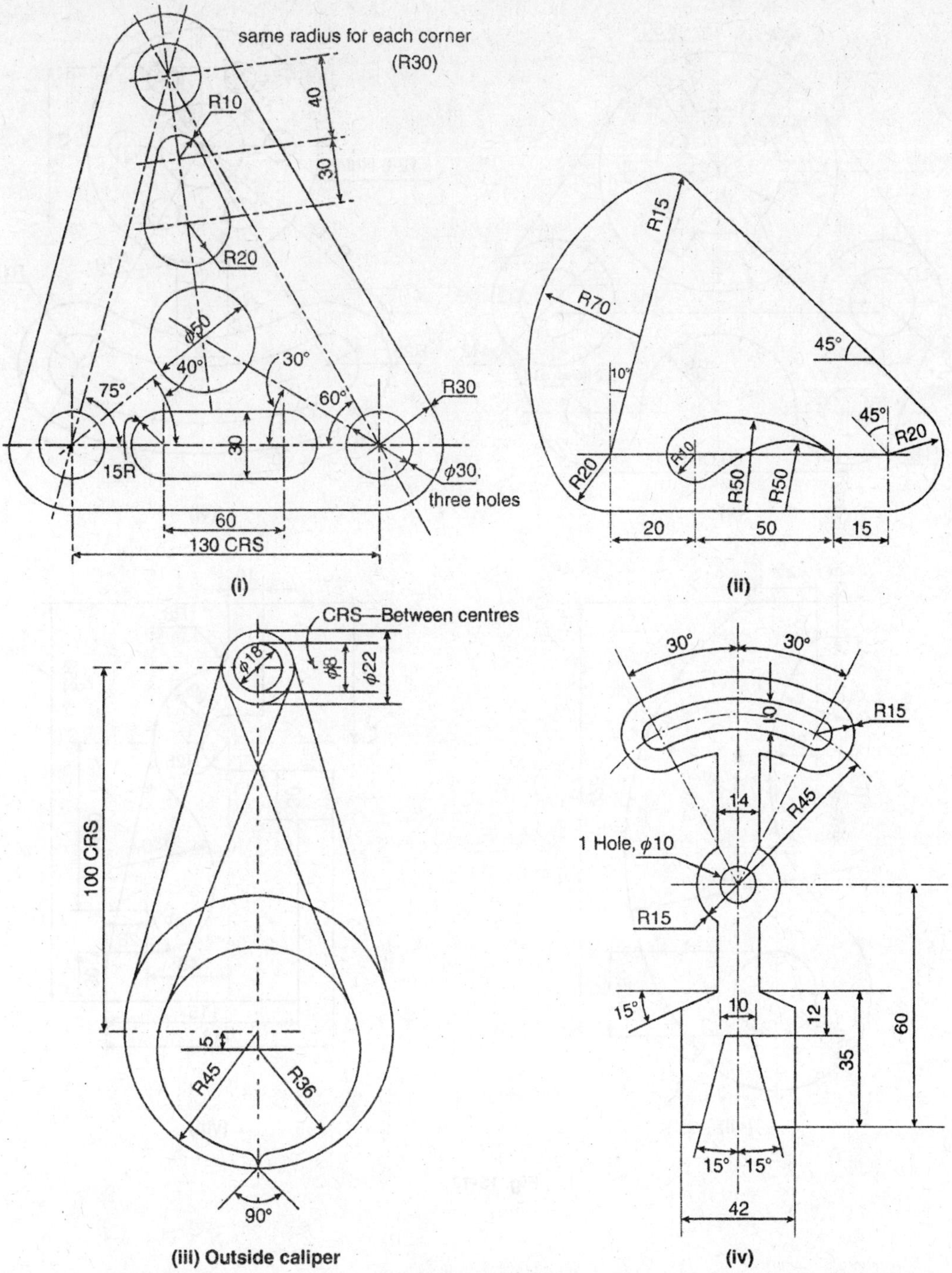
same radius for each corner
(R30)
R10
40
30
R20
ϕ50
40°
30°
75°
60°
R30
30
15R
ϕ30,
three holes
60
130 CRS
(i)
R15
R70
45°
10°
45°
R20
R10
R20
R50
R50
20
50
15
(ii)
CRS—Between centres
ϕ18
ϕ8
ϕ22
100 CRS
5
R45
R36
90°
(iii) Outside caliper
30°
30°
10
R15
14
R45
1 Hole, ϕ10
R15
15°
10
12
60
35
15°
15°
42
(iv)

Fig. 15-13

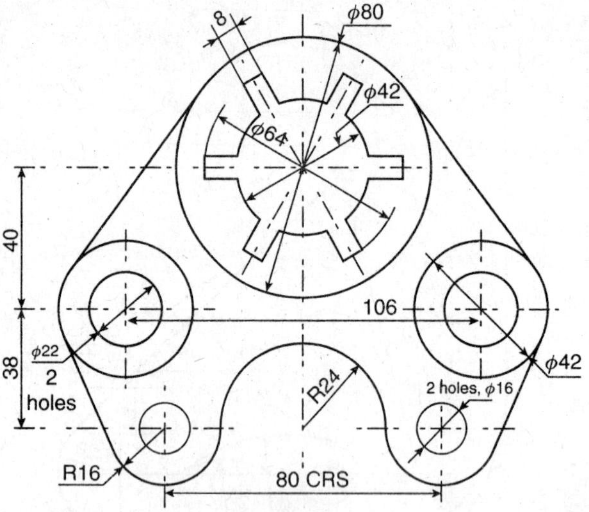

(v)

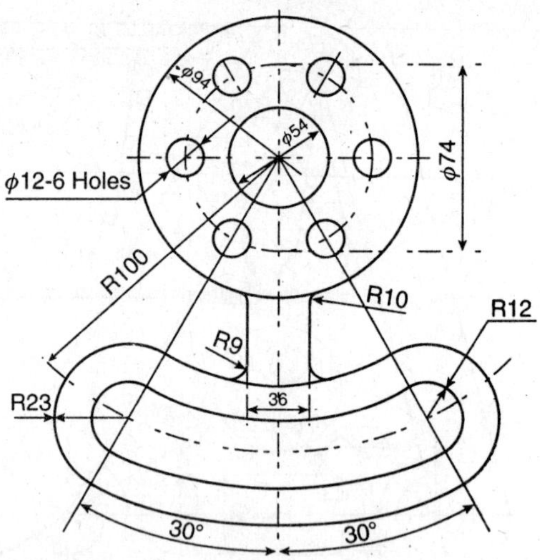

(vi)

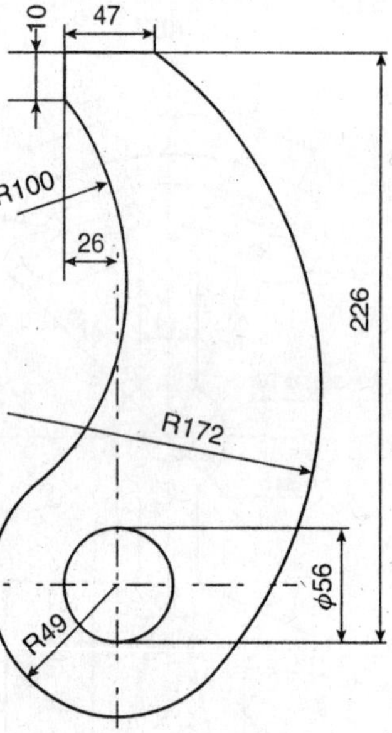

(vii)

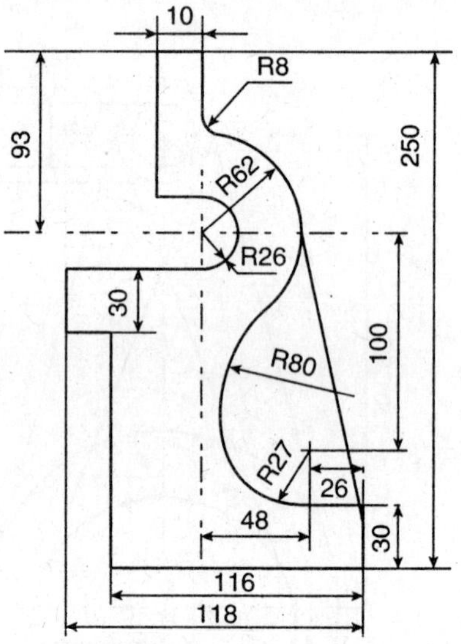

(viii)

Fig. 15-13

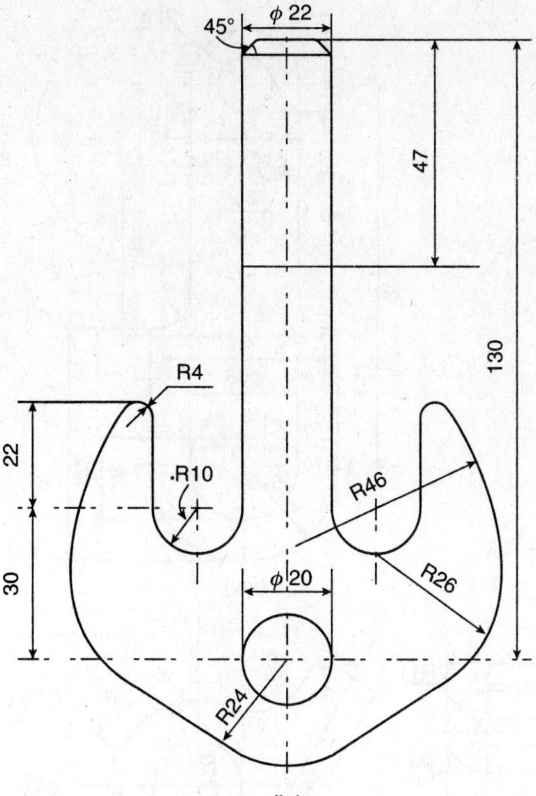

(ix)

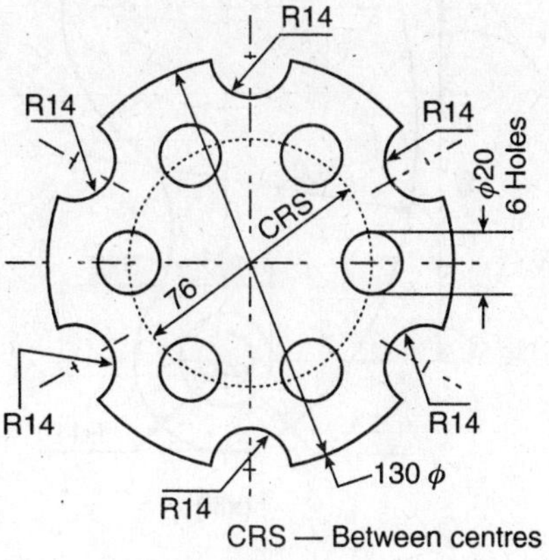

(x)

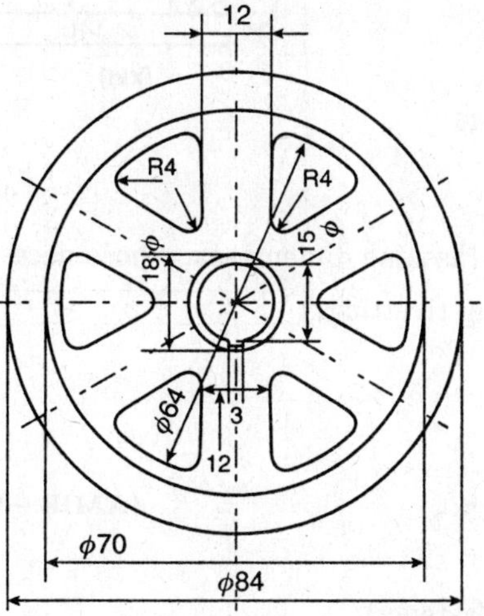

(xi)

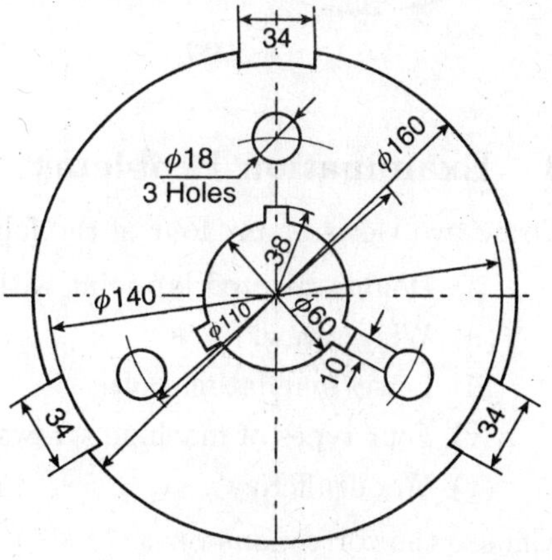

(xii)

Fig. 15-13

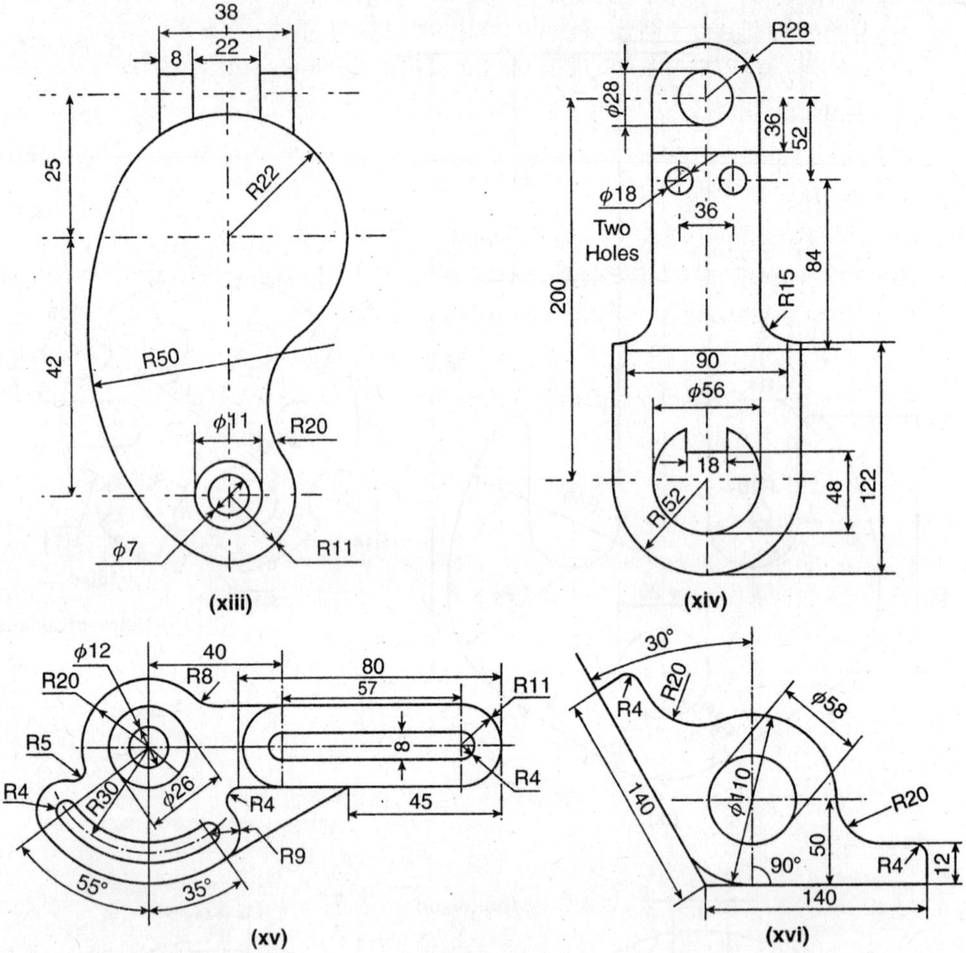

Fig. 15-13

15.11 Examination Problems

1. Draw two views of any four of the following (assume dimensions, where necessary)
   - (i) Double-rivetted lap joint with zig-zag rivetting,
   - (ii) Wing screw,
   - (iii) Lewis foundation bolt,
   - (iv) Four types of machine screws
   - (v) Woodruff Key **(AMIE—Winter, 2000)**
2. Choose the correct answer :
   - (i) Which one of these is a permanent fastener?
     - (a) Nut-bolt (b) Stud (c) Rivet. **(AMIE—Summer, 2001)**

     Ans. (c)

(ii) Commonly used thread-profile for a load-lifting screw-jack is

(a) BSW thread (b) Square thread (c) acme thread. **(AMIE—Summer, 2001)**

Ans. (b) and (c)

3. (a) Draw any two of the following threads, showing their dimensional proportions in terms of the pitch p :

(i) BSW (ii) ACME (iii) Sellers.

(b) For structural rivets, draw any two of the following rivet-heads, showing their dimensions in terms of the rivet-diameter d :

(i) Snap (ii) Pan (iii) Flat counter-sunk. **(AMIE—Summer, 2001)**

(c) Draw two views of any four of the following (assume dimensions, where necessary) :

(i) Eye-foundation bolt with nut

(ii) Double-rivetted lap joint (chain rivetting)

(iii) Gib-head Key

(iv) Hexagonal-headed bolt

(v) Wing nut. **(AMIE—Winter, 2001)**

(d) Draw the sectional front view and top view of double rivetted double-strap butt joint with zig-zag rivetting. Take thickness of main plate $t = 10$ mm, thickness of cover plate, $t_c = 0.625t$, diameter of rivet, $d = 6\sqrt{t}$, pitch of rivet, $p = 3$d. **(AMIE—Summer, 2002)**

(e) Draw any four of the following (assume suitable dimensions, whenever necessary) :

(i) Two views of a single-rivetted butt joint with double-cover plate

(ii) Two views of a square-headed bolt and nut

(iii) Three views of a hexagonal nut

(iv) Acme thread (show two consecutive teeth)

(v) Two views of a double-headed Key

(vi) Pictorial view of a simple journal bearing. **(AMIE—Winter, 2002)**

(f) State four parameters of free-hand sketches.

Ans. See article 15.2.

4. Fill in the blanks with appropriate words :

(i) Nuts and bolts are used to join two or more metal-plates ______.

(ii) Rivet-heads are used to join two or more metal-plates ______.

(iii) Threads are cut externally on ______.

(iv) Internal threads are usually cut on a ______.

(v) The three common materials for free-hand sketching are ______, ______ and ______.

Ans. (i) Temporarily (ii) Permanently (iii) Bolts, Screws or pipes (iv) Nut or pipe (v) Pencil, Eraser, Sketch-pad.

Chapter 16

Building Drawing

16.1 Introduction

This chapter deals with the conventional type simple, single-storied residential building and details of furniture-arrangements, fixture and their standard dimensions. A conventional type of simple, single-storied building consists of a wide living-room with a verandah, bedroom, kitchen, bathroom, latrine etc., without stair.

To draw the plan for the said type of house, the plan-maker should know the size and shape of the land, width of road, how many persons will stay there, their functional requirements, profession, taste of the owner, capacity of finance, etc. Lastly, the rules and regulations of the local corporations or municipalities are to be followed to design the foundation of the building. The plan-maker should know the load-bearing capacity of the soil and, if required, the soil is to be tested in a laboratory, as per Bureau of Indian standards (BIS) specifications.

The details of building materials. (brick, stone, cement, steel, aluminium, sand, glass, wood, stone-chips etc.) may be obtained from the books on building materials.

16.2 Main Portion of Single-Storied Building

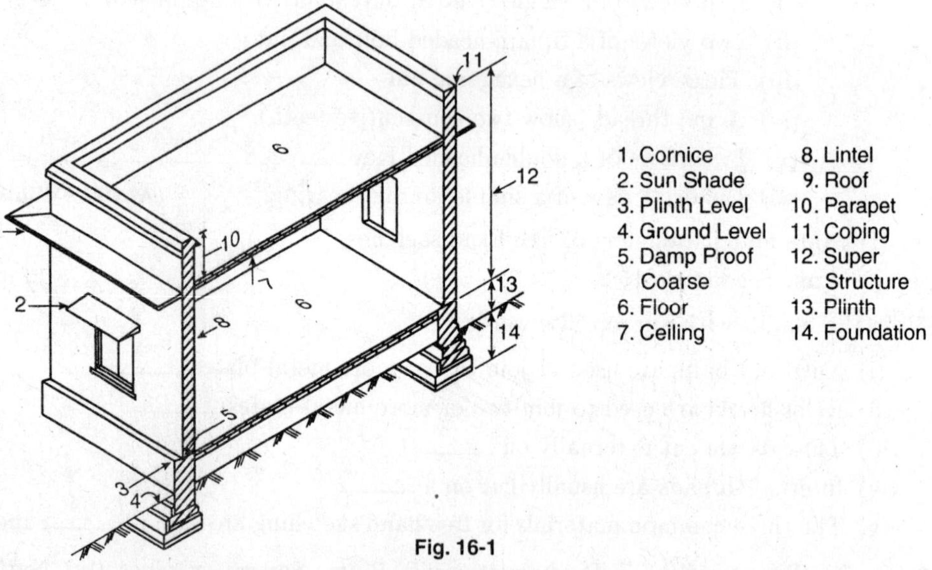

Fig. 16-1

Refer to Fig. 16-1 (Page 198). Main portion of a single-storied building is shown. Only study the names and commit to memory for future guidance.

Study the figures (a), (b) and (c) of Fig. 16-2. In Fig. (a), isometric view, of a single-storied building, is shown with a verandah.

In Fig. (b), sectional plan through window-sill-level of the same building of (a), is shown in isometric view.

In Fig. 16-2(c), sectional elevation of the same building of Fig. 16-2(a), through section A-A, is shown in isometric view. The sectional line A-A is shown in Fig. (a).

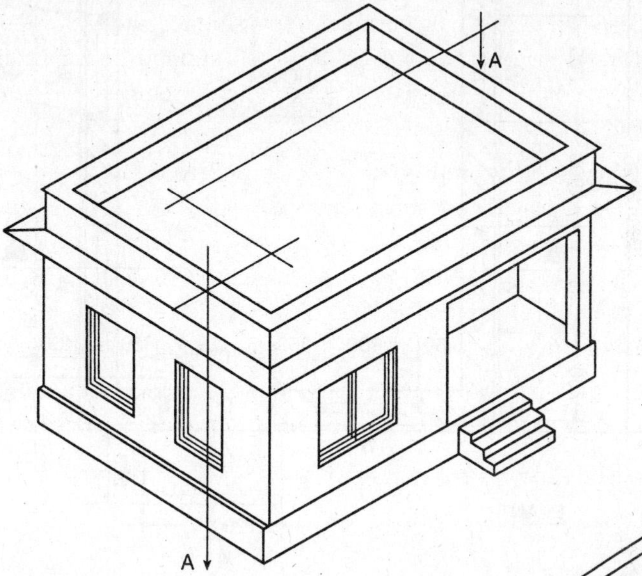

(a)

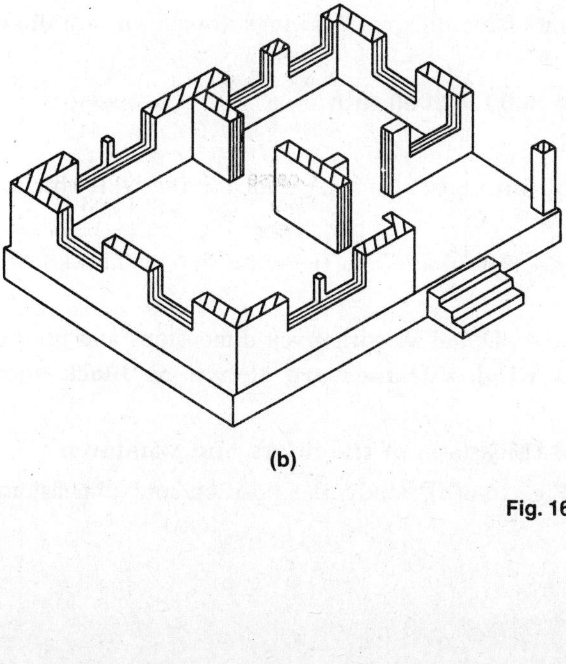

(b)

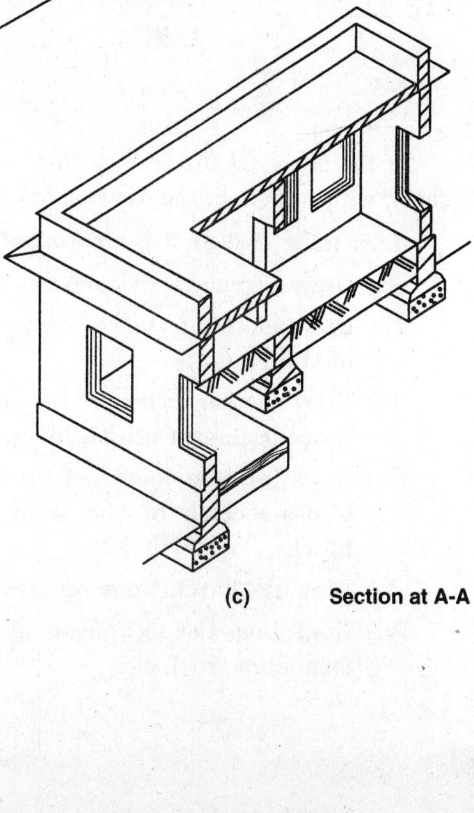

(c) Section at A-A

Fig. 16-2

Now, study and draw, step by step, the following drawings and acquire the skill of building drawing gradually.

Refer to Fig. 16-3.

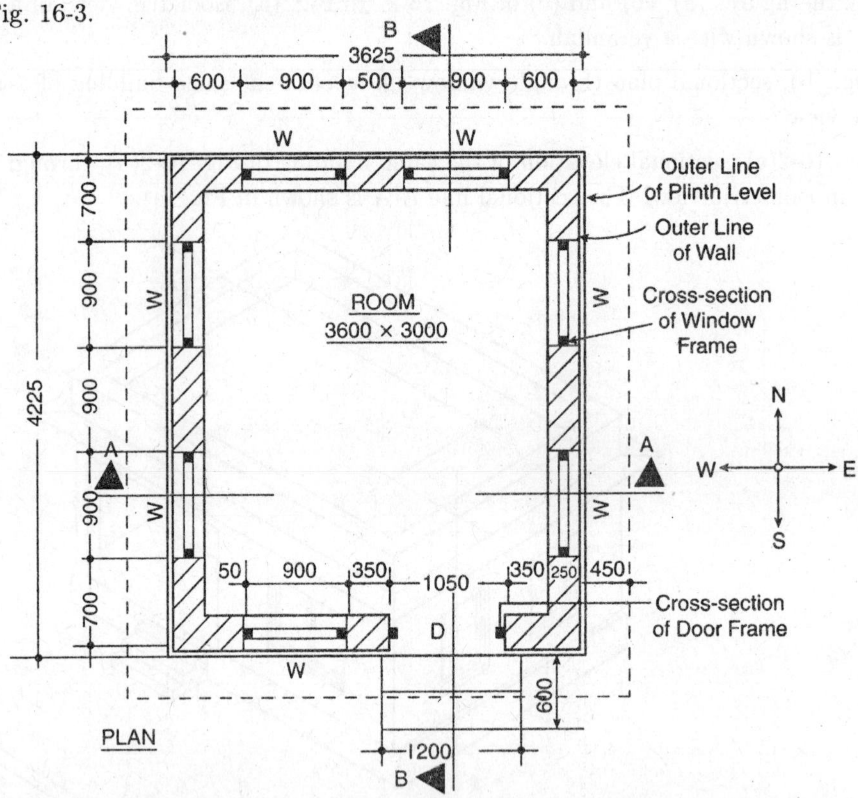

Fig. 16-3(a)

In the figure (c) of the Fig. 16-3, position of the different drawings are shown in a drawing sheet, with respect to the drawing (a).

Refer to fig 16-3(a), it is a **room of size 3600 × 3000 mm.**

(i) Draw a rectangle of same size as above.

(ii) Draw another rectangle, keeping **250 mm** apart from (i), which is the **wall-thickness of the room.**

(iii) Draw another rectangle, keeping **62.5 mm** apart from the outer line of the wall, which is **outer line of plinth-level.**

(iv) Now, draw the doors and windows, i.e., D and W with given dimensions and positions. **Cross-section of the door and window-frames are shown as black square blocks.**

(v) Draw the **hatch-lines on the wall-thickness, of the doors and windows.**

(vi) Now, from the section on 'AA', Fig. 16-3(b), study the position and dimensions of foundation, roof, etc.

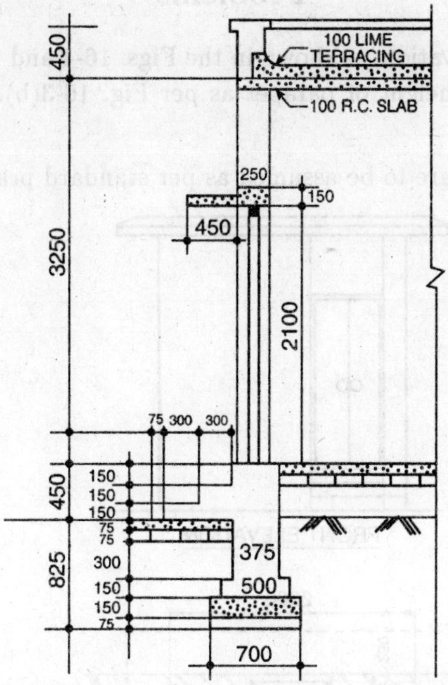

Fig. 16-3(b) Section on AA of Fig. (a) on page 200

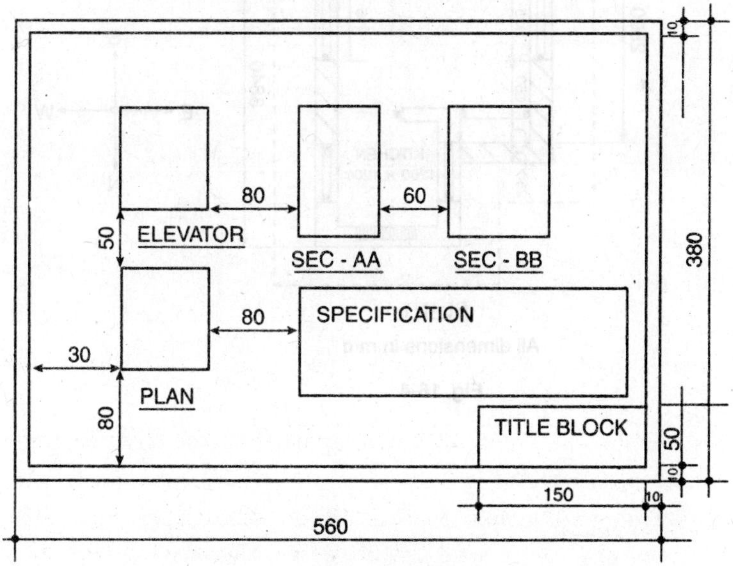

(c) Drawing Sheet arrangement

Fig. 16-3(c)

Problems

1. Draw plan and front elevation, as shown in the Figs. 16-4 and 16-5, assuming plinth-height, height of the building, height of parapet as per Fig. 16-3(b). Take window-sill height as 600 mm.

   Door and window-sizes are to be assumed as per standard practice.

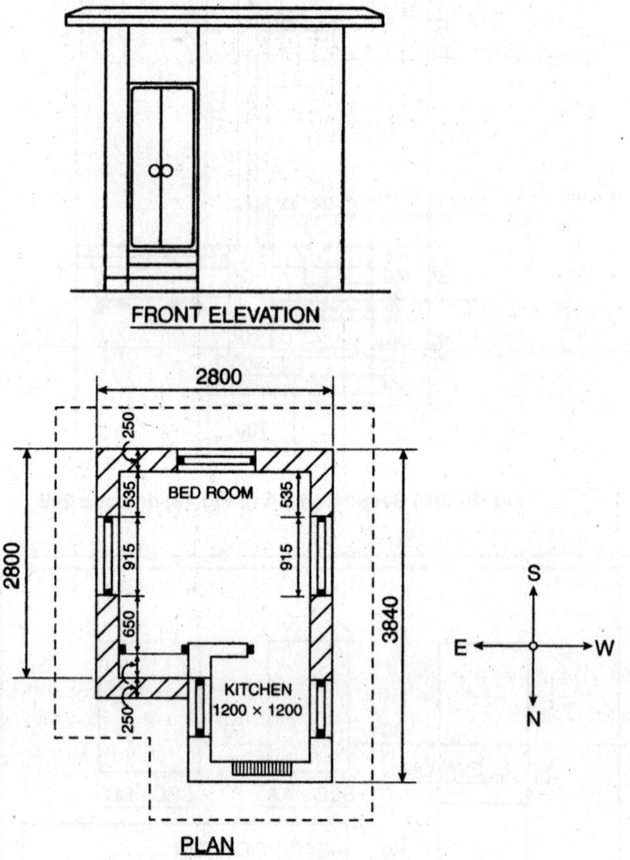

All dimensions in mm

Fig. 16-4

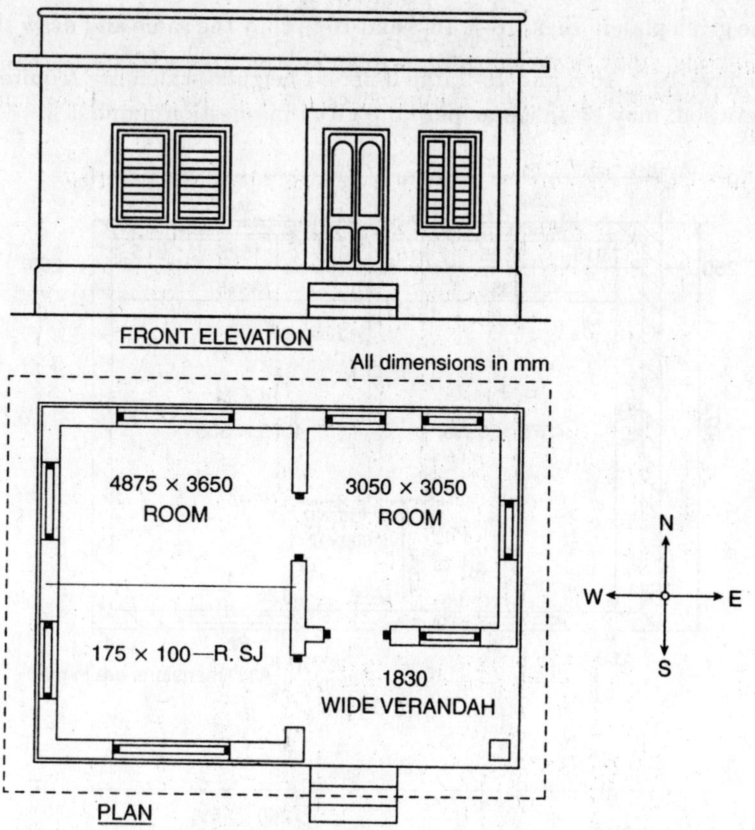

Fig. 16-5

2. From the given plan in Fig. 16-6 (in line diagram), draw plan and front elevation. Given thickness of super-structure wall as 250 mm. Assume plinth-height, height of the building, height of parapet as per Fig. 16-3(b). Take window-sill height as 600 mm. Door and window-sizes are to be assumed as per standard practice.

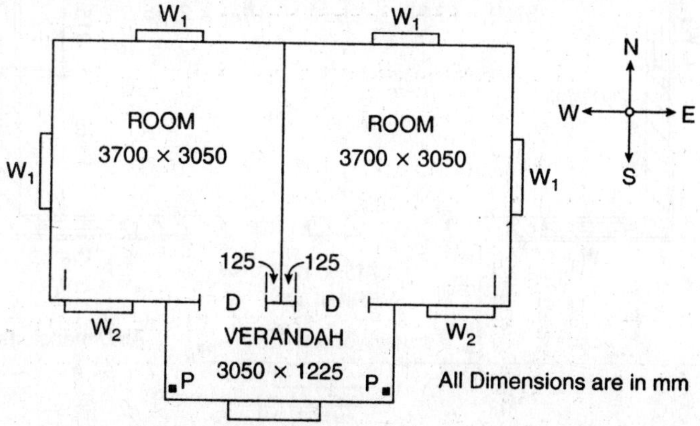

Fig. 16-6

3. From the given plan in Figs. 16-7, 16-8 and 16-9, copy the same and draw the front elevation.

In the figures 16-7, 16-8 and 16-9, the different heights, which are required for drawing the front-elevation, may be taken as per note given in question number 2.

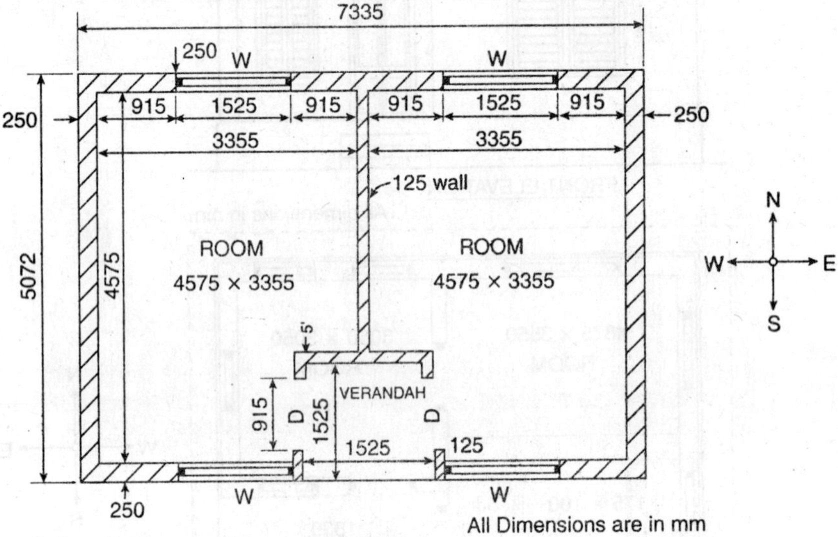

Fig. 16-7

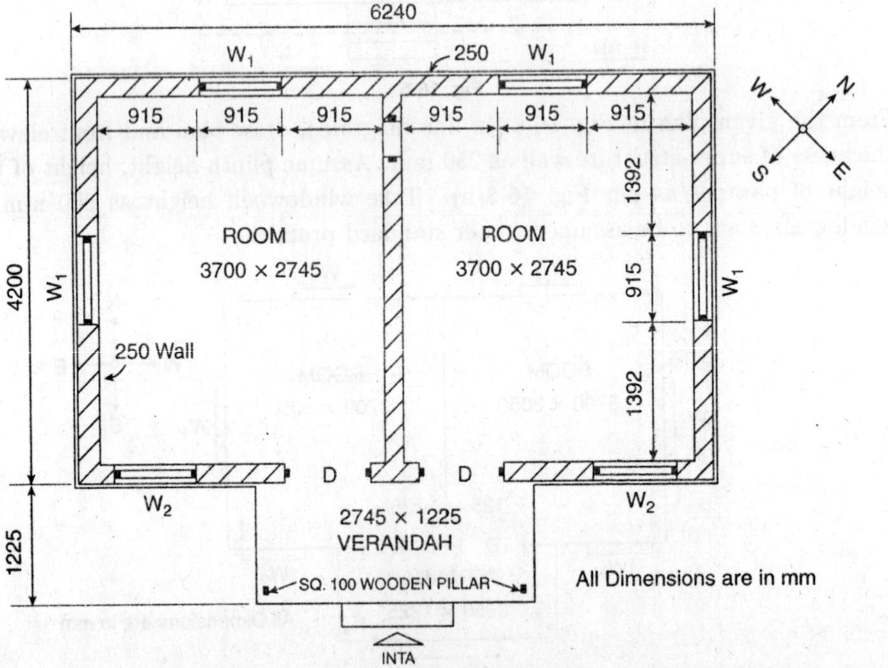

Fig. 16-8

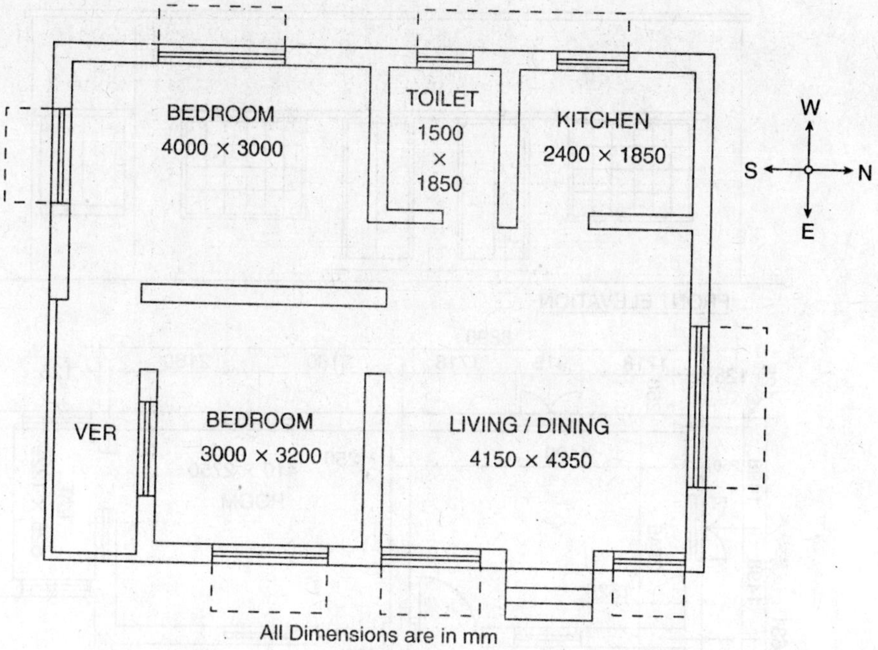

Fig. 16-9

4. In the Fig. 16-10, an isolated 'Kitchen' is shown for study and practise the same. Show in it the position of gas-cylinder with oven, kitchen–sink and space for storage of food.

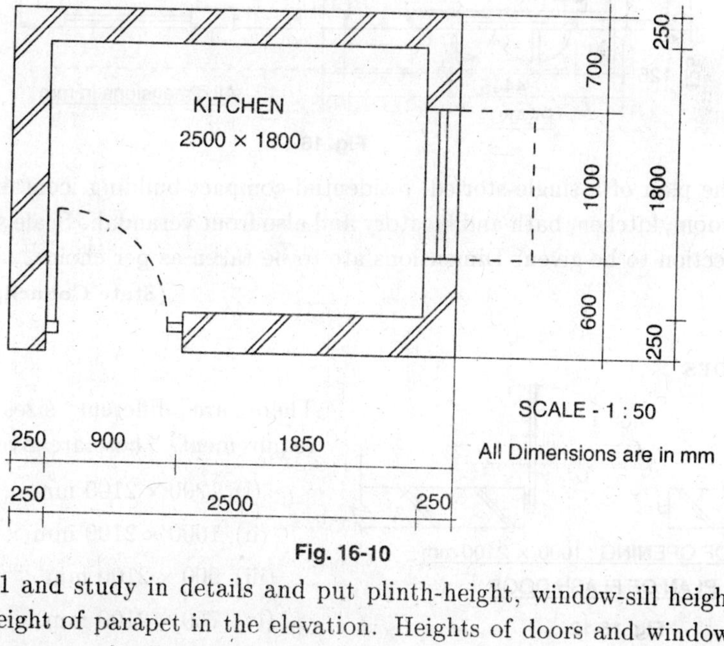

Fig. 16-10

5. Copy Fig. 16-11 and study in details and put plinth-height, window-sill height, height of the building, height of parapet in the elevation. Heights of doors and windows are to be incorporated as per standard practice.

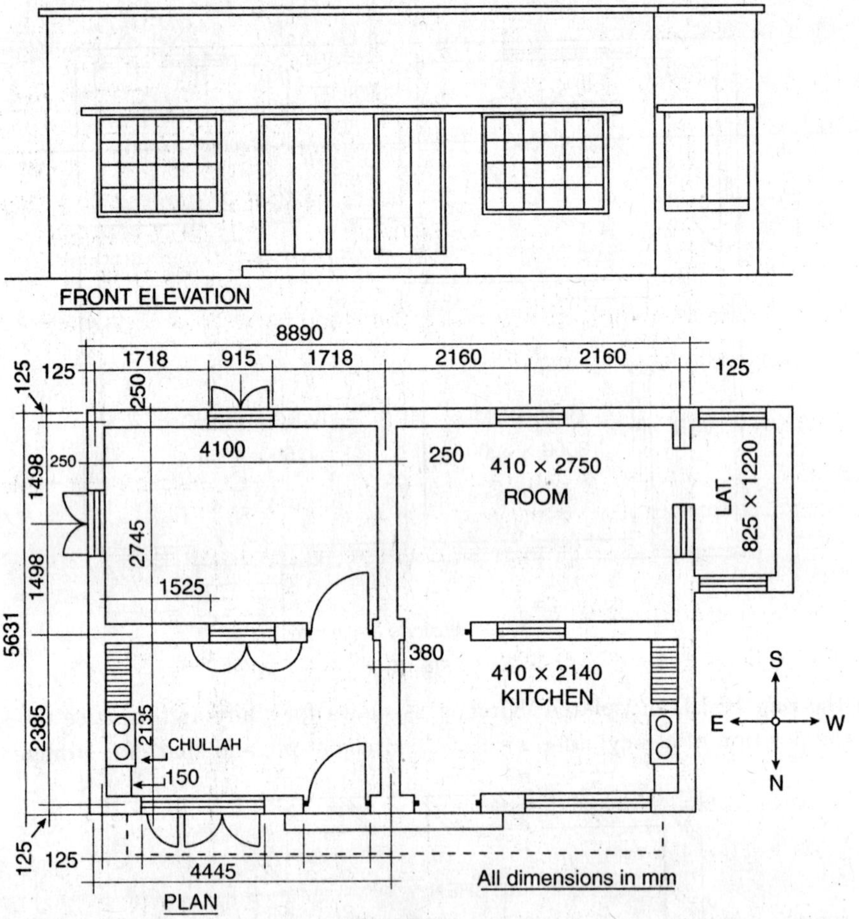

Fig. 16-11

6. Draw the plan of a single-storied, residential-compact-building, containing two bedrooms, dining room, kitchen, bath and lavatory and also front verandah. Scale should be mentioned and direction to be given. Dimensions are to be taken as per choice.

(State Council, W.B.—June, 2006)

16.3 Doors

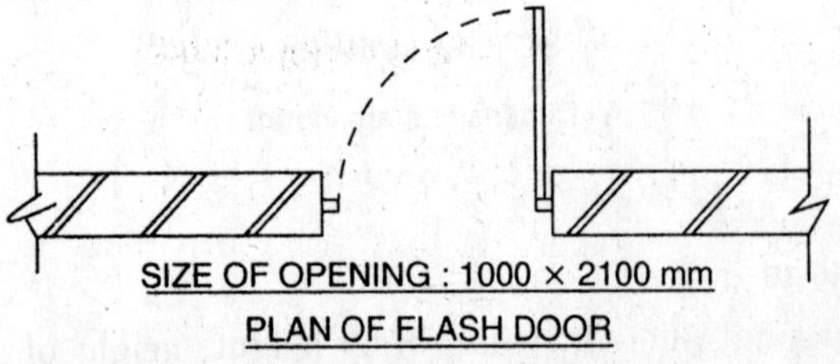

Fig. 16-12

There are different sizes of doors as per requirement. These are given below :

(i) 1200 × 2100 mm

(ii) 1000 × 2100 mm

(iii) 900 × 2100 mm

(iv) 750 × 2100 mm.

In Fig. 16-12, the plan of the Flash-Door is shown. Study the same and state its advantage.

16.4 Windows

Sizes of Windows are given below :

(i) 2000 × 1350 mm

(ii) 500 × 1350 mm

(iii) 1000 × 1350 mm

(iv) 500 × 900 mm

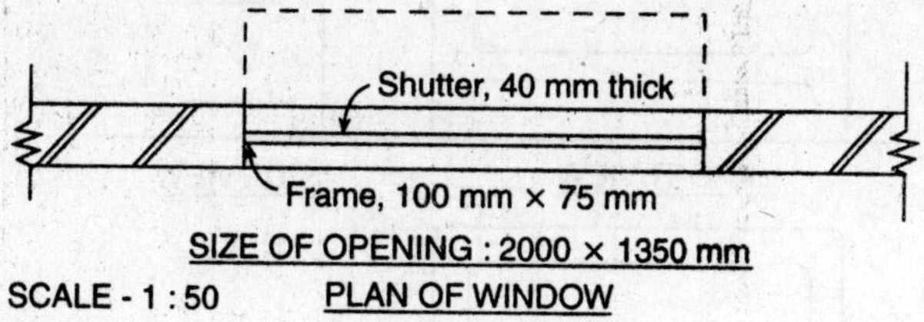

Fig. 16-13

Windows are of different types. In Fig. 16-13, the plan of a window is shown. Redraw it to a larger scale as per choice.

16.5 Furniture

Plan, elevation and side view of one each of double-bed, chair, sofa-set and table are shown in Figs. 16-14(a), (b), (c) and (d) respectively.

Essential furnitures in a bed-room and their sizes are shown below :

(i) Single-bed—1100 × 2100 mm

(ii) Double-bed—1800 × 2100 mm

(iii) Bed-side table—450 × 450 mm

(iv) Almirah—900 × 450 mm

(v) Dressing Table—350 × 750 mm

(vi) Chair—450 × 450 mm

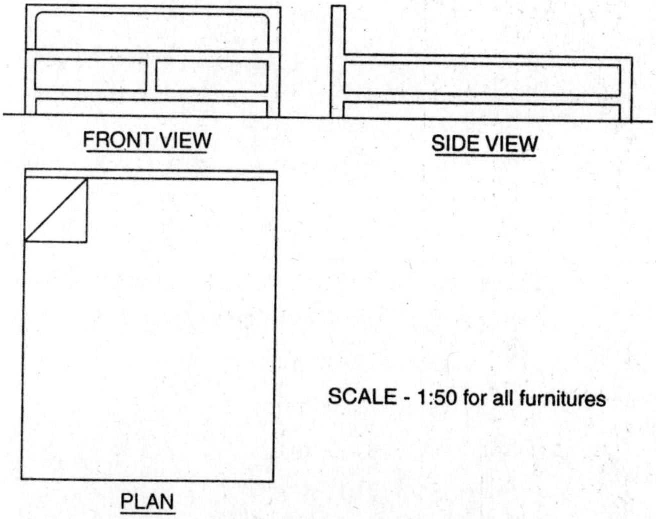

Fig. 16-14(a) A DOUBLE-BED

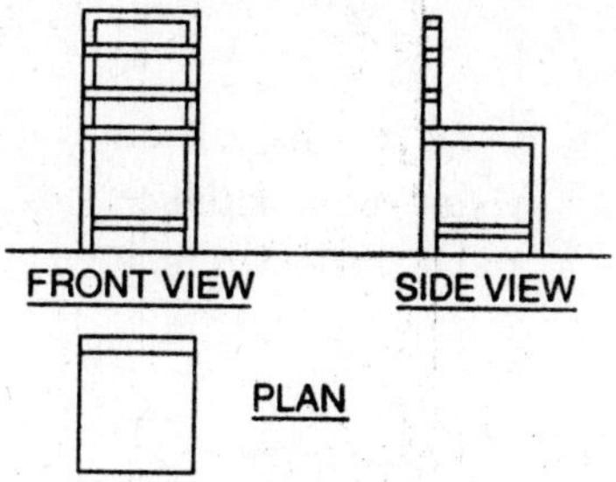

Fig. 16-14(b) A CHAIR

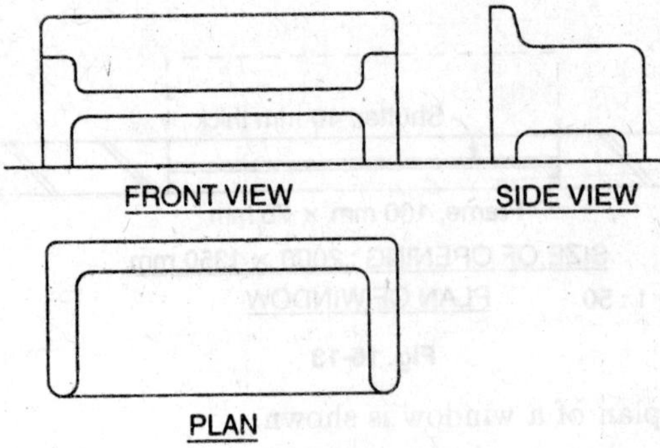

Fig. 16-14(c) A SOFA-SET

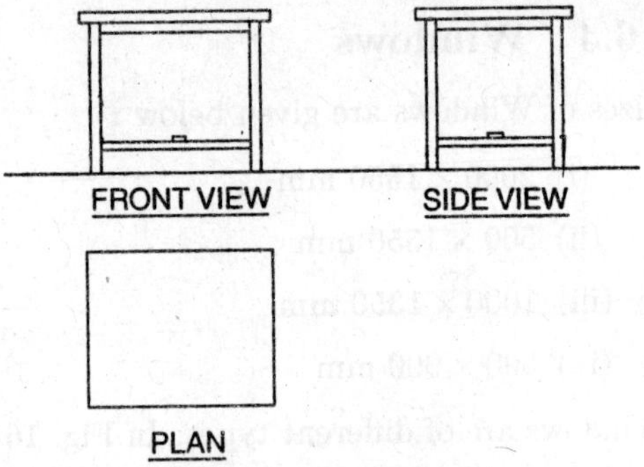

Fig. 16-14(d) A TABLE

We should select the furniture, according to the size of the room and the requirements.

Chapter 17

Intersection of Surfaces

17.1 Introduction

(i) This chapter deals with the intersection of two surfaces only.

(ii) The nature of the surfaces are **straight and curve.**

(iii) **The intersection line of**

(a) Plane-surface to plane-surface create straight lines,

(b) Plane-surface to curved-surface produce curved lines,

(c) Curved-surface to curved-surface result curved lines.

(iv) Please recapitulate the following chapters :

(a) Projection of solids

(b) Surface-development and

(c) Section of solids, as these are interrelated.

(v) This chapter is also known as '**interpenetration of solids**',

(vi) Here, we have dealt with the following regular solids : (a) Prism (b) Pyramid (c) Cylinder (d) Cone and (e) Sphere

17.2 Aim

The designers have to design many types of boilers, smoke-stack,(chimney)', different ducts, ventilation devices, metal containers, tanks, machine-castings, machine tools, equipment, ships, automobiles, aircraft, etc. in various industries. These involve simple and complicated curves. Our aim is to guide the students, so that they can learn how to draw these curves.

17.3 Methods

There are two methods to draw the curves of intersection of surfaces :

(i) **Line Method :**

(a) Here, we will draw some imaginary lines on the solid (e.g., cylinder). Generally, we divide the base-circle into 12 equal parts and draw lines from those points, as in the case of surface-development. We imagine the cylinder to be a 12-sided prism.

(b) Similarly with other solids.

(c) We find out the points of intersection with corresponding lines.

(d) There we will get the required curves from meeting points of those lines.

(ii) **Cutting-Plane Method :**

This system is similar to the line method. Students have to study this system for their physical conception. Here we will follow the line method.

We will explain the cutting plane method later.

17.4(a) Intersection of Prisms

Problem 1 : Two square prisms, one placed vertically and the other horizontally. The horizontal one is smaller in size. One of the sides of both the prisms are parallel to V.P. The smaller prism is completely penetrated into the bigger prism, and their axes intersect. Draw the lines of intersection, when the axis of the smaller prism in parallel both to H.P. and V.P.

Solution : Here, penetrating lines are straight lines. This is the simplest form of penetration,

(i) Draw the isometric view of vertical square prism [Fig. 17-1(i)].

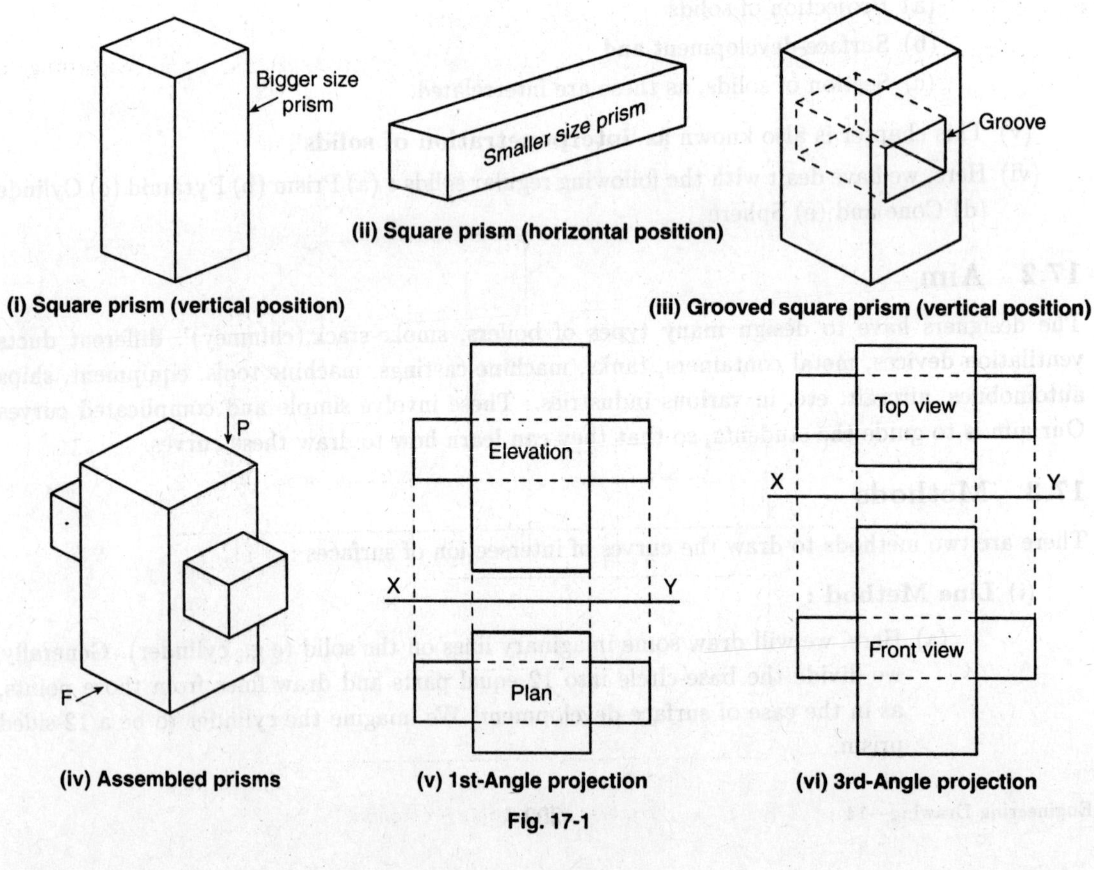

(i) Square prism (vertical position)

(ii) Square prism (horizontal position)

(iii) Grooved square prism (vertical position)

(iv) Assembled prisms

(v) 1st-Angle projection

(vi) 3rd-Angle projection

Fig. 17-1

(ii) Draw the isometric view of horizontal square prism of smaller dimension, [Fig. 17-1(ii)].

(iii) Draw a groove on (i) to suit the prism (ii), [Fig. 17-1(iii)]

(iv) Insert the prism (ii) into the grooved prism [Fig. 17-1(iv)]

(v) Draw the plan and elevation (1st Angle projection) according to Fig. 17-1(v).

Elevation (looking along the arrow F) see Fig. 17-1(iv)

Plan (looking along the arrow P)...... "

(vi) Draw the top and front view (3rd-Angle projection) according to Fig. 17-1(vi).

Front view (looking along the arrow F) see Fig. 17-1(iv).

Top view (looking along the arrow P)...... "

N.B. Now, we can rotate both the prisms at angles of 30°, 45°, 60° or any other angles.

Note : We will proceed with regular prism and cylinder simultaneously, as we imagine the cylinder to be a twelve-sided prism. So, their construction-procedures are same.

Problem 2 : Two square prisms, base-sides of them 45 mm each, penetrated at right-angles and their axes meeting at a point. The axes are parallel to V.P. All the sides of two prisms are equally inclined to the V.P. Draw the penetrating curves, assuming suitable lengths of the prism.

Solution : First angle projection method is applied here.

(i) Draw side-view first, and then the elevation and plan of horizontal prism according to the given conditions (Refer Fig. 17-2).

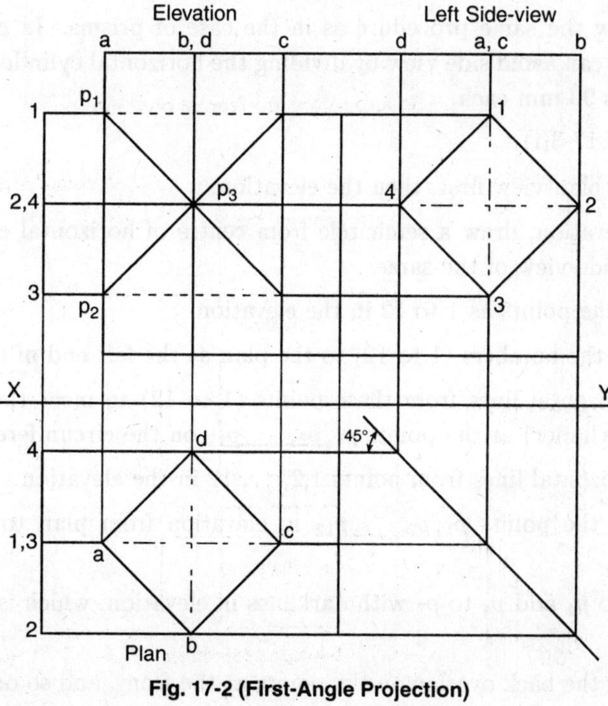

Fig. 17-2 (First-Angle Projection)

(ii) Mark the corners as 1, 2, 3 and 4.

(iii) Draw the plan, elevation and side view of the vertical prism according to given conditions with faint lines.

(iv) Mark the vertical one as a, b, c and d.

(v) Lines 1 and 3 cut the line a at p_1 and p_2 at the elevation.

(vi) Lines 2 and 4 cut the line b and d at p_3

(vii) Join p_1, p_3 and p_2 with straight lines, which is the line of intersection.

(viii) Draw 1-p_1, (2, 4)-p_3, $p_2$3 Dark

(ix) From p_1 and p_2 draw dotted line towards right.

(x) Complete the same at the right side of the vertical axis in elevation.

Note : We will solve the same problem with cylinders of same diameters so that the student can relate with the prisms. (When the number of sides of a regular prism is infinite, the prism becomes a cylinder.)

17.4(b) Intersection of Cylinders

Problem 3 : Two cylinders, of base-diameters 45 mm each, penetrated at right-angles so that their axes bisect each other. The axes are parallel to V.P. Draw the penetrating curves. Assume suitable lengths of the cylinders.

(1st-Angle projection)

Solution : Follow the same procedure as in the case of prisms. In case of a cylinder, we get twelve points. We can avoid side view by dividing the horizontal cylinder-base. We assume length of the cylinders as 90 mm each.

Refer to figure 17-3(i).

(i) Draw the plan-view first, then the elevation.

(ii) In the elevation, draw a semicircle from centre of horizontal cylinder, which represents the half side-view of the same.

(iii) Number the points as 1 to 12 in the elevation.

(iv) Transmit the numbers (1 to 12) to the plan at the left end of the horizontal cylinder.

(v) Draw horizontal lines from these points (1 to 12) to meet the circle (plan-view of the vertical cylinder) at the points $p_1, p_2, \ldots, p_{12}$ on the circumference.

(vi) Draw horizontal lines from points 1,2,..., 12 in the elevation.

(vii) Transmit the points $p_1, p_2, \ldots, p_{12}$ in elevation from plan to the respective points in elevation.

(viii) Join p_1 to p_4 and p_4 to p_7 with darklines in elevation, which is the required penetrating lines.

(ix) Points on the back overlap on the points at the front, and so only one darkline is drawn.

(x) A similar penetrating line will be formed at the right-hand side in elevation (not shown here).

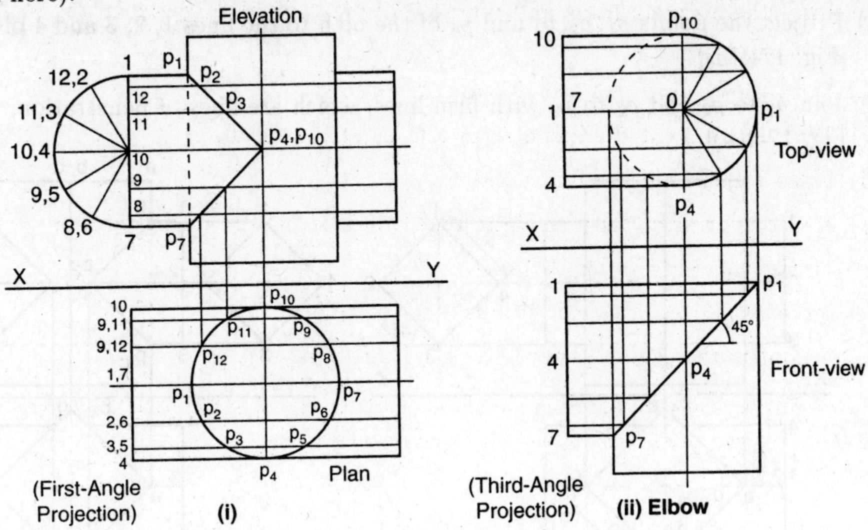

Fig. 17-3

Elbow-Joint

An elbow joint can be drawn, as shown in Fig. 17-3(ii). Procedure is the same as in Fig. 17-3(i). In the manufacturing process, a metal pipe is cut into two pieces at an angle of 45° to the axis. Two parts are turned at an angle of 90° (one side only for each part), and then welded by joining the 45°-edges of the two metal parts.

Problem 4 : A vertical square prism, of end-sides 60 mm each, is completely penetrated by a horizontal square prism of sides 40 mm each, their axes intersect and are parallel to V.P. One of the sides of the two prisms are making an angle 45° to V.P. Draw the projections, showing lines of intersection. Choose any suitable length of the prisms.

(1st-Angle projection)

Solution : Refer to Fig. 17-4(i). Draw the projections, without penetrating curves. We assume the length of each prism as 100 mm.

(i) Draw a square of sides 40 mm each making an angle of 45° with V.P. in the side view, where only the smaller prism is shown.

(ii) Project the elevation and plan of the smaller prism.

(iii) Number the same in one base 1, 2, 3 and 4 in the side-view.

(iv) Draw the plan of vertical prism in the middle position of the plan of the horizontal prism and number it as a, b, c and d.

(v) Draw the figure 17-4(ii).

(vi) Locate the points of intersections in plan.

(vii) Edges, 1, 2, 3 & 4 of the horizontal prism, cut the vertical prism at p_1, p_2, p_3 and p_4, [see plan of Fig. 17-4(ii)]

(viii) Project the points p_1, p_2, p_3 and p_4 of the plan to the lines 1, 2, 3 and 4 of the elevation, [Fig. 17-4(ii)]

(ix) Join p_1 to p_2 and p_2 to p_3 with firm lines, which are lines of penetration, [see elevation, Fig. 17-4(ii)]

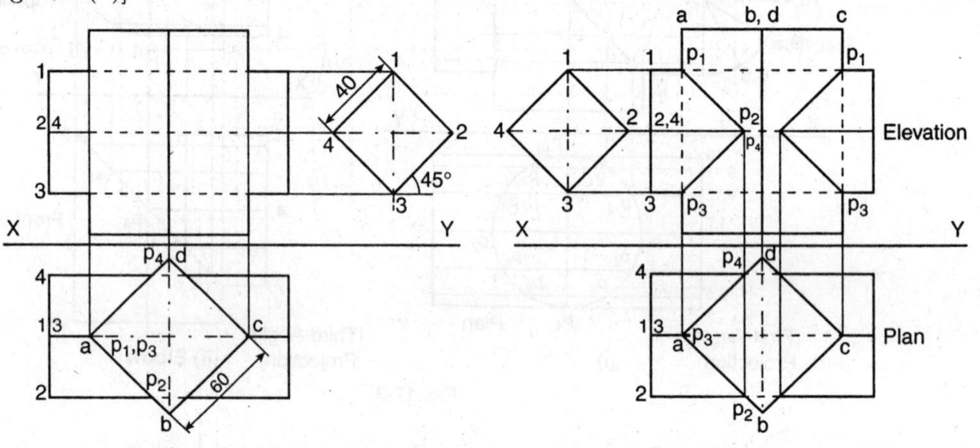

(i) (First-Angle Projection) (ii)

Fig. 17-4

(x) The lines p_1 to p_2 and p_2 to p_3 are hidden portions.

(xi) Draw dotted lines on hidden portions of the horizontal prism, both in plan and elevation, as shown.

(xii) Back-side and front side are symmetrical about the axis. So, penetrating lines at the back coincide with those at the front.

Note : Here, we can relate the above problem with cylinders of unequal diameters, by imagining the cylinder to be a 12-sided prism. The imaginary lines are known as **generators**.

Problem 5 : A vertical cylinder of 75 mm diameter is completely penetrated by another cylinder of 50 mm diameter. Their axes are perpendicular and bisecting to each other, and parallel to V.P. Draw projections, showing curves of penetration.

Solution : We will solve the problem in two methods: (a) Line Method and (b) Cutting Plane Method. Any one method may be used. We are assuming the length of each cylinder as 100 mm.

(a) Line Method : Refer to Fig. 17-5(i).

(i) Draw side view of the two cylinders first.

(ii) Divide the horizontal cylinder (i.e., circle) in 12 equal parts with set-square and number the same as 1, 2, 3,. . . ,12.

(iii) Draw the plan and elevation.

(iv) Transmit the imaginary points 1, 2, 3,. . ., 12, to the horizontal cylinder in plan and elevation as shown.

(v) Lines 1, 2, 3,. . ., 12 cut the vertical cylinder at $p_1, p_2, p_3, \ldots, p_{12}$ in the plan.

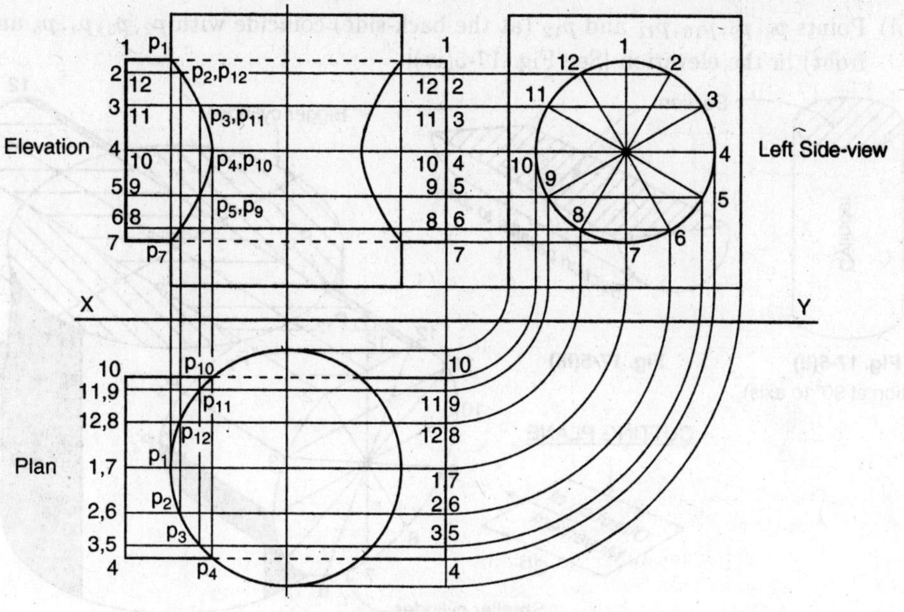

Fig. 17-5(I) (First-Angle Projection)

(vi) Transmit the points $p_1, p_2, p_3, \ldots, p_{12}$ to the elevation to cut '1' with p_1, 2 with p_2 and so on, which indicate the points of intersections.

(vii) Join $p_1, p_2, \ldots, p_7$, (as the figure is symmetrical , so points $p_8, \ldots, p_{12}$ overlap with the points $p_1, p_2, \ldots, p_7$), which is the required curve of penetration.

(viii) Transmit the points to the right side, and draw the curve. See elevation.

(b) Cutting-Plane Method :

(i) There is basically no difference between the line and cutting-plane methods.

(ii) It helps the students to achieve the physical conception of the same.

(iii) It is useful in solving complicated and hidden portion of two solids.

Note : We know that, the cross-section of a cylinder is a circle, when the cutting plane is perpendicular to the axis. See Fig. 17-5(ii). Again, when the cutting plane is parallel to the axis of the cylinder, the section is a rectangle. See Figs. 17-5(iii). Fig. 17-5(ii) and (iii) are shown in isometric views.

(iv) From the Fig. 17-5(iv), we assume a cutting plane (CP) passing through the points 12-2 of the horizontal cylinder to cut the vertical cylinder at p_{12}–p_2, Fig. 17-5(iv) is an isometric view.

(v) Similarly, when the cutting plane passes through 11-3, 10-4, 9-5 and 8-6, we get the points p_{11}–p_3, p_{10}–p_4, p_9–p_5 and p_8–p_6 respectively on the vertical cylinder.

(vi) Join $p_1, p_2, \ldots, p_{12}$, which is the required curve of penetration.

(vii) Points p_8, p_9, p_{10}, p_{11} and p_{12} (at the back-side) coincide with p_2, p_3, p_4, p_5 and p_6 (at the front) in the elevation [See Fig. 17-5(iv)].

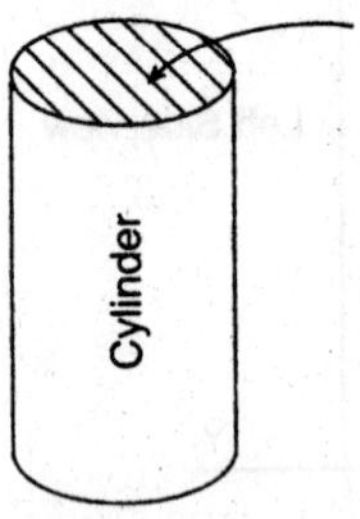

Fig. 17-5(ii)
(Section at 90° to axis)

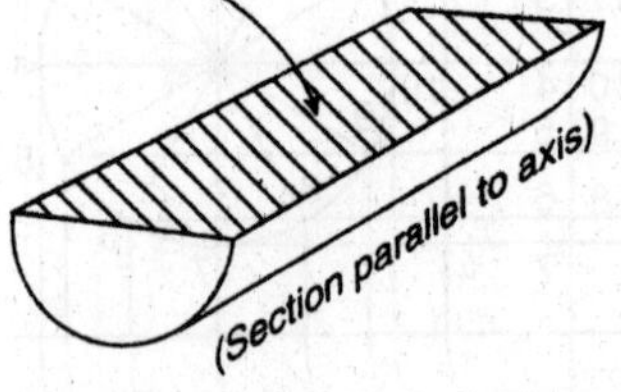

Fig. 17-5(iii)

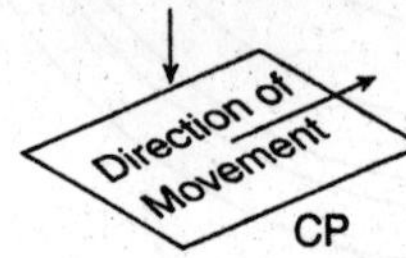

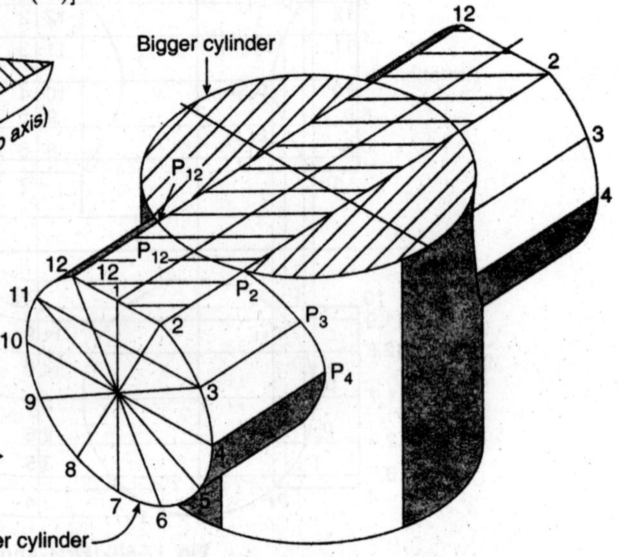

Fig. 17-5(iv)

Problem 6 : One vertical square prism is penetrated by another horizontal square prism, of end-sides 50 mm and 35 mm each respectively. Their axes are 6 mm apart and parallel to V.P. All the faces of the prisms are equally inclined to V.P. Draw the projections. Show lines of intersection.

Solution : This problem is solved by third-angle projection.

Refer to Fig. 17-6, draw the following, step by step :

(i) Draw the plan, elevation and side-view of vertical prism a, b, c and d.

(ii) Draw the side view of the horizontal prism on the side view of the vertical prism. Here we will get the true cross-sectional view of the horizontal prism.

(iii) Draw the axes of the prism 6 mm apart according to the given condition.

(iv) Mark e and f, cutting by the vertical line a in the side-view.

(v) Draw the horizontal prism on the plan and elevation, projecting from the side-view of the same.

(vi) Number points p_1, p_2, p_3 and p_4, meeting lines 1, 2, 3 and 4 in the plan.

(vii) Transmit the points p_1, p_2, p_3 and p_4 to the elevation on the respective lines 1, 2, 3 and 4.

(viii) Transmit the points e and f to the vertical line a of the elevation.

(ix) Join p_1 to p_2 and p_2 to p_3 with firm lines in the elevation.

Note : It is clear that points e, f and 4 are on the back side. So the lines p_1 to e, e to p_4, p_4 to f and f to p_3 will be dotted in the elevation.

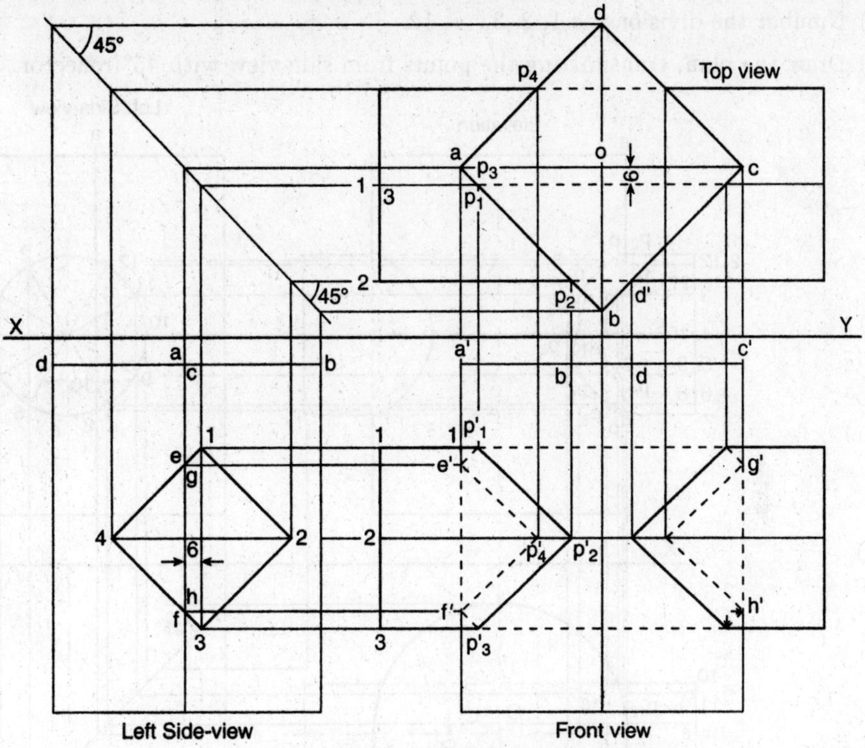

Fig. 17-6 (Third-Angle Projection)

(x) Draw dotted lines from p_1 to e, e to p_4, p_4 to f and f to p_3 in elevation.

(xi) Lines between the points p_1, p_2, p_3, f, p_4, e and p_1 are the penetrating lines in the elevation.

(xii) Draw dotted lines from p_4, p_3 and p_2 towards right to the next boundary line of vertical prism in plan.

(xiii) Draw dotted lines from the points p_1, p_2 and p_3 towards right to the next penetrating points in elevation.

(xiv) Complete the right-hand side of elevation, which is similar to the left-hand side of the axis, as explained above.

Problem 7 : Two cylinders of diameters, 80 mm and 60 mm penetrated each other at right angles. The bigger one is vertical and the other horizontal, and their axes are parallel to V.P. The two axes are 10 mm apart. Draw the projections showing curves of intersection.

Solution : This problem is solved in first-angle projection.

With reference to Fig. 17-7, draw the following, step by step :

(i) Draw side-view of the combination, placing the horizontal cylinder with its centre 10 mm apart.

(ii) Divide the end circle of horizontal cylinder into 12 equal parts with set-square, (See side-view)

(iii) Number the divisions as 1, 2, 3,. . . , 12.

(iv) Draw the plan, transmitting the points from side-view with 45° reflector.

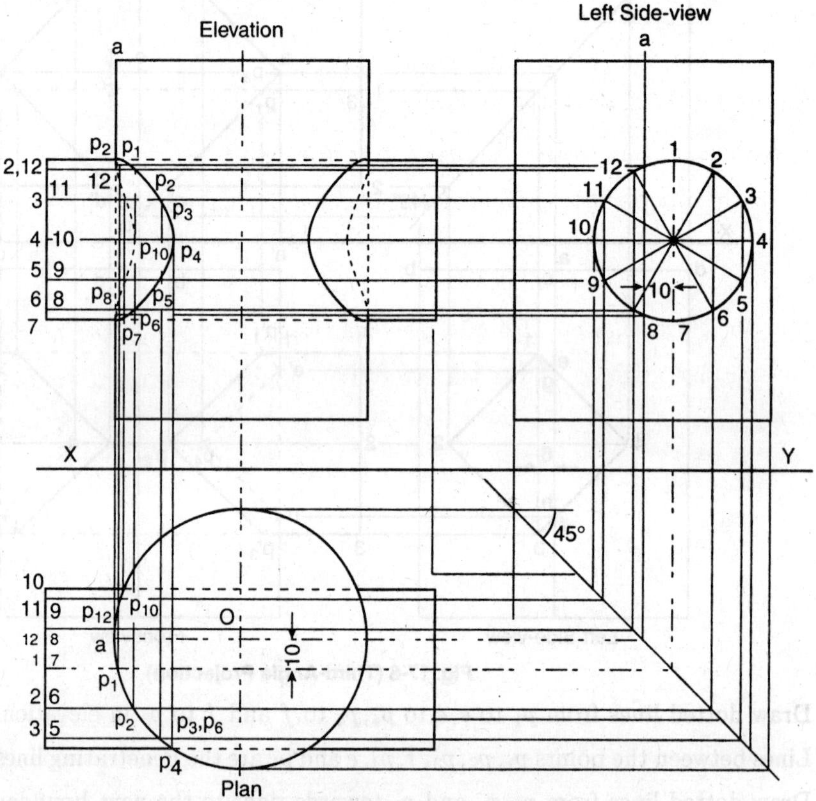

Fig. 17-7 (First-Angle Projection)

(v) Transmit the numbers from side-view to plan.

(vi) Draw the elevation and transmit the numbers of the horizontal cylinder from plan to elevation.

(vii) Points 1, 2, 3,. . . , 12 cut the vertical cylinder in the plan at the points $p_1, p_2, p_3, \ldots, p_{12}$, (See the plan)

(viii) Transmit the points $p_1, p_2, p_3, \ldots, p_{12}$ to the elevation in the respective points to the horizontal cylinder.

(ix) Points p_1 to p_7 are visible and hence join them by firm line. Others (8 to 12) are hidden. So the hidden portion will be shown by in dotted lines.

(x) Complete the right-hand side, which is similar to the left-hand side.

(xi) We get the required curves of penetration (See Fig. 17-7).

17.5 Intersection of Right Cone and Right Cylinder

Problem 8 : A cone, of diameter 70 mm and 90 mm long, is placed on the ground with its base. A cylinder, axis parallel to H.P. and V.P. and of diameter 40 mm, is completely penetrated and their axes meet at a height of 25 mm above the ground. Draw the projections of the solids, showing curves of intersection.

Solution : (Line Method) (First angle projection)

Referring to Fig. 17-8, draw the following, step by step :

(i) Draw the side-view of the combination at the right side.

(ii) Divide the circle of horizontal cylinder into twelve equal parts with set-square, (See side-view).

(iii) Number the division-points as 1, 2, 3,. . ., 12.

(iv) Draw the elevation and plan of the combination, following proper procedure as shown.

(v) Transmit the numbers at the left end of the cylinder in elevation to its plan.

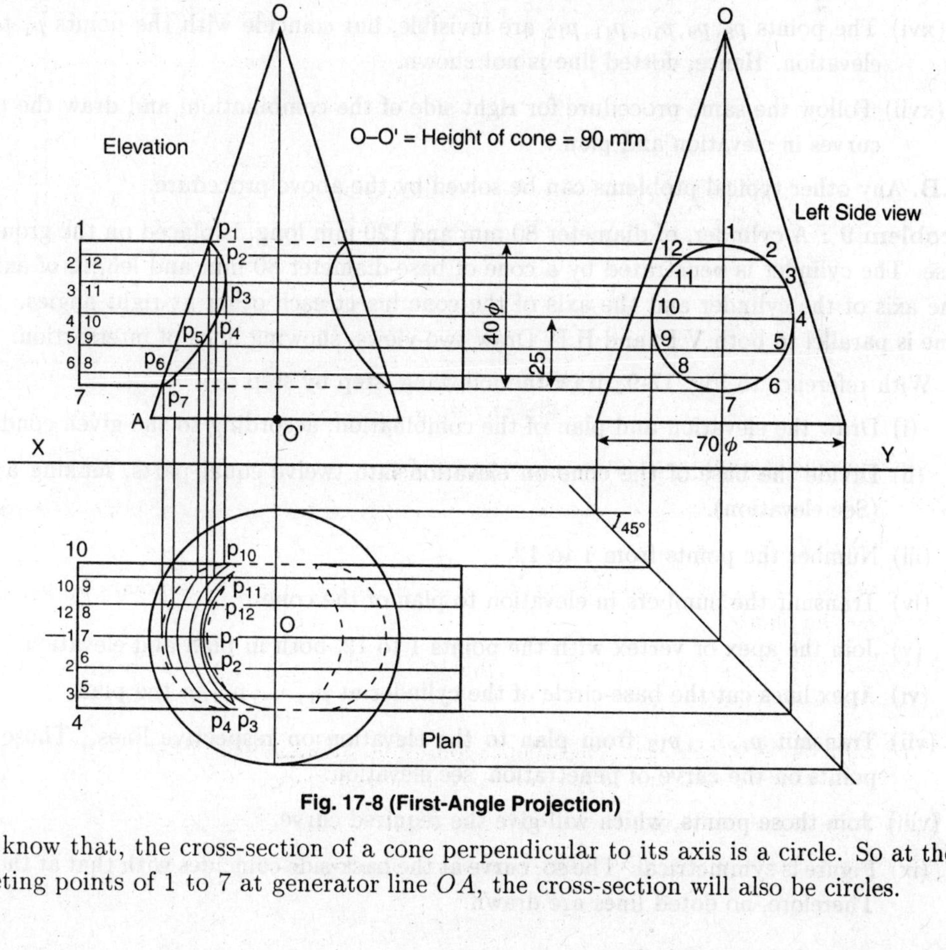

Fig. 17-8 (First-Angle Projection)

(vi) We know that, the cross-section of a cone perpendicular to its axis is a circle. So at the meeting points of 1 to 7 at generator line OA, the cross-section will also be circles.

(vii) Measure the distance from the axis of the cone to its generator-line OA at points 1, 2, 3,. . ., 7 in elevation.

(viii) The above distances are the radii of cross-sections of the cone at the respective points (1,2,3,. . .,7). See elevation.

(ix) Draw the circles with these radii in the plan with O as centre.

(x) Radius on line '1' in elevation will cut the 1, 7 lines at point p_1 on the plan.

(xi) From radii of the different lines from 2 to 12, we will get the points $p_2, \ldots, p_{12}$ on the plan.

(xii) Draw a firm curved line, joining the points $p_{10}, p_{11}, p_{12}, p_1, p_2, p_3$ and p_4, as it is visible from the top.

(xiii) Draw dotted curved line by joining the points p_4 to p_{10}, (See plan).

(xiv) Transmit the points p_1 to p_{12} from the plan to the elevation up to the lines 1 to 12.

(xv) Join the points $p_1 \cdots p_7$ with a firm curved line, (See elevation).

(xvi) The points $p_8, p_9, p_{10}, p_{11}, p_{12}$ are invisible, but coincide with the points p_1 to p_7 in the elevation. Hence, dotted line is not shown.

(xvii) Follow the same procedure for right side of the combination, and draw the penetration curves in elevation and plan.

N.B. Any other typical problems can be solved by the above procedure.

Problem 9 : A cylinder, of diameter 80 mm and 120 mm long, is placed on the ground with its base. The cylinder is penetrated by a cone of base-diameter 80 mm and length of axis 120 mm. The axis of the cylinder and the axis of the cone bisect each other at right angles. Axis of the cone is parallel to both V.P. and H.P. Draw two views, showing lines of intersection.

With reference to Fig. 17-9, draw the following, step by step :

(i) Draw the elevation and plan of the combination, according to the given conditions.

(ii) Divide the base of the cone on elevation into twelve equal parts, making a semi-circle (See elevation).

(iii) Number the points from 1 to 12.

(iv) Transmit the numbers in elevation to plan of the cone.

(v) Join the apex or vertex with the points 1 to 12, both in plan and elevation.

(vi) Apex lines cut the base-circle of the cylinder at $p_1, \ldots, p_{12}$ in the plan.

(vii) Transmit $p_1, \ldots, p_{12}$ from plan to the elevation on respective lines. Those points are points on the curve of penetration, see elevation.

(viii) Join those points, which will give the required curve.

(ix) Figure is symmetrical. The so, curve at the back-side coincides with that at the front-side. Therefore, no doted lines are drawn.

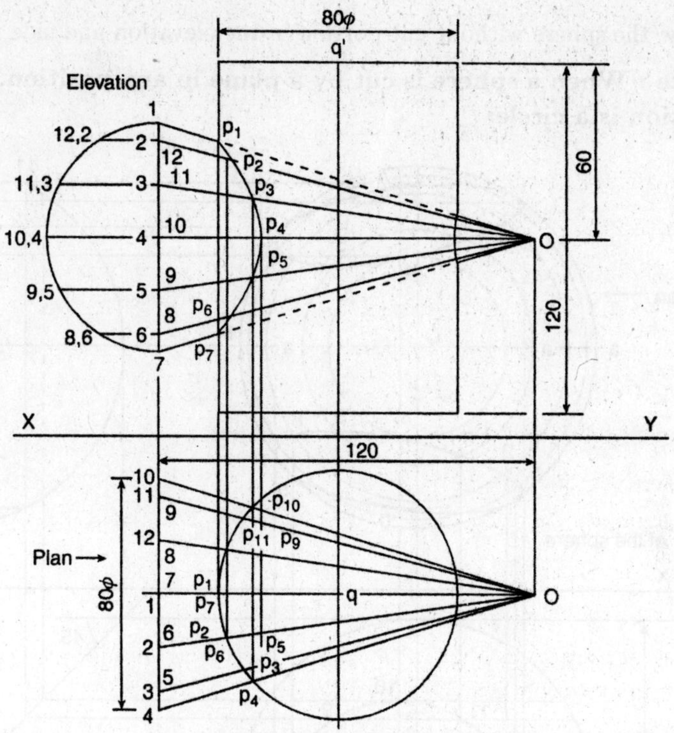

Fig. 17-9 (First-Angle Projection)

(x) We will get the required curve near the apex or vertex in the same manner. (See right side of elevation).

17.6 Intersection of Sphere and Cylinder

Problem 10 : On a sphere of 80 mm diameter, a hole of 46 mm diameter is drilled. The axis of the hole is 10 mm away from the centre of the sphere and axis of the hole is inclined at 60° to the V.P. Draw three views of the sphere with the drilled hole.

Solution: (First angle projection). With reference to the figure 17-10, draw the following, step by step : This problem is solved by the cutting plane method.

(i) Draw the sphere in the plan view with the diameter of circle as 80 mm, which is shown as a-b passing through the centre O.

(ii) With centre at 0 and radius 10 mm, draw an arc. From centre 0, draw a line making an angle of 60° with a-b such that $\angle a - 0 - 11 = 60°$. Produce the line 11–0 so that it cuts the previous arc at 0′. Thus, 0′ is the centre of the hole, 46 mm diameter. With centre at 0′ and radius = 23 mm, draw a circle, which is the plan-view of the drilled hole.

(iii) Divide the circle into twelve equal parts, and number them from 1 to 12.

Note : Total penetrating curve will be **visible in the elevation** (see Fig. 17-10), as the hole is in front of axis of the sphere.

(iv) Draw the sphere without cut-portion in the elevation and side view.

Note : When a sphere is cut by a plane in any position, the true-shape of the section is a circle.

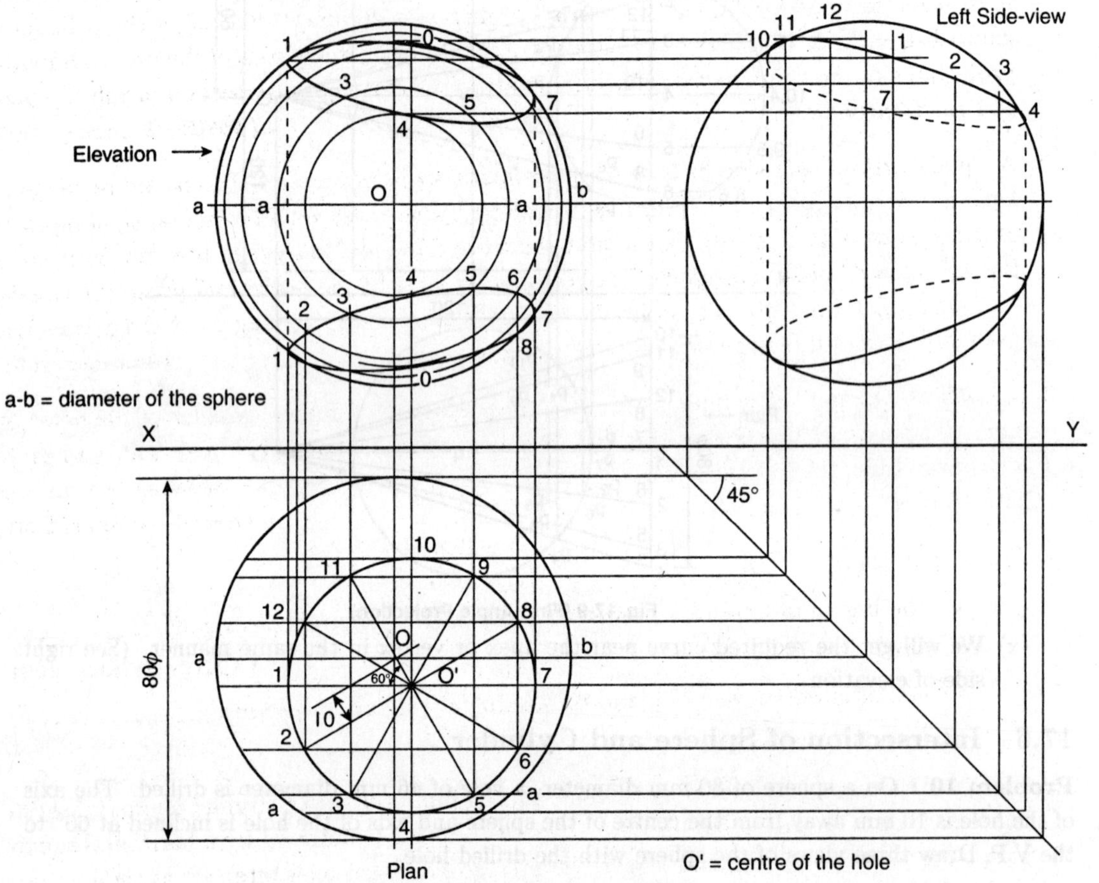

Fig. 17-10 (First-Angle Projection)

(v) Assume a cutting plane parallel to V.P. on point 4. Similarly, on 3–5 (extended as a–a), 2–6, 1–7, 12–8, 11–9 at the point 10.

(vi) Draw the circles on elevation.

(vii) Transmit the points 1 to 12 from plan on circles of cutting plane at elevation on upper and lower portion, i.e., 1 to 1 circle, 2 to 2–6 circle and so on.

(viii) Those points are on the penetration curve. Join the points from 1 to 12.

(ix) Hole is cut by straight line on elevation, which is invisible, marked by dotted line.

(x) Project the points from elevation to the side view.

(xi) Draw the curve and draw the hole by dotted line (See side-view).

Problems

1. A cone, of base diameter 60 mm, axis 70 mm long, stands vertically with its base on H.P. It is penetrated by a cylinder (horizontal) of diameter 26 mm. The axis of the cylinder is parallel to V.P., 20 mm above the base and 5 mm in front of the axis of the cone. Draw the projections of the combination of solids, showing the curves of intersection in the top view and the front view.

   (AMIE—Summer, 2000)

2. A square prism, each side of base 45 mm, stands with its base on HP and its rectangular faces are equally inclined to V.P. It is completely penetrated by a horizontal square prism, of sides of base 35 mm, each such that the axis of the vertical prism is 8 mm in front of the axis of the horizontal prism. The rectangular faces of the horizontal prism are equally inclined to H.P. and its axis is parallel to V.P. Assume suitable lengths of the prisms and draw the lines of intersection of surfaces of both the prisms. **(Summer, 2001)**

3. A vertical cylinder, of diameter 80 mm is penetrated by another cylinder of diameter 50 mm. The axis of the penetrating cylinder is parallel to both H.P. and V.P. and is 10 mm away from that of the vertical cylinder. Draw the projections, showing the lines of intersection.

   (AMIE—Winter, 2001)

4. Fill in the blanks :

   (a) The line of intersection between a prism and a pyramid consists of ____ _____.

   Ans. Straight lines. **(AMIE—Winter, 2001)**

   (b) The intersection line of plane-surface to curved-surface produce _____ _____.

   Ans. Curved lines.

5. A regular hexagonal pyramid, with base-edges 30 mm each and altitude 85 mm, is resting on its base on H.P. with a base-edge parallel to V.P. It is penetrated by a horizontal square prism, of base-edge 20 mm and length 120 mm, so that their axes intersect at right angles. The base-edges of the prism are equally inclined to H.P. and its axis is 30 mm above H.P. Draw the projections of the combination of solids, showing the curves of interpenetration.

   (AMIE—Summer, 2002)

6. A pipe, of diameter 25 mm, is joined to another vertical pipe of diameter 50 mm, at an angle of 45°. Their axes are parallel to V.P. and are 12.5 mm apart. Draw the projections of the pipes, showing the curves of intersection. Develop the surface of the smaller (branch) pipe.

   (AMIE—Winter, 2002)

Chapter 18*

Perspective Projection

18.1 Introduction

In perspective projection, an object is shown at a particular point, as may be seen by the human eyes or a camera. Only from the orthographic projections (plan, elevation and side view), it is often difficult to appreciate the actual appearance of any object, e.g., a building, an aeroplane, a car or a ship. But accurate perspective drawings of the said objects are able to present, as nearly as possible, the actual appearance of the objects in terms of lines on a two-dimensional plane. That is why, perspective drawings are very important for the work of architects, engineers, industrial designers, interior designers, etc. Perspective projection is also used by the Advertising agencies in order to popularise their products or services through advertisements in T.V., newspaper, etc.

A number of methods are used for the construction of perspective projections. But for convenience, only two methods have been explained in this chapter.

In order to study the perspective projections, we should know the principles of orthographic projections, particularly the third-angle projection.

For the purpose of general reference, some basic definitions of perspective projections are explained here.

18.2 Some Terms

18.2.1 Station Point (SP)

Station point is the point, where the eye of the observer is placed at the time of looking the object. In other words, station point is the point of observation, at which the rays from the eye to the object in the scene meet [See Fig. 18-1(a)].

18.2.2 Picture Plane (PP)

The vertical plane, on which the view is formed by the piercing (intercepting) points of the visual rays from the object, is called the picture plane. The piercing points reproduce the scene (view), the size of which depends upon the location of the picture plane (PP).

Usually, the picture plane is placed between the object and the station point, that is, the picture plane is placed in front of the object [See Fig. 18-1(a)].

*This chapter has been contributed by Mrs. Rita Sarkar, Junior Lecturer, Department of Architecture, North Calcutta Polytechnic.

It may be noted that, objects of the same height intercept a greater distance on the picture plane (PP), when placed close to it, (i.e., PP) than when farther away.

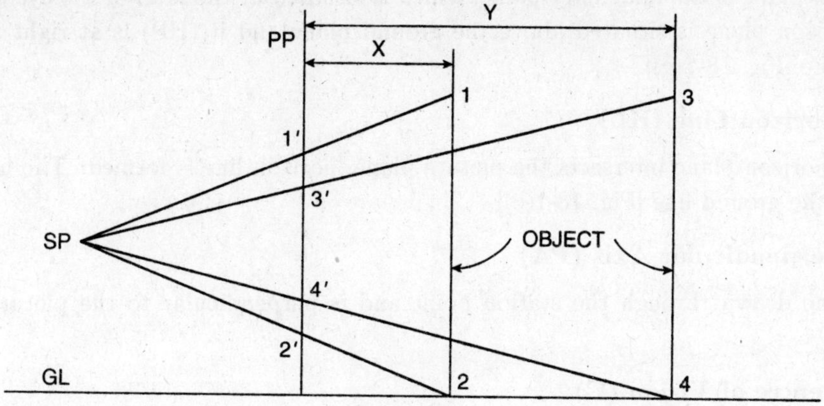

Fig. 18-1(a)

Referring to Fig. 18-1(a), rays from the object 1–2 (placed at a distance x from PP) intercepts a distance 1′–2′ on PP. But, the rays from the object 3–4 (same height as 1–2 and placed at a distance y from PP) intercepts a distance 3′–4′ on PP. We find that 3′–4′ is less than 1′–2′, because the distance Y is greater than x.

When the picture plane touches the face of the object, then the size of the view is the same as that of the object. (i.e., true size of the object).

Similarly, the picture plane may be placed behind the object. In this case, the size of the view is greater than that of the object. [see Fig. 18-1(b)]. Here 1–2 is the object. Size of the view is 1′–2′, which is greater than the height 1–2.

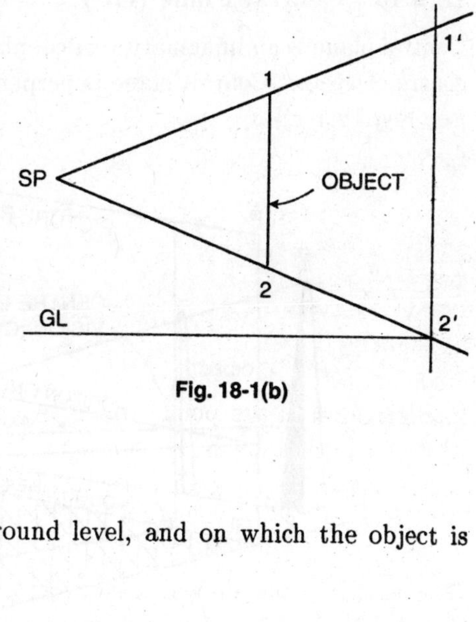

Fig. 18-1(b)

18.2.3 Ground Plane (GP)

Ground plane is a horizontal plane situated on the ground level, and on which the object is assumed to be located. See Fig. 18-1(c).

18.2.4 Ground Line (GL)

The line of intersection of the picture plane with the ground plane is called the ground line [see Fig. 18-1(c)].

18.2.5 Auxiliary Ground Plane (AGP)

On this auxiliary ground plane, the top view of the object and the perspective view are projected. It is the horizontal plane placed above the ground plane.

18.2.6 Horizontal Plane or Horizon Plane (HP)

The horizon plane is the imaginary plane, which is situated at the level of the eye or the station point. Horizon plane is situated above the ground plane and it (HP) is at right angles to the picture plane [Fig. 18-1(c)].

18.2.7 Horizon Line (HL)

When the horizon plane intersects the picture plane, horizon line is formed. The horizon line is parallel to the ground line [Fig. 18-1(c)].

18.2.8 Perpendicular Axis (PA)

It is the line drawn through the station point and is perpendicular to the picture plane [Fig. 18-1(c)].

18.2.9 Centre of Vision (C)

The meeting point of the axis of vision (perpendicular axis) and the picture plane is known as the centre of vision [Fig. 18-1(c)].

18.2.10 Central Plane (CP)

Central plane is an imaginary vertical plane, which passes through both the station point and the centre of vision. Central plane is perpendicular both to the picture plane and the ground plane [see Fig. 18-1(c)].

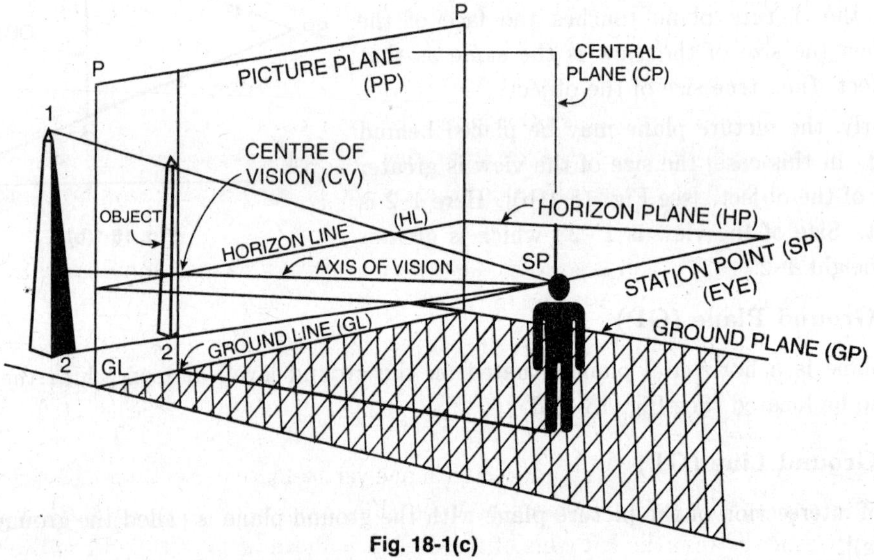

Fig. 18-1(c)

Problems

1. A cube *abcdefgh* is placed with one of its vertical faces *abfe* touching the picture-plane and parallel to it. The station point (SP) is assumed to be situated at the left (i.e., at a

distance of x from the left edge of the object, y ahead of of PP and z above GL). Draw the perspective.

Solution : We shall now draw the perspective in the following way, step by step :

(i) PP and GL are drawn parallel to each other at any convenient distance apart (see Fig. 18-2).

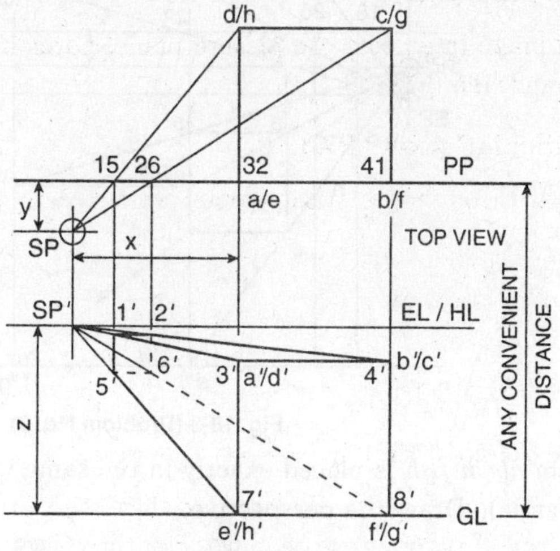

Fig. 18-2 (Problem No. 1)

(ii) In the top view, the object is placed as per problem and according to the required scale (i.e., one face parallel to PP and touching it).

(iii) HL/EL is established parallel to GL and at a distance z above it.

(iv) In the top view, the station point (SP) is located as per problem (i.e., x from left edge and y ahead of PP. Its front view SP′ is projected on HL. The piercing points 1, 2, 3, 4, 5, 6, 7 and 8 are located on the PP by drawing rays from the corners of object to SP.

(v) The front view of the object's closest face is drawn as per some scale, from SP′ rays are connected to its corners.

(vi) Vertical projectors are drawn perpendicular to picture plane to intersect the rays drawn in step (v) at $1', 2', 3', 4', 5', 6', 7'$ and $8'$.

(vii) Join the points to obtain the perspective (Fig. 18-2).

2. A cube, a, b, c, d, e, f, g and h is placed with one vertical face a–b–f–e at a distance w behind the PP and parallel to it. The station point (SP) is assumed to be situated at the left end (i.e., distance x from the left edge of the object, y ahead of the PP and z above GL). Draw the perspective.

Solution : We shall draw the perspective in a similar way, as in the previous problem (Fig. 18-2). The perspective is shown in Fig. 18-3.

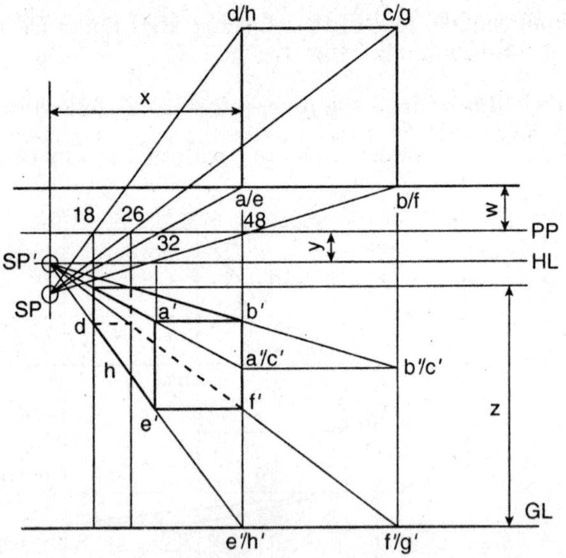

Fig. 18-3 (Problem No. 2)

3. A square prism *abcdefgh* is placed exactly in the same way, as in the previous problem (in the case of a cube). Draw the perspective.

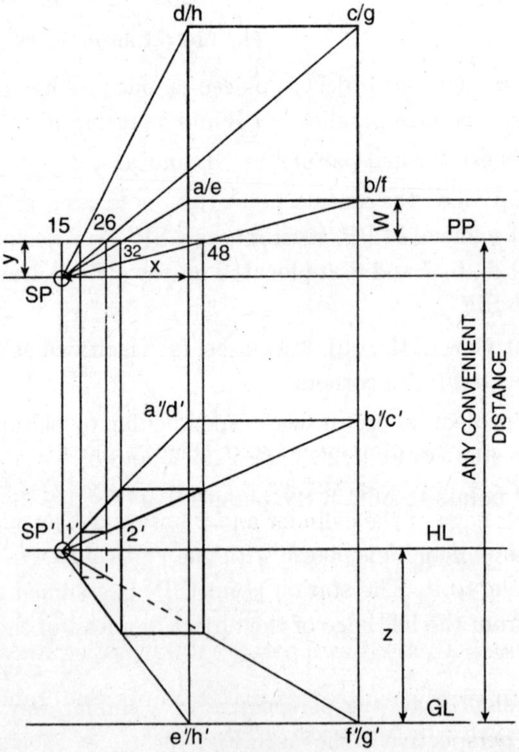

Fig. 18-4 (Problem 3)

Solution : The procedure of drawing the perspective is exactly the same, as in the previous problem See Fig. 18-4 for the perspective.

4. A square pyramid $0, a, b, c$ and d is placed with base edge ab parallel to PP and touches it. The station point (SP) is located as in the case of a cube (see Fig. 18-3). Draw the perspective.

Solution : The procedure of drawing the perspective is same as explained in the seven steps in case of the cube. (refer to Fig. 18-2). The perspective of the pyramid is shown in Fig. 18-5.

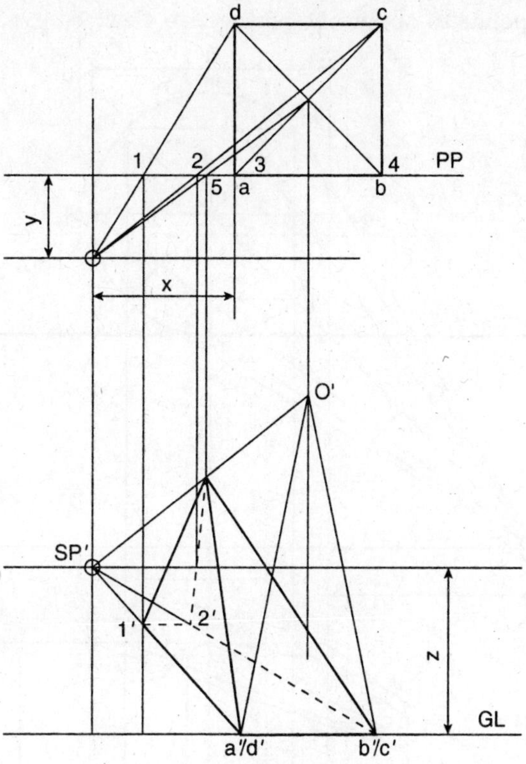

Fig. 18-5 (Problem 4)

5. A right cylinder, of base-diameter d and height h, rests on ground plane, touching the picture plane (PP) with its axis vertical. The station point is y unit ahead of PP, x unit left of the extreme left edge of the cylinder and z unit above ground line. Draw the perspective projection of the cylinder.

Solution : Refer to Fig. 18-6. The steps are as under :

(i) The PP and GL are drawn parallel to each other at any convenient distance apart.

(ii) In the top view, the object is placed as per the problem and to the required scale (i.e., upright cylinder touching the PP).

(iii) HL/EL is established parallel to GL and z unit above it.

(iv) In the top view, the station point (SP) is located as per the problem (i.e., x unit left from left edge of the object and y ahead of it). The front view SP' is projected on HL, the piercing points are located on the PP by drawing rays from the marked points a, b, c, etc. to the station point.

(v) The front view of the closest face to PP is drawn to the same scale, from SP′, rays are connected to a', b', c', d', etc.

(vi) The vertical projectors are drawn perpendicular to the PP in order to intersect the rays drawn in steps (v) at a', b', c', etc.

(vii) Join the points to obtain the perspective view.

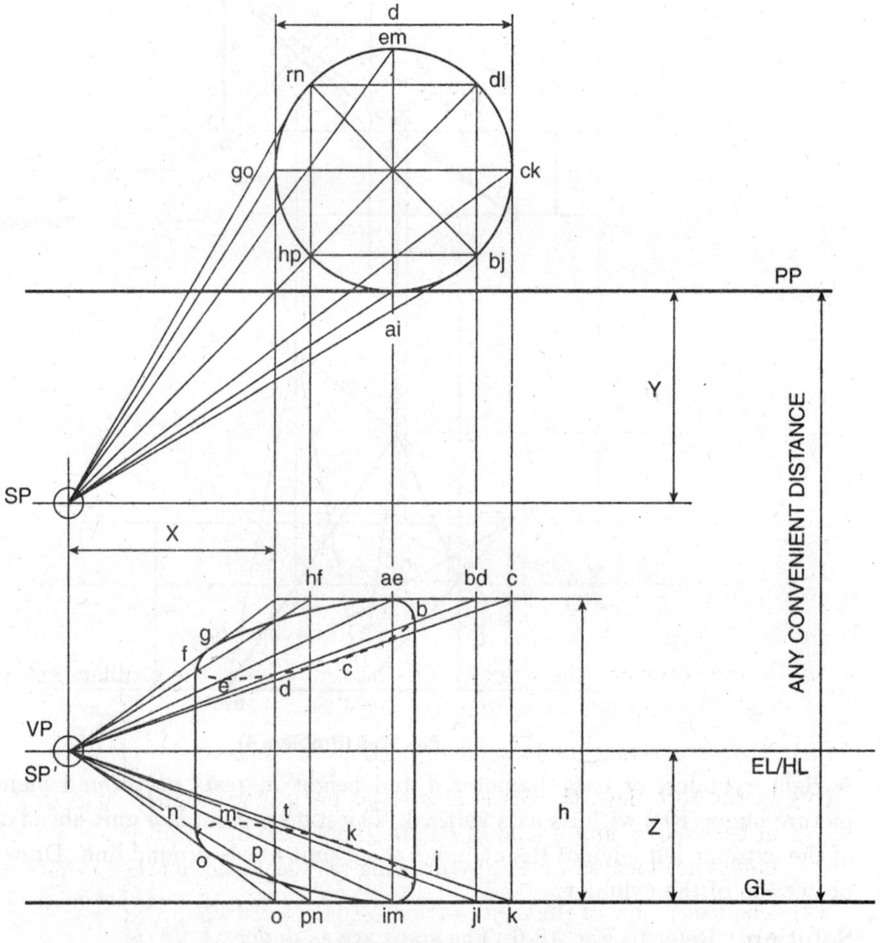

Fig. 18-6 (Problem 5)

6. A right cone, of base-diameter d and height h, rests on ground plane, touching the picture plane (PP) with its axis vertical. The station point (SP) is y unit ahead of PP and z unit above GL. The extreme right point of the base of the cone is x unit left of the SP. Draw the perspective view.

Solution : Refer to Fig. 18-7. The steps are as under :

(i) The PP and GL are drawn parallel to each other and any convenient distance apart.

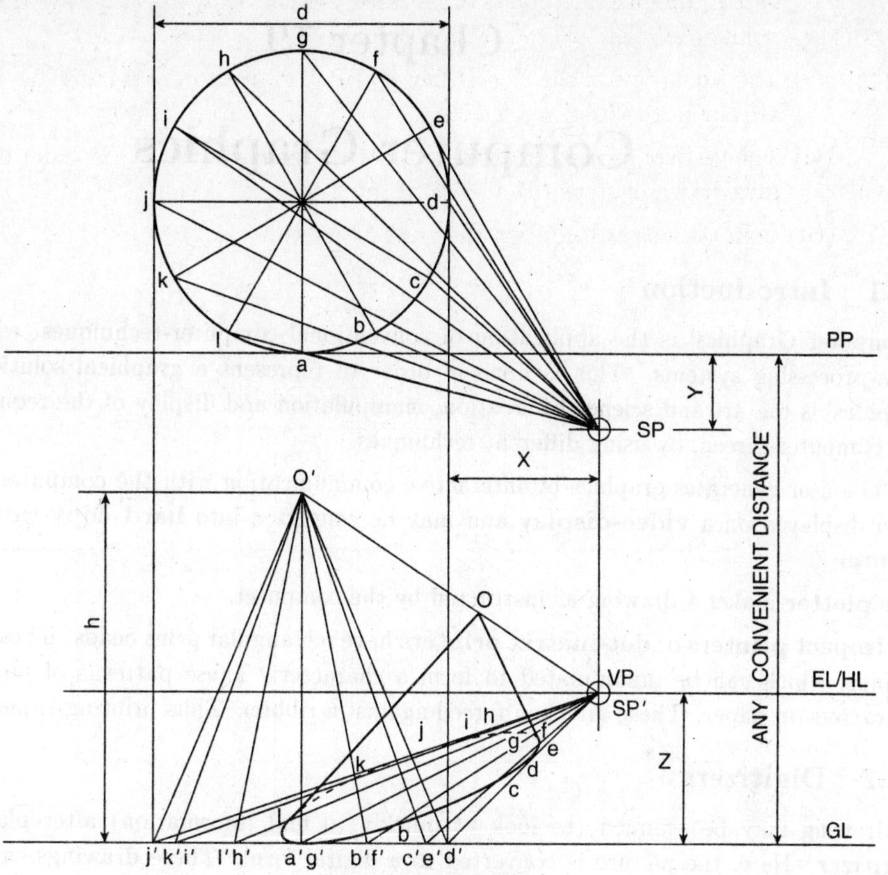

Fig. 18-7 (Problem 6)

(ii) In the top view, the object is placed according to the problem and to the required scale (i.e., upright cone touching the PP).

(iii) HL/EL is established parallel to GL and z above it.

(iv) In the top view, the SP is located as per the problem (i.e., x unit from the right of base and y unit ahead of PP, its front view SP′ is projected on HL, the piercing points are located on PP by drawing rays from the marked points a, b, c, etc.

(v) The front view of the object's nearest face to PP is drawn to the same scale. From SP′, rays are connected to the corresponding marked points a', b', c', etc.

(vi) The vertical projectors are drawn perpendicular to PP in order to intersect the rays drawn in step (v).

(vii) Join the points to obtain the required perspective (Fig. 18-7).

Chapter 19

Computer Graphics

19.1 Introduction

'Computer Graphics' is the application of conventional computer-techniques, with the help of data-processing systems. This is done in order to represent a graphical solution, 'Computer graphics' is the art and science of creation, manipulation and display of the required images on the computer-screen, by using different techniques.

The user generates graphics by interactive communication with the computer. Graphics are then displayed on a **video-display** and may be converted into **hard copy** with a **plotter** or **printer**.

The **plotter** makes a drawing as instructed by the computer.

Impact printers or **dot-matrix printers** have rectangular print heads. These are composed of pins, which can be manipulated to form a character. These patterns of pins make dotted characters on paper. These are then forced against a ribbon. Thus printing is performed.

19.2 Digitizers

A drawing may be scanned (to look at quickly to find information) after placing it on the **digitizer**. Here, the picture is converted to a digital form. These drawings can be stored for future use or correction or modification, if necessary. Digitizers are also called **digitizer-tablet** or **digitizer-Board**.

19.3.1 Locators

The **locators** give the positional information. From a locator, the computer receives the co-ordinates (x and y) for a point.

19.3.2 Selectors

The **selectors** are used to choose (pick) a particular object, but no information is furnished about its screen position (i.e., light pen).

19.3.3 Keyboard

'Keyboard' generally looks like a typewriter, and usually contains many function-keys. This permits to communicate through a set of English alphabet (A to Z) and numerals (1 to 9 and 0) and other function-keys.

19.4 Cathode-Ray Tube (CRT)

It is a video-display device, which consists of a phosphor-coated screen and an electron gun. The electron gun throws a beam and sweeps out raster lines on the screen.

Each raster line consists of a number of dots, called **Pixels**. By turning the pixels on and off, the images are generated on the screen.

19.5.1 Terminal

It consists of a Keyboard, a Cathode-Ray tube (CRT) and interconnections with the computer. With the help of a terminal, the user is able to communicate with the system.

19.5.2 Auto CAD

'CAD' is the abbreviation of 'Computer Aided Design'. 'CAM' is the abbreviation of 'Computer Aided Manufacturing'.

Basically, 'CAD' is a software, by which we can draw easily many types of Engineering Drawings (e.g., Mechanical, Civil, Architectural, Electrical, etc.)

19.5.3 Use of the Mouse

The '**mouse**' of a computer is a point-picking device, and has three buttons (Fig. 19.1).

The **Left button** (Pick button) is used to specify points on the screen. The **Right button** is equivalent to 'ENTER' in the Key board. The **middle button** generally displays the object snap menu or activates the real time panning (to follow a subject).

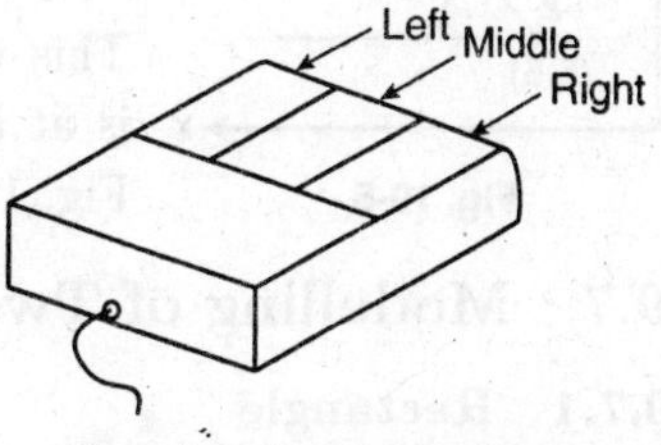

Fig. 19-1 Computer Mouse

19.6.1 Plotting a Point on the Screen

In an Auto-CAD, we shall refer to any drawing-object as an entity (a thing which exists separately from other things). The most simple entity is a point. To plot a point on the drawing, 'Point Command' is used.

19.6.2 Command Point ↵

("↵" refers to press ENTER Key of the Keyboard).

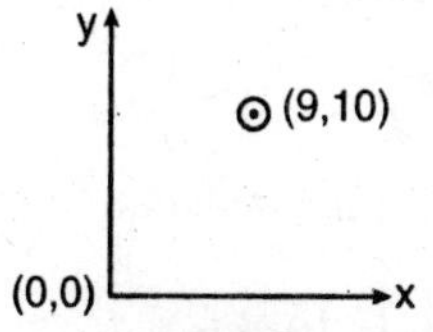

Fig. 19-2

19.6.3 Point (9,10)

This will draw a point, whose coordinates are (9,10), i.e., x-coordinate is 9 and y-coordinate is 10. See Fig. 19-2. (0,0) is the origin.

Both the x and y-axes are known as 'Cartesian Coordinates', after the French mathematician Rene Descartes.

19.6.4 Line Drawing Displays

By the use of three types of coordinate system, the plotting of a line on the screen is performed. These are Absolute Coordinate, Relative Coordinate and Polar Coordinate systems.

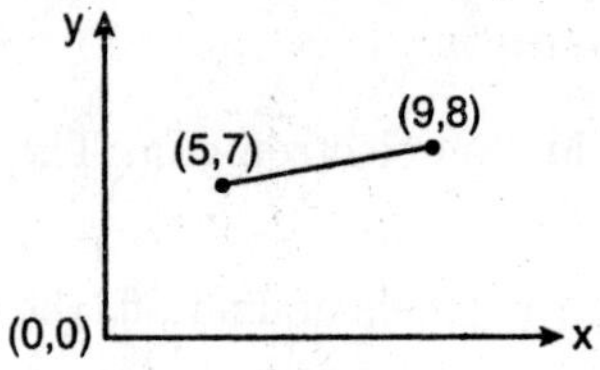

Fig. 19-3

14. Absolute Coordinate System

Command : LINE ↵

From point : 5, 7 ↵

To point 9, 8 ↵

It will draw a line from the point (5,7) to the point (9,8). Refer to Fig. 19-3, where x and y-axes are shown.

15. Relative Coordinate System

Command : LINE ↵

From point : 5, 7 ↵

To point 7, 5 ↵

This will draw a line from a point (5, 7) to point (7, 5), using Relative Coordinate System, see Fig. 19-4.

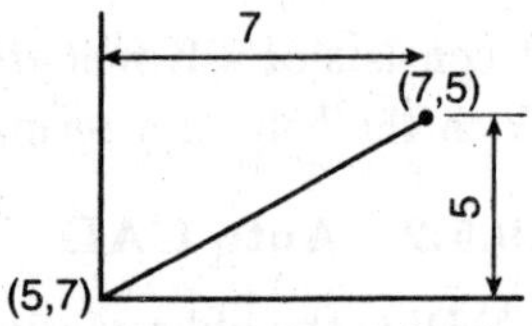

Fig. 19-4

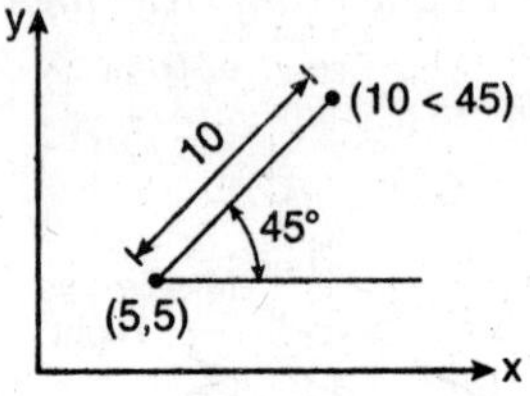

Fig. 19-5

16. Polar Coordinate System

Command : LINE ↵

From point : 5, 5 ↵

To point 10 < 45

This will draw a line, where initial point is (5, 5), and the final point is at a distance of 10 units at an angle of 45° to the horizontal. See Fig. 19-5.

19.7 Modelling of Two-Dimensional (2D) Objects

19.7.1 Rectangle

In order to draw a rectangle, we are to select its two diametrically opposite corners.

Command : RECTANG

First Corner : 5, 5

Second Corner : 15, 15 (See. Fig. 19-6)

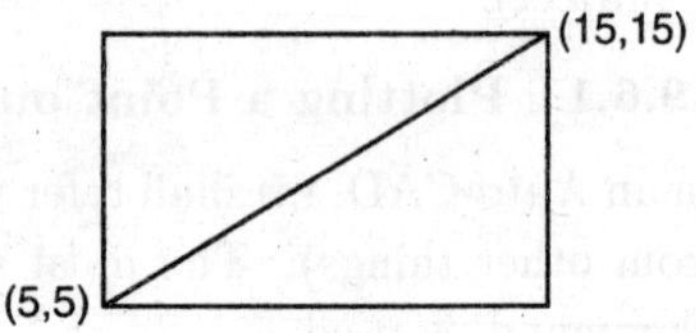

Fig. 19-6

19.7.2 Circle

We can draw a circle in many options, as detailed below :

19.7.2.1 Given the coordinates of the centre-point and the radius

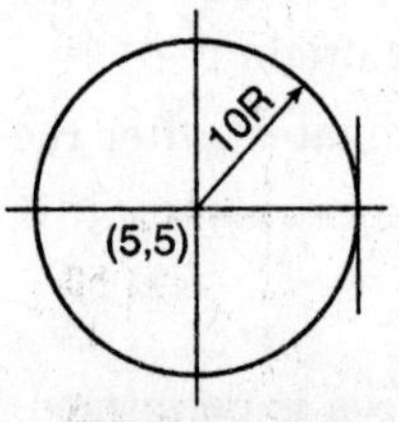

Fig. 19-7

Command : CIRCLE

3P/2P/Ttr/<Centre-point> : 5, 5

Diameter/<Radius> :10 (In this case, it is radius, which is 10 mm).

N.B. "Ttr" Stands for "tan tan radius" (See. Fig. 19-7).

19.7.2.2 Given the coordinates of the centre-point and the diameter

Command : CIRCLE

3P/2P/Ttr/<Centre-point> 5, 5

Diameter/<Radius> : D

Diameter : 10 (See Fig. 19-8)

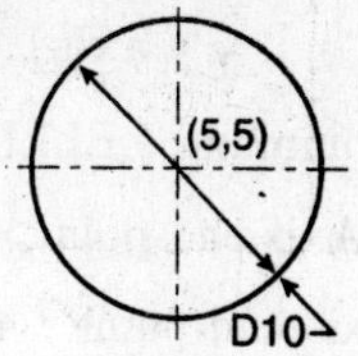

Fig. 19-8

19.7.2.3 Given the three distinct point-coordinates

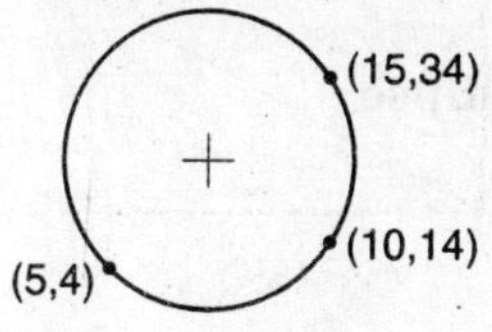

Fig. 19-9

Command : CIRCLE

3P/2P/Ttr/<Centre-point> : 3P

First Point : 5, 4

Second point : 15, 34. (See Fig. 19-9)

Third point : 10, 14.

19.7.2.4 Given the two points. In this case, the distance between the two points is taken as the diameter

Command : CIRCLE

3P/2P/Ttr/<Centre-point> : 2P

First Point on diameter : 5, 19 (See Fig. 19-10)

Second point on diameter : 15, 19

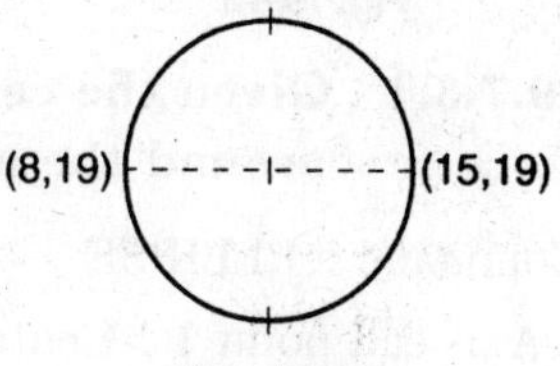

Fig. 19-10

19.7.2.5 Given the tangent points of two other objects and radius of that circle

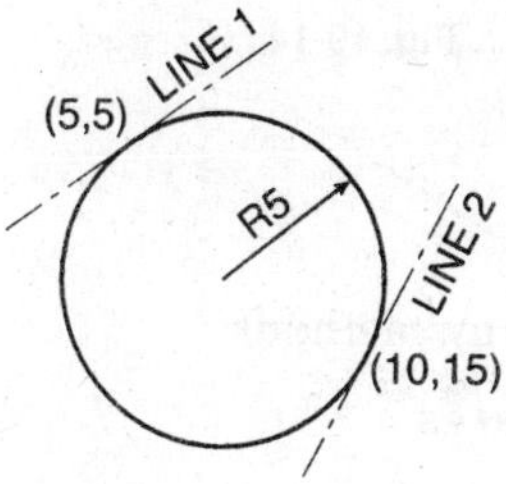

Fig. 19-11

Command : CIRCLE

3P/2P/Ttr/<Centre-point> : Ttr

Enter first tangent specification : 5,5

Enter second-tangent specification : (10,15)

Radius : 5

In this case, we can draw the circle of radius 5 mm, which lies between the two tangents, where the touching points are (5, 5) and (10, 15). See Fig. 19-11.

19.7.3 Ellipse

In order to draw an ellipse, four methods are given below :

19.7.3.1 Axis end-points of the ellipse are given

Command : ELLIPSE

<Axis end-point 1 >Centre : 5,5

Axis end-point 2 : 50,50

<Other axis distance(Minor axis)>rotation : 40,40 (See Fig. 19-12)

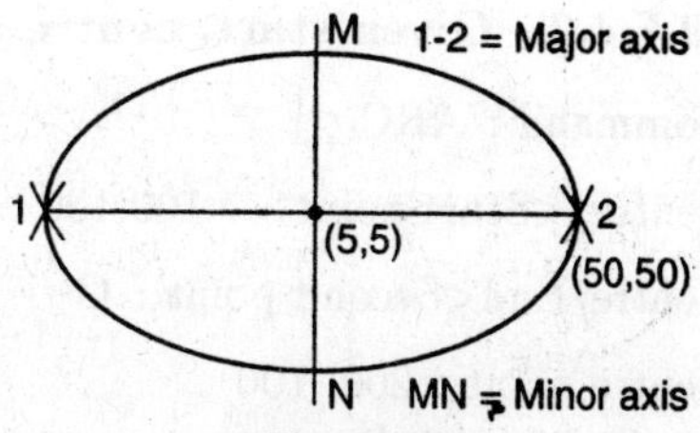

Fig. 19-12

19.7.3.2 Given the centre, major axis end-point, and the length of other axis (Minor axis)

Command : ELLIPSE

<Axis end-point>/Centre : C

Axis end-point : 40, 50

<Other axis distance>/Rotation : 30°

The figure will be almost similar as in Fig. 19-12.

19.7.3.3 Given major axis end-points and rotation-angle of the ellipse

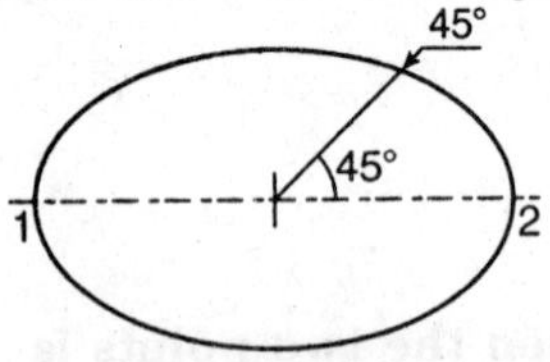

Fig. 19-13

Command : ELLIPSE

< Axis end-point 1 >/Centre : 40,40.

Axis end-point 2 : 90,90 ↵

< Other axis distance>/Rotation : *R* ↵

Rotation around major axis : 45° (See Fig. 19-13)

19.7.3.4 Given the centre, end-point-coordinates and rotation-angle of ellipse around the axis

Command : ELLIPSE

<Axis end point 1 >Centre : C

Centre of ellipse : 25,25

<other axis distance >/Rotation : R

Rotation around major axis : 45° (See Fig. 19-14)

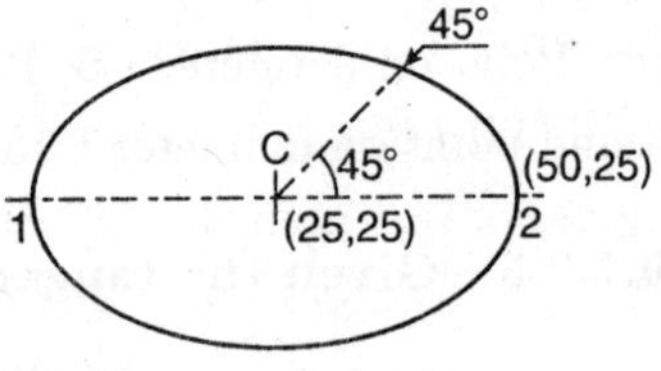

Fig. 19-14

19.7.4 ARC

It is known that, an arc is a part of circumference of an entity.

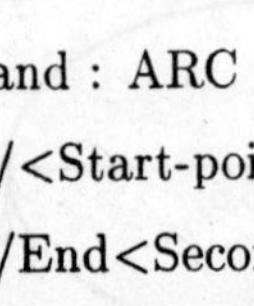

Fig. 19-15

For drawing an arc in auto CAD, there are many methods.

19.7.4.1 Given the three point-coordinates

Command : ARC

Centre/End/<Second point> : 50,50

End-point : 90,90 (See Fig. 19-15)

19.7.4.2 Given start, centre and end-point coordinates (S.C.E.)

Command : ARC ↵

Centre/<Start-point> : 100,100

Centre/End<Second point : C

Centre point : 200, 100

Angle/Length of Chord/<End-point (200,50)

19.7.4.3 Given the start, centre-point coordinates and included angle value (S.C.A.)

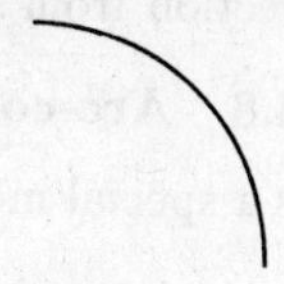

Fig. 19-16

Command : ARC ↵

Centre<Start-point> : 25, 25

Centre/End/<Second point> C

Centre : 100,100

Angle/Length of chord/<End-point>: A

Included angle : 45°. (See Fig. 19-16)

N.B. Included angle may be regarded as **positive** or **negative**, based on clockwise or anti-clockwise direction respectively.

19.7.4.4 Given the start and centre-point-coordinates and length of chord (S.C.L.)

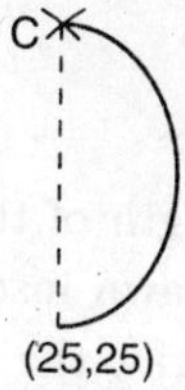

Fig. 19-17

Command : ARC

Centre/<Start-point>: 25, 25

Centre/End/<Second point> : C

Centre : 100, 200

Angle/Length of chord/<End-point>: L

Length of chord : 80 (See Fig. 19-17)

19.7.4.5 Given start, end and included angle (S.E.A.)

Command : ARC

Centre/<Start-point> : 200,100

Centre/End/<Second point> : E

End-point : 100,200

Angle/Direction/Radius/<Centre point> : A

Included angle : 45° (See Fig. 19-18)

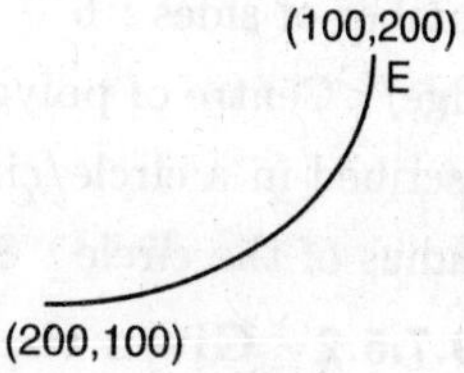

Fig. 19-18

19.7.4.6 Given the start and end-point-coordinates and the length of radius (S.E.R.)

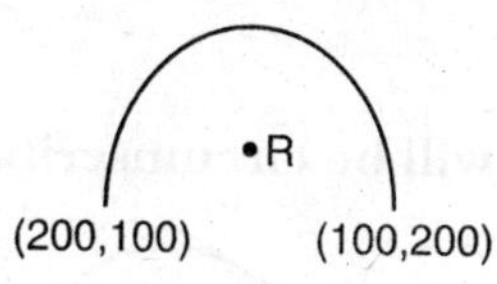

Fig. 19-19

Command : ARC

Centre/<Start-point> : 100,200

Centre/End/<Second point>: E

End point : 200,100

Angle/Direction/Radius/<Centre point> : R

Radius : 50 (See Fig. 19-19)

19.7.4.7 Given start-point end-point-coordinates and starting direction (S.E.D.)

Command : ARC

Centre/<Start-point> : 50,50

Centre/End/<Second point> : E

End-point : 50,75

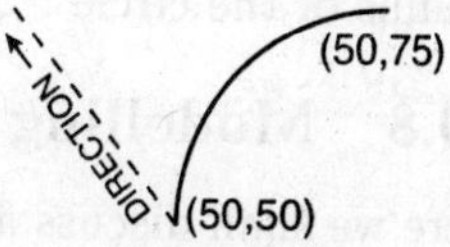

Fig. 19-20

Angle/Direction/Radius/<Centre-point> : D

Direction from start-point : 135 (See Fig. 19-20)

19.7.4.8 Arc-continuation method

This is a special method of drawing an arc. In this case, the starting point of the arc is assumed to be the last point of the existing drawn object on the screen.

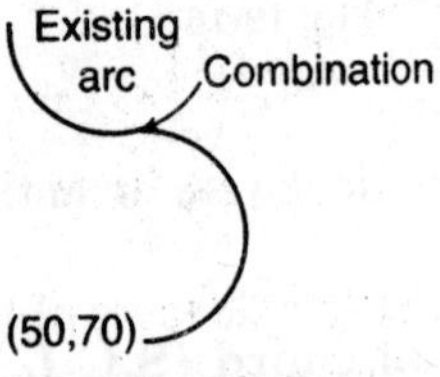

Fig. 19-21

Then, with the given parameters, the arc can be drawn, We suppose that the coordinates of a point arc (90,90). In order to draw an arc from this point, we proceed as follows :

Command :ARC

Centre/<start-point> : ↵ (Enter)

End-point : 50,70 (See Fig. 19-21)

19.7.5 Polygons

The "POLYGON" command is used to draw a regular polygon for a given length of the side or edge. We can draw polygons of 3 to 1024 sides. The polygon may also be drawn inscribing or circumscribing a circle of given radius. Only three methods arc given here to draw the polygons.

19.7.5.1 Given the radius of a circle, where the regular polygon will be inscribed

Command : ARC

Number of sides : 6

Edge/<Centre of polygon> : 50,50

Inscribed in a circle/circumscribed about a circle (I/C) : I

Radius of the circle : 80 (See Fig. 19-22)

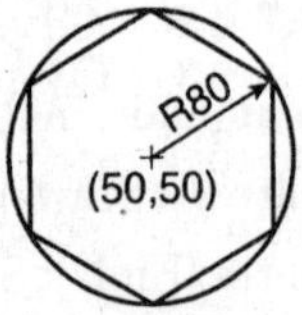

Fig. 19-22

19.7.5.2 Given the number of sides of a regular polygon

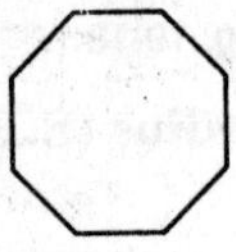

Fig. 19-23

Command : POLYGON

Number of sides : 8

First end-point of edge : 50,50

Second end-point : 100,120 (See Fig. 19-23)

19.7.5.3 Given the radius of a circle, about which the polygon will be circumscribed

Command : POLYGON

Number of sides : 6

Edge/<centre of polygon> : 75,95

Inscribed in a circle/Circumscribed about a circle (I/C) : C

Radius of the circle : 50 (See Fig. 19-24)

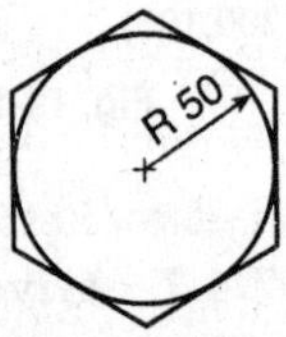

Fig. 19-24

19.8 Modelling of Three-Dimensional (3D) Objects

Here we shall discuss about three basic three-dimensional (3D) models, namely box (rectangular solid or cube), cone and cylinder.

In order to draw the 3D objects, we have to consider the *Z*-axis as the height, along with *X-Y* plane (i.e., both length and breadth).

19.8.1 BOX—Given the Corners of a Box and its Height

Command : BOX ↵

Centre/<Corner of Box> (0,0,0):5,5,5 ↵

Cube/Length/<other Corner> : 10,10,10 ↵

Height : 14 (See Fig. 19-25)

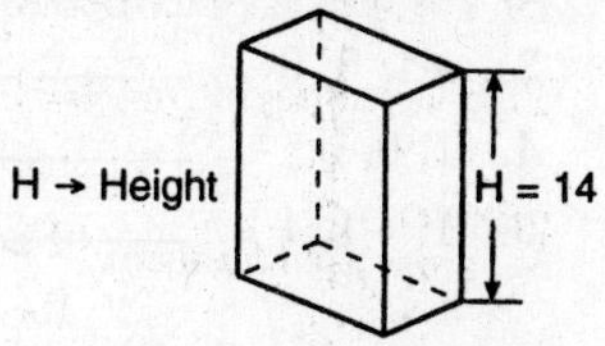

Fig. 19-25

19.8.2 Cone

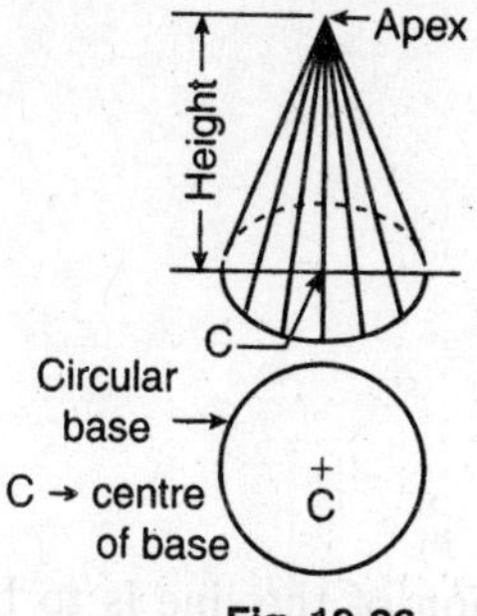

Fig. 19-26

In order to draw a cone, at first we have to consider whether the base of the cone will be circular or elliptical in nature. The options are centre-point and elliptical (default).

Command : CONE ↵

Elliptical/<Centre point><0,0,0>:5,5,5 ↵

Diameter/<Radius> : 3 ↵

Apex/<Height>: 5 ↵ (See Fig. 19-26)

19.8.3 Cylinder

In this case, we have to consider whether the base of the cylinder is circular or elliptical in nature. The options are–elliptical and centre-point (default).

Command : CYLINDER ↵

Elliptical/<Centre-point>< 0, 0, 0 >: 5, 5, 0 ↵

Diameter/<Radius>: 10,10,0 ↵

Centre-point of other end/<Height>: 5 (See Fig. 19-27)

C_1 = centre of top end, C_2 = centre of bottom end

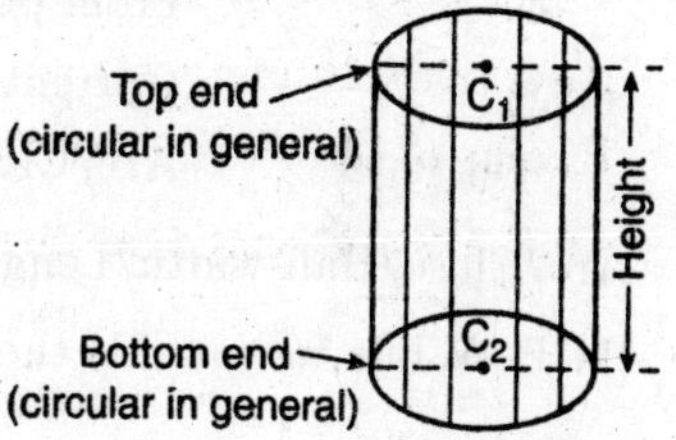

Fig. 19-27

19.9.1 Display Solid Objects

While drawing the views (elevation, plan, side-view or isometric view), we may frequently need to observe the object. Again, we may need to change the current views, if necessary. We can do this, using the "view command".

command : VIEW ↵

?/Delete/Restore/Save/Withdraw

If we put a question mark (?), it shows a list of all saved views.

19.9.2 V-POINT Command

The V-POINT command is used to set a viewing direction for a three-dimensional visualisation of an object.

The *X*, *Y* and *Z* coordinate-values are given below to specify the view-point direction.

V-point Coordinates	View Displayed
1. 0, 0, 1	TOP VIEW
2. 0, −1, 0	FRONT VIEW
3. −1, 0, 0	LEFT-SIDE VIEW
4. 1, 0, 0	RIGHT-SIDE VIEW
5. 1, −1, 1	TOP, FRONT, RIGHT-SIDE VIEW (ISOMETRIC VIEW)
6. $-1-1,1$	TOP, FRONT, LEFT-SIDE VIEW (ISOMETRIC VIEW)

We can set the view-point in three ways as follows :

(i) using coordinates (as explained above)

(ii) using axes

(iii) using the rotate option.

19.10 Polyline

Very often, we need a polyline in order to draw a tool. Here, the direction of the line is to be changed according to necessity. To create a polyline, we have to use the PLINE Command.

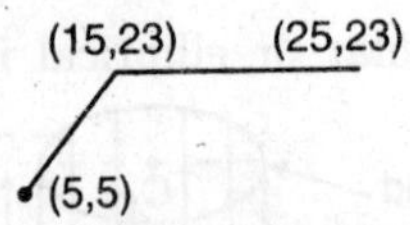

Fig. 19-28

Command : PLINE ↵

From point : 5,5

Current line width : 1.0000

Arc/Close/Half-width/Length/Undo/Width/<End-point of line> : 15,23 ↵

Arc/Close/Half-width/Length/Undo/Width/<End-point of line>: 25, 23 (See Fig. 19-28)

[**N.B.** While we provide the 'C' option, the ployline closes]

19.11.1 ERASE Command

Using this command, we can erase the total or part of the object from the screen.

In order to erase an object, we have to select the object first. This may be done by just pressing the left button of the mouse. Then giving the ERASE command, the selected portion of the object will be removed from the screen.

Command : ERASE ↵

Select objects : <Select the object>

19.11.2 TEXT (Dimensioning Command)

It is the most important part of the drawing. Because by dimensioning, the size of the object is determined. This TEXT command provides the facility to plot the text in the drawing.

Command : TEXT ↵

19.11.3 Hard Disk

A **hard disk** is a high-speed, high-capacity, rotational storage device. Usually, hard disks cannot be removed from the computer. Hence, hard disks were sometimes called 'fixed disks'.

A hard disk consists of a rigid **ferro-magnetic material or aluminium-alloy** disk, which is coated with a magnetic oxide material like that of a diskette.

Because the disk are rigid, they can be spun much faster than a floppy. Maximum speed of a hard disk may be attained up to 10,800 r.p.m.

19.11.4 UNION Command

UNION command is used to join a number of solids or regions to create a single solid or region.

19.11.5 SUBTRACT Command

This command is used to subtract one or more solids or regions from another solid or region.

19.11.6 UNDO Command

The word 'Undo' means 'to nullify or to render void'. UNDO command is used to reject a previous command or more than one command at a time.

19.11.7 REDO Command

This command brings back the objects that were removed previously, using the UNDO command.

19.11.8 AutoCAD EDITing

A drawing can be modified during its preparation or at the time of revising the design. This is known as EDITing of a drawing.

19.11.9 EDITing Commands

There are two types of EDITing :

(a) Modification of a drawing or entity. Modification is done with the help of ERASE, UNDO, REDO, SELECT, CHANGE—Commands.

(b) To make more number of copies of a drawing. More copies of a drawing are produced by using COPY, ARRAY, OFFSET—Commands.

19.12 Some Other Commands

19.12.1 SNAP Command

This command forces the cursor to step a specific distance. If we want to maintain accuracy while entering distances with the cursor 'SNAP' command is useful in that case. In AutoCAD, there are two SNAP modes—GRID SNAP and POLAR SNAP.

19.12.2 EXTRUDE Command

'Extrude Command' is used in AutoCAD in solid-modelling.

A solid object may be created by adding thickness (definite amount) in a direction perpendicular to the x-y plane of a certain area, i.e., in the z-axis direction. For example, a **square** area may be converted to a **cube**, when the thickness equals to one side of the square. Solid objects may be created by extruding the following figures : Polygons, circles, ellipses, closed splines and regions of different shapes.

19.12.3 OFFSET Command

OFFSET Command creates an object, which is parallel to and is situated at a definite distance from its original position. We are to use OFFSET Command to draw parallel lines, polylines, concentric circles, arcs etc.

19.12.4 ZOOM Command

In a camera, the ZOOM-lens can enlarge or reduce the size of an image, without changing the focus. Similarly, the 'zoom command' can enlarge or reduce the apparent size of an object (or entity). The size of the object remains the same.

When very small circuits (in the fields of electronics or computer) are enlarged with the help of ZOOM command, it is easier to visualize the details in the circuits. While working on other commands, ZOOM Command can be used simultaneously.

19.12.5 HATCH Command

When a solid or hollow engineering component is cut by an imaginary section-plane, the material of the object (excluding the hollow or void space) can usually be shown by thin continuous lines, inclined at an angle of 45° to the horizontal. These lines are called "Hatch or hatching lines". In the AutoCAD, the hatching process is done with the help of "HATCH" Command".

19.12.6 Utility Commands

The basic functions of AutoCAD are controlled by the 'Utility Commands'. In order to provide assistance to set the different variables and options in the AutoCAD, utility commands are used.

The basic functions of some utility commands are explained below.

19.12.7 HELP Command

All the AutoCAD commands can be listed by the 'HELP' command. HELP command may be used, when a person forgets any of the commands.

19.12.8 END Command

When we want to return to the main menu and update the drawing file, END command can be used.

19.12.9 QUIT Command

When we want to return to the main menu, without updating the drawing file, QUIT command can be used.

19.13 SAVE Command

When we want to save the current drawing and try to remain in the drawing editor screen, for further editing, SAVE command is used.

19.13.1 LIMITS Command

While working on a drawing, LIMITS command can change the upper and lower limits of the drawing area.

19.13.2 Some MS-DOS* Commands

19.13.3 DIR

Its function is to list the files and subdirectories.

19.13.4 DIR *

Its function is to list the subdirectories only.

19.13.5 CD

Its function is to set the current directory.

19.13.6 DIR *.Ext

Its function is to list the files with some extension.

19.13.7 DIR A*.*

Its function is to list the files and filename starting with 'A'.

19.13.8 COPY A:*.*C

Its function is to copy files from the floppy to the hard-disk.

19.13.9 COPY C:*.*A

Its function is to copy files from the hard-disk to the floppy.

19.14 TYPE A.Ext

Its function is to display contents of file A.Ext.

N.B. For a complete list of DOS commands (from Wikipedia, the free encyclopedia), please refer the following website :

**http://en.wikipedia.org/wiki/List_of_DOS_commands

Website : It is a collection of web-pages grouped by theme, as indicated above for DOS commands.

Web-page : It is a single document or information on a particular subject, institution, organisation etc.

*DOS—Disk operating system, MS—Microsoft, **http—Hyper text transfer protocol

Software and hardware

Software or program, enables a computer to perform specific tasks, as opposed to the physical components of the system, called **hardware**.

19.14.1 AREA Command

This command can calculate the area and perimeter of the surfaces of different objects.

19.14.2 ABOUT Command

This command displays information about AutoCAD.

19.14.3 BREAK Command

This command erases parts of objects or splits an object into two parts.

19.14.4 CHAMFER Command

This command bevels (make sloping) the edges of objects to specified angles as per requirement.

19.14.5 CLOSE Command

This command closes the current drawing.

19.14.6 COLOR Command

This command defines colour for new objects.

19.14.7 COPY Command

This command duplicates any object.

19.14.8 CAL Command

This command evaluates mathematical and geometric expressions. For example, $(5)^4$, which is equivalent to $5 \times 5 \times 5 \times 5$ or 625.

19.14.9 3DFACE Command

This command creates a three-dimensional face, such as in isometric view.

19.15 DVIEW Command

This command defines parallel projection or perspective views.

19.15.1 DIST Command

This command measures the distance and angle between two points.

19.15.2 INTERSECT Command

This command creates composite solids or regions from the intersection of two or more solids or regions, and removes the areas outside of the intersection.

19.15.3 MENU Command

This command loads a menu file.

19.15.4 MIRROR Command

This command creates a mirror image copy of objects.

19.15.5 MOVE Command

This command displaces objects to a specified distance in a specified direction.

19.15.6 MVSETUP Command

This command sets up the specifications of a drawing.

19.15.7 format Command

This command delete all the files on the disk and reformat it for MS-DOS. This command can potentially erase everything on a computer's hard disk.

19.15.8 mem Command

This command displays memory usage. It is equivalent to the Unix command → free.

19.15.9 msd Command (Microsoft Diagnostic)

This command provides detailed technical information about the computer's hardware and software.

19.16 rem Command

This command is used for remark statement, normally used within a batch file.

19.17 time and date Command

This command sets or displays date and time. It is equivalent to the Unix command → date.

19.18 LEADER Command

This command creates a line, that connects annotation to a feature.

19.19 LSLIB Command

This command maintains libraries of landscape-objects.

19.20 LSNEW Command

This command adds realistic landscape items, such as trees and bushes to our drawings.

Questions

A.M.I.E., Summer, 2006

1. Choose the *correct* answer from the following :

(i) Hard disk of computer is made of

(a) Rigid plastic (b) Flexible plastic (c) Metal.*

Ans. (c)

(ii) Acronym for one binary digit 1 or 0 is

(a) bite (b) bit (c) byte.**

Ans. (b)

(iii) Boolean operation is used in

(a) Boundary representation (b) Constructive solid geometry (c) Cellular decomposition.

Ans. (b)

(iv) Colour monitor has three electron guns, each gun emits one of primary colours

(a) Red, yellow or blue (b) Red, blue or green (c) Red, white or blue.

Ans. (b)

(v) Drawing a line from point (3,3) to a point 5 units in X-axis and 8 units in Y-axis is drawn by using the method of

(a) relative co-ordinates (b) absolute co-ordinates (c) polar co-ordinates.

Ans. (a)

(vi) Digitizer† is an

(a) Output device (b) Input device (c) A visual display unit.

Ans. (b)

(vii) First corner (10,10), second corner (30,20) is used in AutoCAD to draw

(a) Line (b) Rectangle (c) Square.

Ans. (b)

(viii) Three-dimensional modelling refers to the modelling of

(a) Solids (b) Surfaces (c) Circles.

Ans. (a)

---

*Ferro-magnetic material or Aluminium alloy

**1 Byte = 8 bits, 1 bit = 1 binary digit, 0 or 1.

†'Digitizer' is also called 'digitizer tablet'

(ix) Wireframe* modelling is used to model

(a) Prism (b) Surface (c) both.

Ans. (b)

(x) EXTRUDE command in AutoCAD is used in

(a) Solid modelling (b) Wireframe modelling (c) Surface modelling.

Ans. (a)

A.M.I.E., Winter, 2006

2. Fill in the blanks :

(i) The number of points to be specified on conics in computer graphics is ______.

(ii) The full form of CSG for solid modelling is ______.

(iii) The full form of GUI in computer graphics is ______.

(iv) The symbol of positional tolerance is ______.

(v) For defining circle in computer graphics, the minimum number of points is ______.

Ans. (i) 3 (Three) (ii) Constructive Solid Geometry (iii) Graphical User Interface (iv) ⌖ (v) 2 (Two)

A.M.I.E., Summer, 2005

3. Choose the *correct* answer among the three answers suggested for each question :

(i) Joystick is

(a) an output device in a computer (b) a storage device in a computer

(c) an input device in a computer.

Ans. (c)

(ii) Digitizer is

(a) used to plot a drawing (b) used to store the drawing files

(c) a graphic input device.

Ans. (c)

(iii) Operating system is

(a) a computer with peripherals, which do not become obsolete

(b) a set of computer programs which coordinate the operation of the computer

(c) a software exclusively used in biomedical engineering.

Ans. (b)

---

*'Wireframe modelling' is also called 'Edge-Vertex' or 'Stick-figure' modelling.

(iv) Centre, radius : (5, 12); R; 20 is the command used in AutoCAD to draw

(a) an ellipse (b) a circle (c) an arc.

Ans. (b)

(v) SNAP is a command used in AutoCAD to

(a) set increments for cursor movement (b) set the grid spacing

(c) set the precision of the decimal place.

Ans. (b)

(vi) The command VPOINT is used in AutoCAD, to

(a) move an object on the screen (b) split the screen into more than one view

(c) view a 3-dimensional object in the screen.

Ans. (c)

(vii) To join several solids to create a single solid, the command option in AutoCAD is

(a) UNION (b) SUBTRACT (c) INTERSECTION.

Ans. (a)

(viii) The following command option is used to select a group of objects lying within a window region and passing through the window

(a) Laser (b) Window (c) Crossing.

Ans. (c)

(ix) The drawing prepared in AutoCAD can be permanently stored in

(a) ROM (b) RAM (c) Auxiliary memory.

Ans. (c)

A.M.I.E., Winter, 2005

4. Choose the *correct* answer among the three answers suggested for each question :

(i) 1 Kilobyte means

(a) *1024 Bytes (b) 1024 KB (c) 1024 MB.

Ans. (a)

(ii) Hardware of a computer consists of

(a) only input unit (b) only input and output unit

(c) input processor and output unit.

Ans. (c)

(iii) QUIT is a command used in AutoCAD to

(a) return to the main menu and update drawing file

*1024 = $(2)^{10}$, 1 Kilo = 1000

(b) return to main menu without updating drawing file

(c) list all the autocad commands.

Ans. (a)

(iv) Input unit of a computer is

(a) Digitizer (b) ROM and RAM (c) Printer.

Ans. (a)

(v) Data can be permanently stored in

(a) ROM (b) RAM (c) Auxiliary.

Ans. (c)

(vi) ORTHO is a command used in AutoCAD to

(a) display dotted lines on the screen

(b) draw either horizontal or vertical lines

(c) locate specific points on the drafting sheet.

Ans. (b)

(vii) The GRID command is used in AutoCAD to

(a) set increments for cursor movement (b) display dotted lines on the screen

(c) move an object on the screen.

Ans. (b)

(viii) MS-DOS is an

(a) autocad software (b) operating system (c) auxiliary memory.

Ans. (b)

(ix) DIR is a command used in MS-DOS to

(a) list the files and subdirectories (b) list the subdirectories

(c) set the current directory.

Ans. (a)

(x) Centre, radius, (2, 8), R, 30 is the command used in AutoCAD to draw

(a) an ellipse (b) an arc (c) a circle.

Ans. (c)

Reference

For MS-DOS Commands, the following book will also be a helpful guide :

MS-DOS Commands.

Website : http//www.computerhope.com/msdos.htm

www—World Wide Web

(b) return to main menu without updating drawing file

(c) list all the autocad commands

Ans. [illegible]

(iv) Input unit of a computer is [illegible]

(a) Digitizer (b) ROM and RAM (c) Printer

Ans. (a)

(v) Data can be permanently stored in

(a) ROM (b) RAM (c) auxiliary

Ans. [illegible]

(vi) ORTHO is a command used in AutoCAD to

(a) display dotted lines on the screen

(b) draw either horizontal or vertical lines

(c) locate specific points on the drafting sheet

Ans. (b)

(vii) The GRID command is used in AutoCAD to

(a) set increments for cursor movement (b) display dotted lines on the screen

(c) move an object on the screen

Ans. (b)

(viii) MS-DOS is an

(a) autocad software (b) operating system (c) auxiliary memory

Ans. (b)

(ix) DIR is a command used in MS-DOS to

(a) list the files and subdirectories (b) list the subdirectories

(c) set the current directory

Ans. (a)

(x) Centre, radius, (2, 3), R 30 is the command used in AutoCAD to draw

(a) an ellipse (b) an arc (c) a circle

Ans. (c)

Reference

For MS-DOS Commands, the following book will also be a helpful guide :

MS-DOS Commands.

Website : http://www.computerhope.com/msdos.htm

---

www—World Wide Web

Annex I

List of *Indian Standards Replacing IS 696:1972
(Bureau of Indian Standards)

1.	IS 9609 (Pt 1) : 1985	Lettering on technical drawings, Part 1 English lettering (First revision)
2.	IS 9609 (Pt 2) : 1985	Lettering on technical drawings, Part 2 Greek characters
3.	IS 10711 : 1983	Sizes of drawing sheets
4.	IS 10712 : 1983	Presentation of item—references on technical drawings
5.	IS 10713 : 1983	Scales for use in technical drawings
6.	IS 10714 : 1983	General principles of presentation on technical drawings
7.	IS 10715 : 1983	Presentation of threaded parts on technical drawings
8.	IS 10716 : 1983	Rules for representation of springs on technical drawings
9.	IS 10717 : 1983	Conventional representation of gears on technical drawings
10.	IS 10718 : 1983	Technical drawings—Dimensioning and tolerancing of cones (first revision)
11.	IS 10719 : 1983	Method of indicating surface—texture on technical drawings
12.	IS 10720 : 1983	Technical drawings for structural metal works
13.	IS 11663 : 1986	Conventional representation of common features
14.	IS 11664 : 1986	Folding of drawing prints
15.	IS 11665 : 1985	Technical drawings—Title block
16.	IS 11666 : 1985	Technical drawings—Item lists
17.	IS 11667 : 1991	Technical drawings—Linear and angular tolerancing—Indication on drawings (First revision)
18.	IS 11669 : 1986	General principles of dimensioning on technical drawings
19.	IS 11670 : 1993	Technical drawings—Abbreviations and symbols for use in technical drawings (First revision)

*For the details of Indian Standards, please write to : The Director, Bureau of Indian Standards, 9 Bahadur Shah Zafar Marg, New Delhi-110002 or contact at the Website : http://www.bis.org.in

References for Further Study

	Author	Title	Publisher
1.	Abbott, W.	Practical Geometry and Engineering Graphics	Blackie
2.	Bhatt, N. D.	Engineering Drawing	Charotar
3.	Bogolyubov, S. and others	Engineering Drawing	Mir Publishers, Moscow
4.	Ghose, D. N.	Engineering Drawing	Dhanpat Rai
5.	Jeyapoovan, T.	Engineering Drawing with autocad 2000	Vikas
6.	Luzadder	Fundamentals of Engineering Drawing	Prentice-Hall
7.	Morris, I. H. and others	Geometrical Drawing	Orient Longman
8.	Narayana, K. L. and Kannaiah, P.	Engineering Graphics	Tata-McGraw-Hill
9.	Omura, G.	Mastering autocad 2004 and autocad I.T. 2004	B.P.B.
10.	Parkinson, A. C.	A first year Engineering Drawing	E.L.B.S.
11.	Sharma, S. C.	Engineering Graphics with autocad 2000	Galgotia
12.	Venugopal, K.	Engineering Drawing and Graphics + autocad	New Age

INDEX

D

E

F

G

H

I

West Bengal State Council Examination Questions

(a) 1995

Engineering Drawing

Time Allowed—4 Hours *Full Marks—80*

Draw a rectangle 80 × 50 *mm on right-hand bottom corner of the Drawing Sheet, touching the border line. Write your Centre of Examination, Subject, Roll No. and Date in the rectangle.*

The figures in the margin indicate marks for each question.

Mention the scale and show all dimensions.

Draftsmanship will carry marks.

For Group A, Group B and Group C—60 marks.

Answer any five questions taking at least one from each group.

For Group D—20 marks.

Answer all the questions in the question paper itself and attach the answered question paper with the Drawing Sheet.

Group A

1. On a map, the actual distance of 5 m is represented by a line of 25 mm long. Calculate the R.F. of the scale. Construct a diagonal scale to read from single decimetre to 30 m at a time. Mark on the scale distances of 19.9 m and 11.1 m. 12

2. Inscribe an ellipse tangential to a rhombus of sides 60 mm each. The shorter diagonal of the rhombus is also 60 mm. Locate the foci and measure the major axis of the ellipse. 12

3. Print the following words in single-stroke block capitals, 30 mm high, in 7 : 4 ratio :

"WEST BENGAL—JULY '95" 12

Group B

4. Draw the projections of the following points on the same ground line, keeping the projectors 15 mm apart : 12

(i) Point A 35 mm below H.P. and 20 mm behind V.P.
(ii) " B 25 mm above H.P. and 40 mm in front of V.P.
(iii) " C 50 mm in front of V.P. and 25 mm below H.P.
(iv) " D on the H.P. and 20 mm behind V.P.
(v) " E on the V.P. and 40 mm below H.P.
(vi) " F on the H.P. and 35 mm in front of V.P.
(vii) " G on both H.P. and V.P.
(viii) " H 45 mm behind V.P. and 15 mm above H.P.

5. A regular hexagonal pyramid, of base-edge 30 mm, height 60 mm, is resting on one of its base-edge on H.P. with a triangular face inclined to H.P. at 60° and the edge on which it is resting on H.P. is inclined to V.P. at 45°. Draw its projections in this position. 12

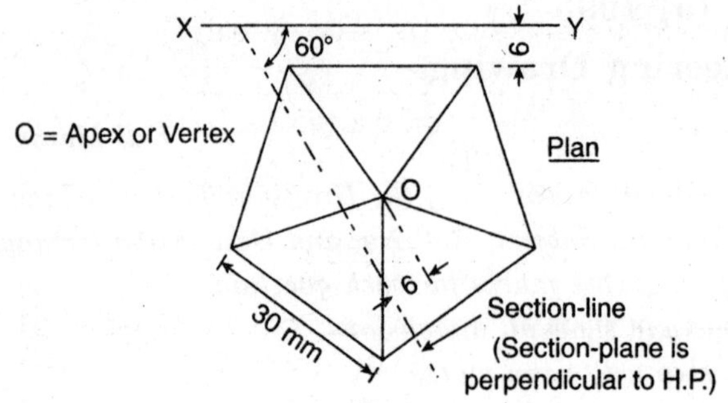

Fig. 1

6. A regular pentagonal pyramid, of base-edge 30 mm, height 70 mm, rests on its base on the ground, with a base-side parallel to V.P. It is cut by a plane inclined at 60° to V.P. as shown in Fig. 1. Draw its plan, sectional elevation and true shape of section. 12

Group C

7. Draw an isometric view of three solids, resting one upon the other with same vertical axis, using ordinary scale. The bottom one is a right cylinder of 70 mm dia. and 30 mm height. The middle one is a cube of sides 60 mm each. A sphere of 50 mm dia. is resting centrally above the cube. 12

8. Develop the surfaces of the pentagonal pyramid that remains (i.e. right-hand side) after section, as given in question No. 6 (Fig. 1) 12

9. Copy the "DOUBLE HOOK" given in Fig. 2 with scale full-size. All dimensions are in mm. Print the name in single-stroke block capitals, 10 mm high. Locate the centre 'O'. 12

Hint. To locate the centre O.

PC = 10 mm

PE = PC + CE = 10 + 24 = 34 mm

OE = 46 mm

OP = OE – PE = 46 – 34 = 12 mm

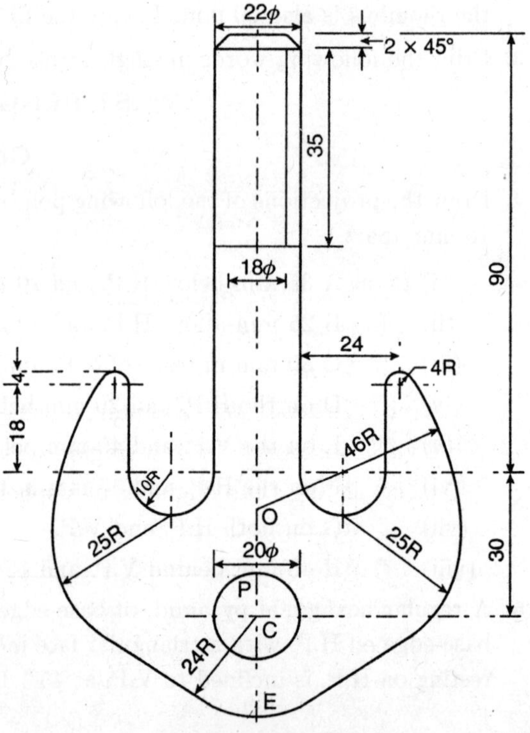

Fig. 2. Double Hook

Or, (For the students of Architecture Department only)

Line diagram of a residential building is given in Fig. 3. Draw, to a scale of 1 : 50, its sectional plan through window-sill level, with the following specifications : 12

Thickness of main walls—250

Thickness of partition walls—125 (in bath-room)

Height of plinth above G.L.—600

Doors : D = 1050 × 2100, D_1 = 900 × 2000, D_2 = 750 × 1800

Windows : W = 1800 × 1200, W_1 = 900 × 1200, W_2 = 600 × 900

All dimensions are in mm. Assume any other data, if necessary.

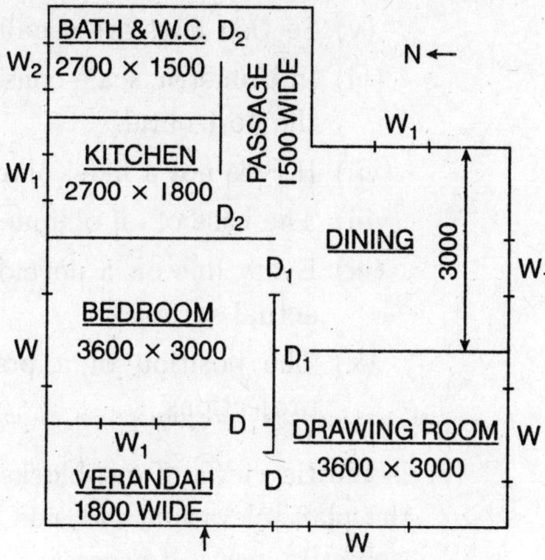

Fig. 3. Plan of a residential building

Group D

10. (a) Fill in the gaps with appropriate words: $\frac{1}{2} \times 10 = 5$

(i) For drawing of small objects, ______ scale is always used.

(ii) Two methods of placing dimensions are ______ and ______.

(iii) If sum of the focal distances from any point on an ellipse is 110 mm, the major axis will be ______.

(iv) ______ projection is a drawing system that uses projector at right angles to produce views of an object.

(v) When ______ of a symmetrical object is imagined to be removed, the section is called a half-section.

(vi) When a sphere is cut by a plane inclined at 30° to H.P., the true shape of section is ______.

(vii) The solid which contains four equilateral triangles is known as ______.

(viii) Isometric view is a single plane ______ projection.

(ix) If 'D' is the diameter of a sphere, its value in isometric projection will be equal to ______.

(x) A right ______ and a ______ is a composite solid, bounded by double-curved-surfaces.

(b) Answer the following statements *(true or false)* : $\frac{1}{2} \times 10 = 5$

(i) Front views of two lines are parallel. Hence the lines must be parallel.

(ii) The distance of a point from the H.P. is marked from XY in the top view.

(iii) When a square prism is resting on its base on V.P., the elevation is a square.
(iv) Sectional-area is never bounded by dotted lines.
(v) Section-lines are usually drawn at an angle 30° to G.L.
(vi) In isometric scale-construction, true lengths are cut on a line inclined at 30° to the horizontal.
(vii) H.P. is not a fixed plane.
(viii) The base of an oblique cylinder is an ellipse.
(ix) Every line on a development must be equal to true length of that line on the actual surface.
(x) The position of a point in space may be determined from any one of its projections.

(c) Isometric views of cut-blocks are given in Fig. 4(a), (b), (c), (d) and (e). Complete the plan, elevation and side view of each block, measuring the dimensions from the isometric views, if necessary. $2 \times 5 = 10$

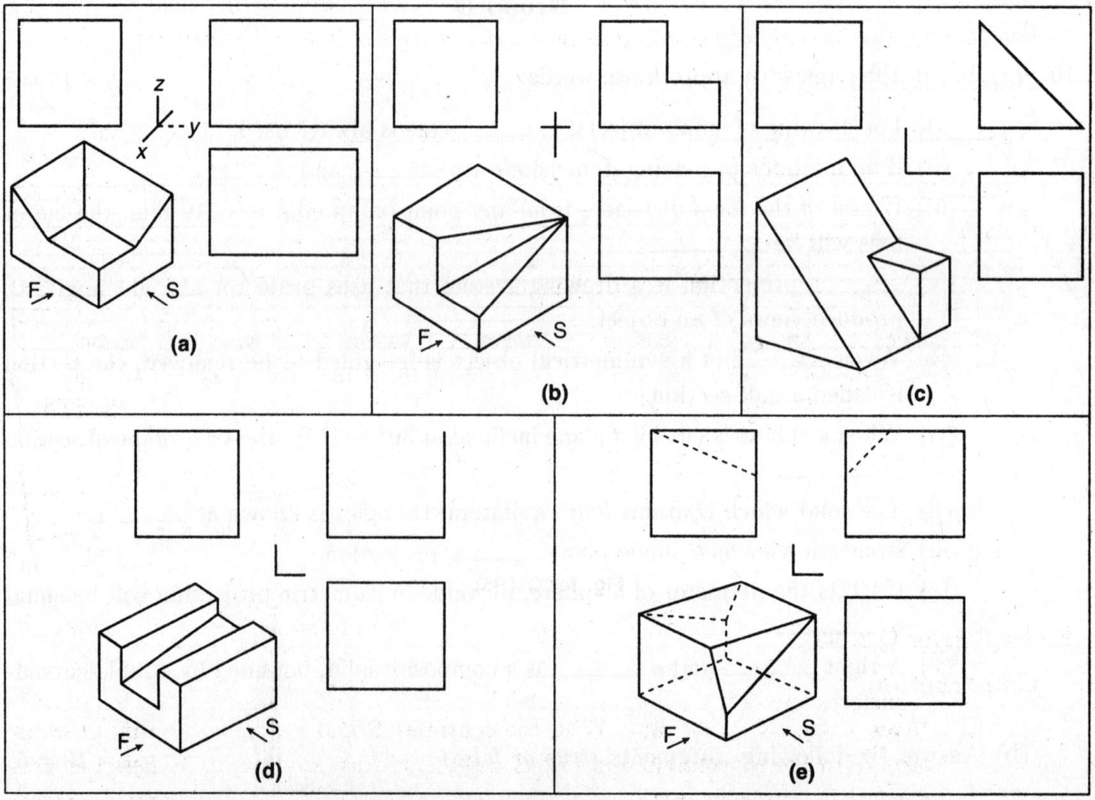

Fig. 4

Question-Answer of West Bengal State Council Examination

(a) 1995

Engineering Drawing

All the Drawings are drawn not to scale.

1. Actual distance = 5 m = 5 × 100 × 10 = 5000 mm. (1 m = 100 cm, 1 cm = 10 mm)

Map-distance = 25 mm.

$\therefore \quad RF = \frac{25}{5000} = \frac{1}{200}.$

Maximim distance to be measured = 30 m.

$\therefore$ length of the scale = RF × max. length to be measured = $\frac{1}{200} \times 30 \times 100 = 15$ cm.

Take a line 15 cm long and divide it into 15 divisions, so that one small-division will represent 2 m (Fig. Q.1('95))

Now, divide the left-hand small-division into 2 equal parts, and each small-division will represent 1 m. Then draw a perpendicular at the left-end of the line and take 30 mm long line. Divide this line into 10 equal divisions. Then, each division = 1 decimetre.

Draw a diagonal line 10-0, as shown in the figure : (Q1 '95)

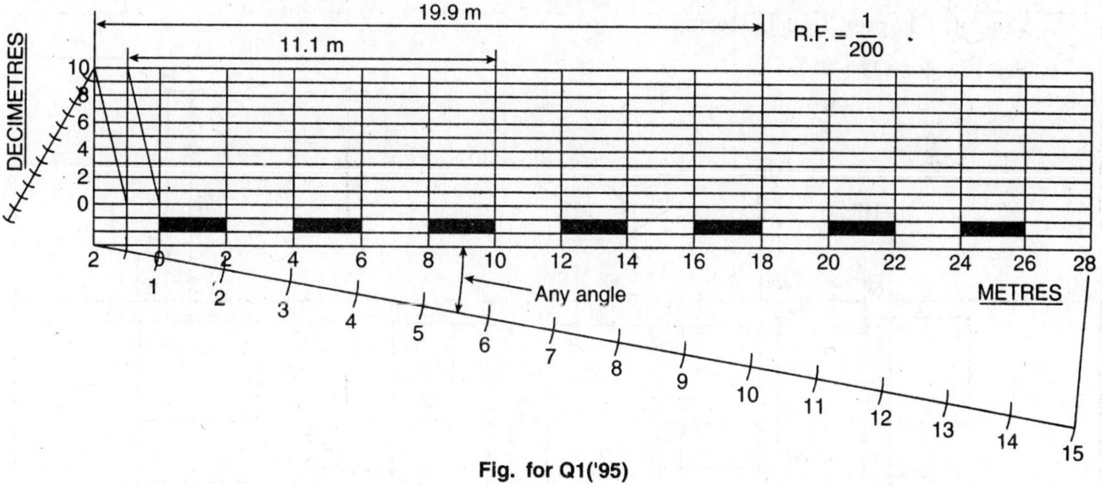

Fig. for Q1('95)

2. See Fig. for Q.2('95).

Construction

(1) Draw a line BC = 60 mm. With the centre at B and radius = 60 mm, draw an arc. With the centre at C and radius = 60 mm, draw another arc. These two arcs meet at D. Similarly, with the centres at B and D, and radius = 60 mm, draw two arcs to meet at A. Thus, ABCD is the rhombus in question.

(2) Bisect AD and BC to get the points G and H respectively. Similarly, bisect AB and DC to get the points E and F respectively.

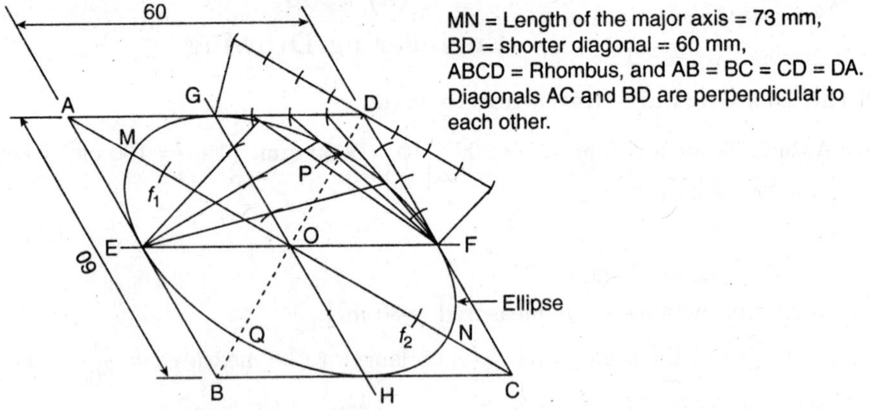

Fig. for Q2('95)

(3) Now, we use the "oblong-method" to draw the ellipse, which touches the rhombus at the four points E, F, G and H.

(4) Join AC, which cuts the ellipse at the points M and N.

(5) PQ = Minor axis. With centre at P and radius OM (half the major axis), draw two arcs, which cut the major axis at f_1 and f_2, which are the foci.

3. Consult Chapter 3 on Lettering.

4. See Fig. for Q4 ('95).

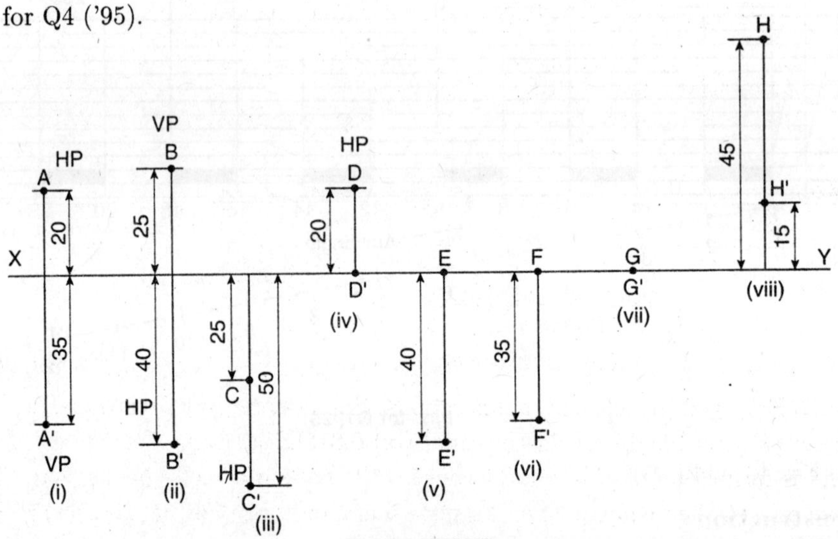

Fig. for Q4('95)

5. See Chapter 11 Problem No. 28.

6. See Chapter 12 Problem No. 8.

7. See also-Chapter 14.

8. See Chapter 16, Problem No. 9.

9. Draw from the question No. 9, one view of 'Double Hook' with scale full-size.

Or, (for the students of Architecture Department only)

See the figure Q.9 or ('95), scale 1 : 50

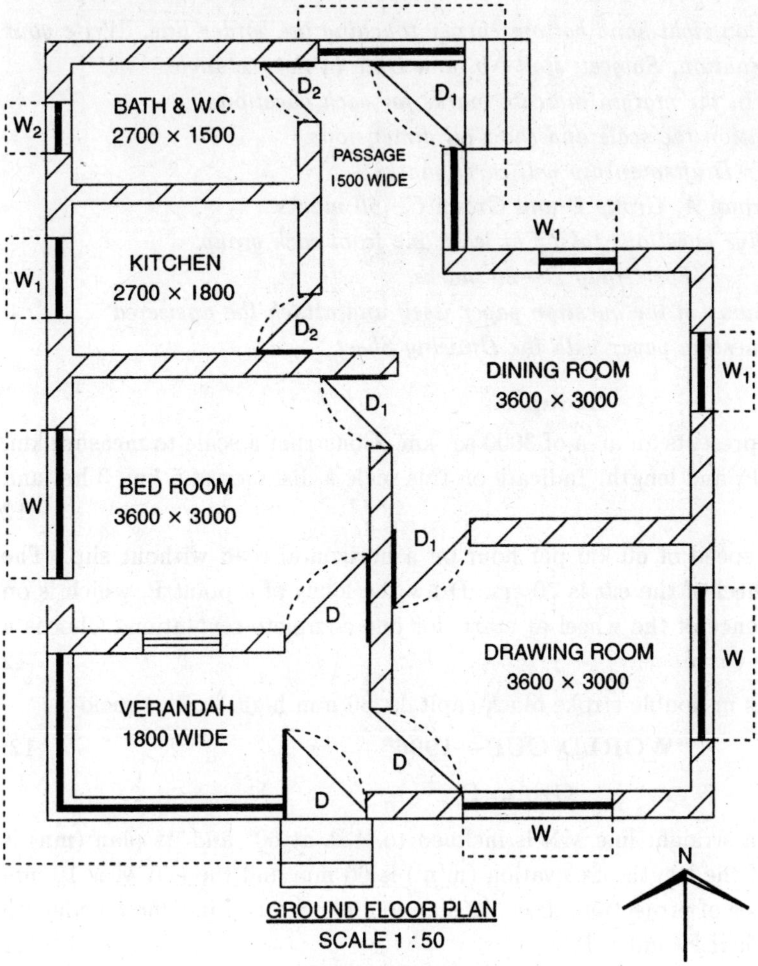

Fig. for Q9 or ('95)

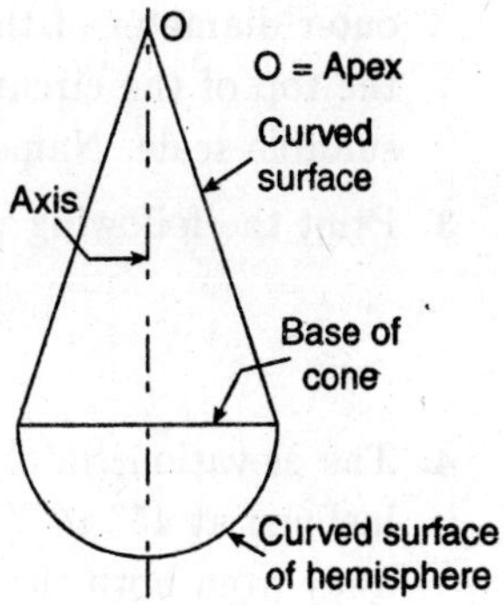

Fig. Q. 10(a)(x)

10. (a) (i) Enlarged (ii) Aligned, unidirectional (iii) 110 mm (iv) Orthogonal (v) One-fourth (vi) Circle (vii) tetrahedron (viii) pictorial (ix) 0.815D (x) Cone, hemisphere. [This composite solid is formed by placing a right cone on the flat-surface of a hemisphere. The base of the cone and the hemisphere have the same diameters. See Fig. Q. 10(a)(x).]

(b) (i) False (ii) False (iii) True (iv) True (v) False (vi) False (vii) True (viii) False (ix) True (x) False.

(c) From the isometric view of each block, the students should try to draw the plan, elevation and side view. Refer Fig. 4(a) to (e) of the Question, 1995.

West Bengal State Council Examination Questions

(b) 1996

Engineering Drawing

Time Allowed—4 Hours *Full Marks—80*

Draw a rectangle 80 × 50 mm *on right-hand bottom corner touching the border line. Write your Centre of Examination, Subject, Roll No. and Date in the rectangle.*

The figures in the margin indicate marks for each question.

Mention the scale and show all dimensions.

Draftsmanship will carry marks.

For Group A, Group B and Group C—60 marks.

Answer any five questions taking at least one from each group.

For Group D—20 marks.

Answer all the questions in the question paper itself and attach the answered question paper with the Drawing Sheet.

Group A

1. A map 120 × 100 cm^2 represents an area of 3000 sq. km. Construct a scale to measure km, hm and dm. Find its R.F. and length. Indicate on this scale a distance of 6 km, 3 hm and 9 dm. 12

2. A car is moving with a speed of 60 km per hour on a horizontal road without slip. The outer diameter of the wheel of the car is 70 cm. Draw the locus of a point P, which is on the top of the circumference of the wheel at start, for one complete revolution. Choose a suitable scale. Name the curve. 12

3. Print the following words in double-stroke block-capitals, 30 mm high, in 7 : 4 ratio :

"WORLD CUP—1996" 12

Group B

4. The elevation (m′n′) of a straight line MN is inclined to H.P. at 60° and its plan (mn) is inclined at 45° to V.P. If the length of elevation (m′n′) is 60 mm and the end M is 10 mm away from both the planes of projection, draw its plan and elevation. Find the true length and true inclinations with H.P. and V.P. 12

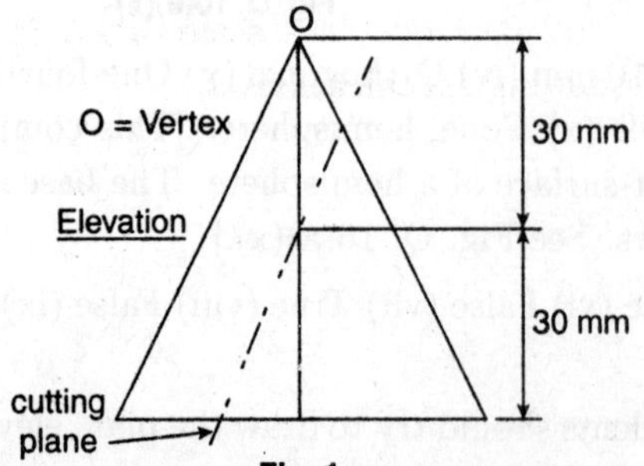

Fig. 1

5. Draw three views (plan, elevation and side view) of a regular pentagonal pyramid, base-edge 30 mm, height 60 mm, when one of its triangular faces is vertical. Measure the inclination of its base-surface with H.P. 12

6. A regular hexagonal pyramid, of base-side 30 mm and height 60 mm rests on ground with two of its parallel base-sides perpendicular to V.P. It is cut by an inclined plane parallel to its slant-surface and cuts the axis at its mid-height as shown in Fig. 1 (Elevation only). Draw its sectional plan and sectional side-view (First-Angle projection). 12

Group C

7. Draw an isometric projection of two solids, resting one upon the other, with same vertical axis, using iso-scale, looking in the direction of arrow 'x'. The bottom one is a regular hexagonal prism of base-side 30 mm and height 50 mm. A hemisphere of 50 mm dia. is resting on the top-surface of the prism, as shown in Fig. 2. 12

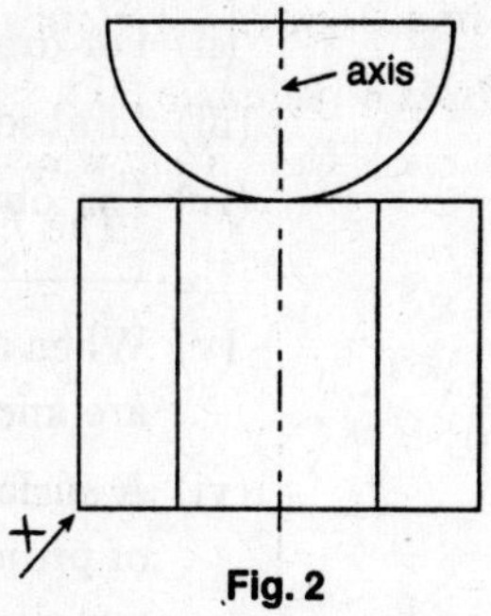

Fig. 2

8. Develop the surfaces of the hexagonal pyramid that remains after section as given in question No. 6 (Fig. 1), i.e. right-hand portion of the pyramid, after cutting. 12

9. Copy the 'WRENCH' as given in Fig. 3 with scale full-size. All dimensions are in mm. Print the name in single-stroke, block-capital, 10 mm high. Locate the unknown centres. 12

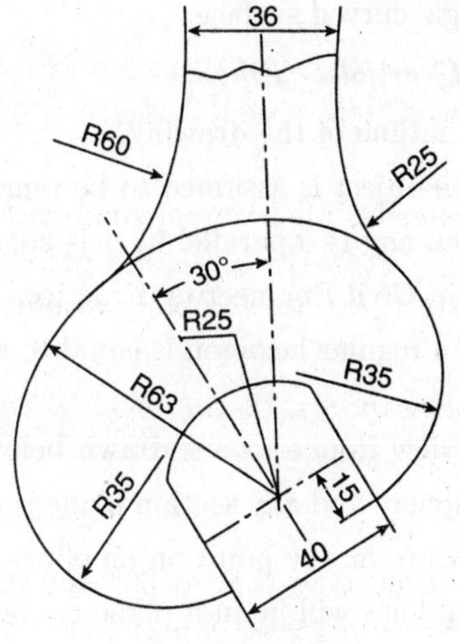

Fig. 3. Wrench

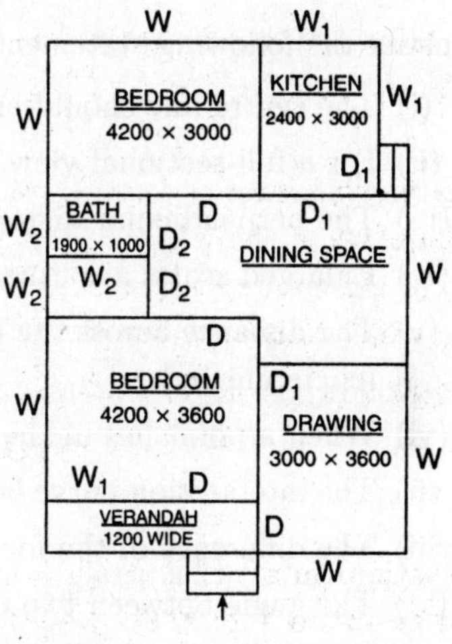

Fig. 4. Plan of a residential building

Or, (For the students of Architecture Department only)

Line-diagram of a residential building is given in Fig. 4. Draw to a scale of 1 : 50, its sectional plan through window-sill level, with the following specifications : 12

Thickness of main walls—250

Thickness of partition walls—125 (in between bathroom & WC)

Height of plinth above G.L.—450

Doors : D = 1050 × 2100, D_1 = 900 × 2000, D_2 = 750 × 1800

Windows : W = 1800 × 1200, W_1 = 900 × 1200, W_2 = 600 × 900.

All dimensions are in mm. Assume any other data, if necessary.

Group D

10. (a) Fill in the gaps with appropriate words : $\frac{1}{2} \times 10 = 5$

(i) Dimension lines should not ______ each other.

(ii) For drawing large objects ______ scale is used.

(iii) In absence of a ______, a scale of chords may be used.

(iv) For obtaining a sectional view, the part of the object between ______ and ______ is assumed to be removed.

(v) When a cone is cut by planes at different angles, the intersection curves obtained are known as ______.

(vi) A surface of an object appears in its true shape when it is ______ to the plane of projections.

(vii) When a point is ______ V.P., its view from above is above XY.

(viii) A straight line is defined as the ______ distance between two points.

(ix) ______ is a solid bounded by single curved surface.

(b) Answer the following statements *[true (T) or false (F)]* : $\frac{1}{2} \times 10 = 5$

(i) The centre line should end at the outline of the drawing.

(ii) For a full-sectional view, half of the object is assumed to be removed.

(iii) The perpendicular distance between any two parallel lines is constant.

(iv) Enlarged scales are usually found in Civil Engineering Practice.

(v) The distance across the corners of a regular hexagon is equal to the diameter of its circumcircle.

(vi) When a point lies in the V.P., its view from above is drawn below XY line.

(vii) The intersection-curve between a sphere and any section plane is always a circle.

(viii) The difference of the focal distances from any point on an ellipse is constant.

(ix) The angle between two intersecting lines will lie in a plane containing the lines.

(x) The base of an oblique hexagonal prism is not a regular hexagon.

(c) (i) Fig. 5(i) shows elevation and side view and Fig. 5(ii) shows plan and elevation of two straight lines. Draw their third views. $1 \times 2 = 2$

(ii) Isometric views of four cut-blocks are given in Fig. 5(iii), (iv), (v) and (vi). Complete the plan, elevation and side view of each block measuring the dimensions from the isometric view, if necessary. $1 \times 2 = 2$

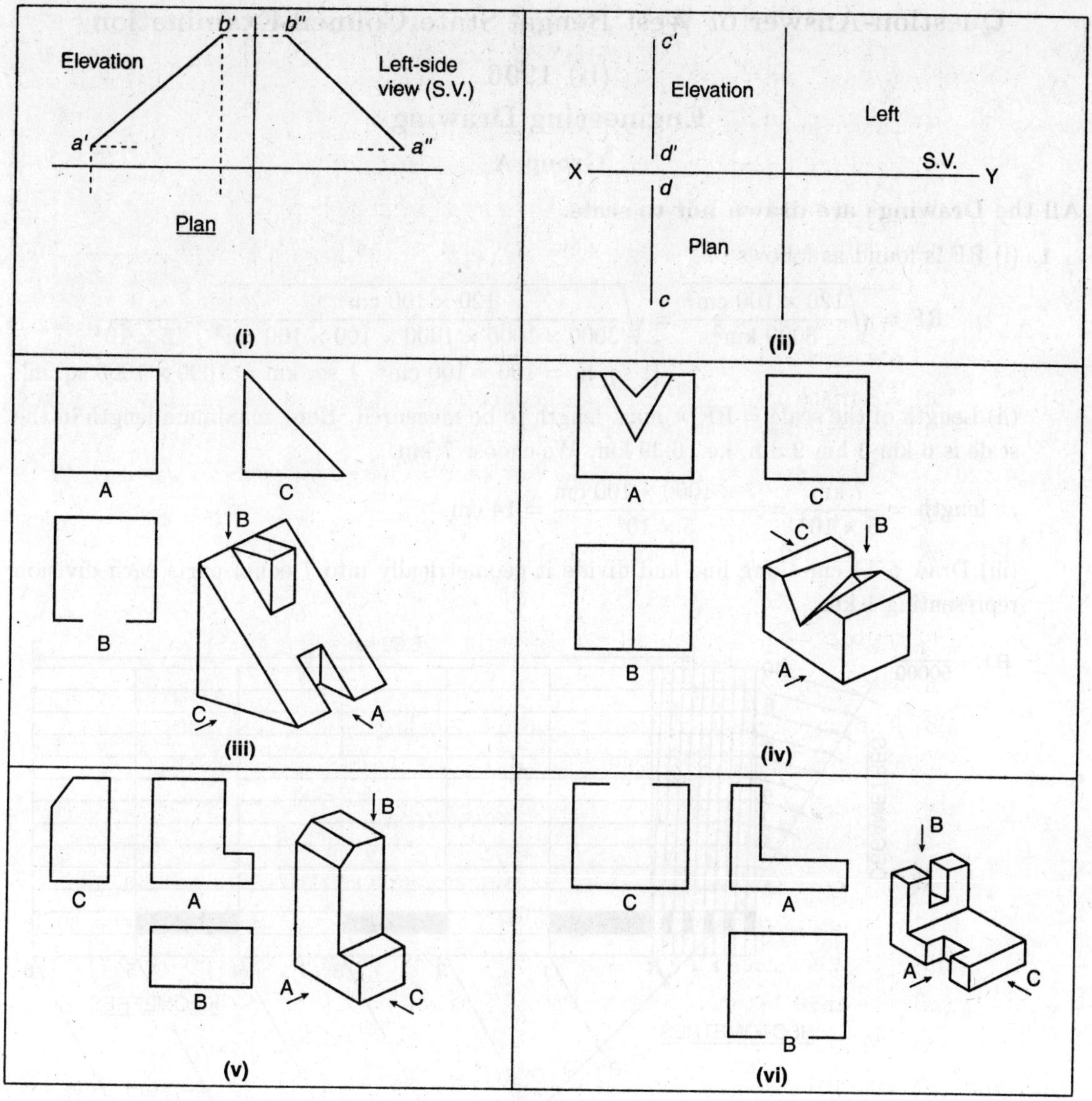
Elevation
b'
b''
Left-side view (S.V.)
a'
a''
Plan
(i)
c'
Elevation
Left
d'
S.V.
X
Y
d
Plan
c
(ii)
A
C
B
B
C
A
(iii)
A
C
B
C
B
A
(iv)
C
A
B
B
A
C
(v)
C
A
B
B
A
C
(vi)

Fig. 5

Question-Answer of West Bengal State Council Examination

(b) 1996

Engineering Drawing

Group A

All the Drawings are drawn not to scale.

1. (i) RF is found as follows :

$$RF = \sqrt{\frac{120 \times 100 \text{ cm}^2}{3000 \text{ km}^2}} = \sqrt{\frac{120 \times 100 \text{ cm}^2}{3000 \times 1000 \times 1000 \times 100 \times 100 \text{ cm}^2}} = \frac{1}{5 \times 10^4}$$

[1 sq. m = 100×100 cm^2, 1 sq. km = 1000×1000 sq. m].

(ii) Length of the scale = RF $\times$ max. length to be measured. Here, maximum length in the scale is 6 km 3 hm 9 dm, i.e., 6.39 km. We choose 7 km.

$$\therefore \text{ length} = \frac{7 \text{ km}}{5 \times 10^4} = \frac{7 \times 1000 \times 100 \text{ cm}}{5 \times 10^4} = 14 \text{ cm}.$$

(iii) Draw a 14 cm. long line and divide it geometrically into 7 equal parts each division representing 1 km.

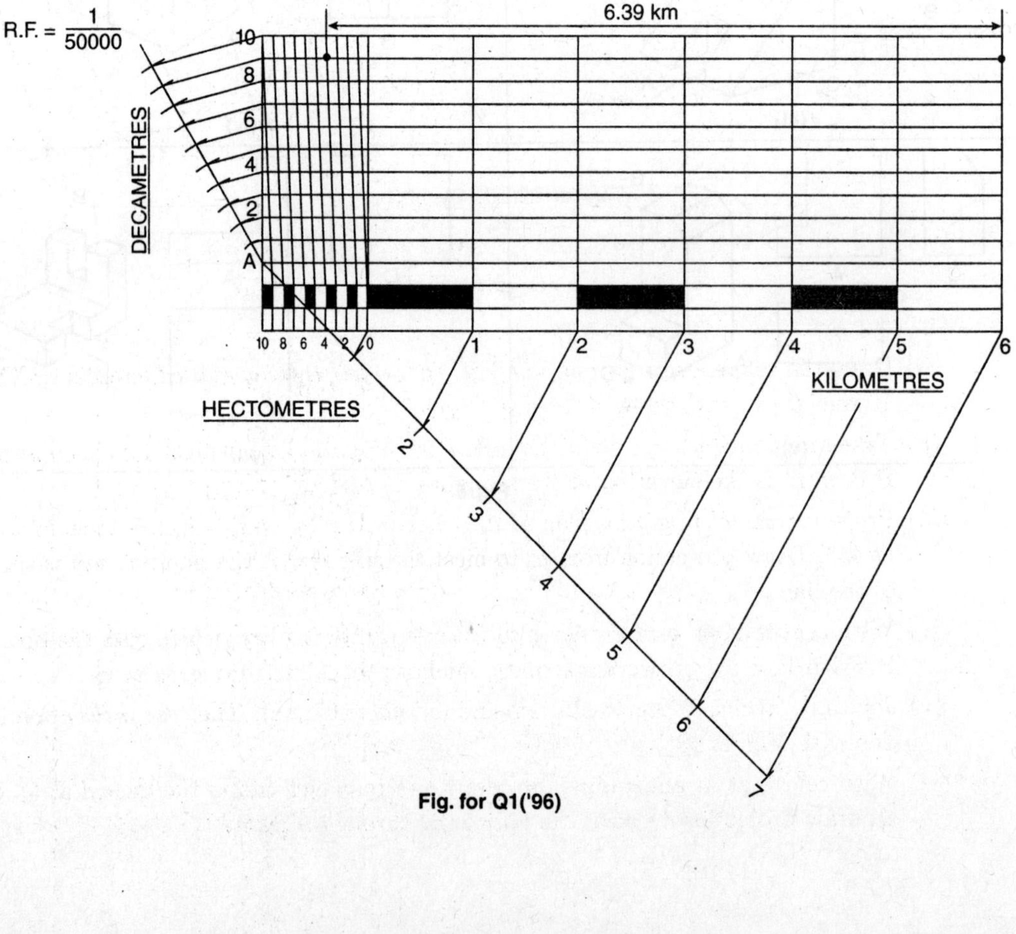

Fig. for Q1('96)

Again divide the small-division geometrically into 10 equal parts. One small-division will represent 100 metres or hm. Now, to get dm, divide 1 hm division diagonally, i.e., the scale is a diagonal scale.

2. Speed of the car is not required here. We choose the scale for drawing the wheel (diameter = 70 cm) as 2 cm = 1 mm, i.e., dia. = 35 mm. For construction, see chapter 6 on curves. Name of the curve is *cycloid.*

Refer Fig. Q2('96).

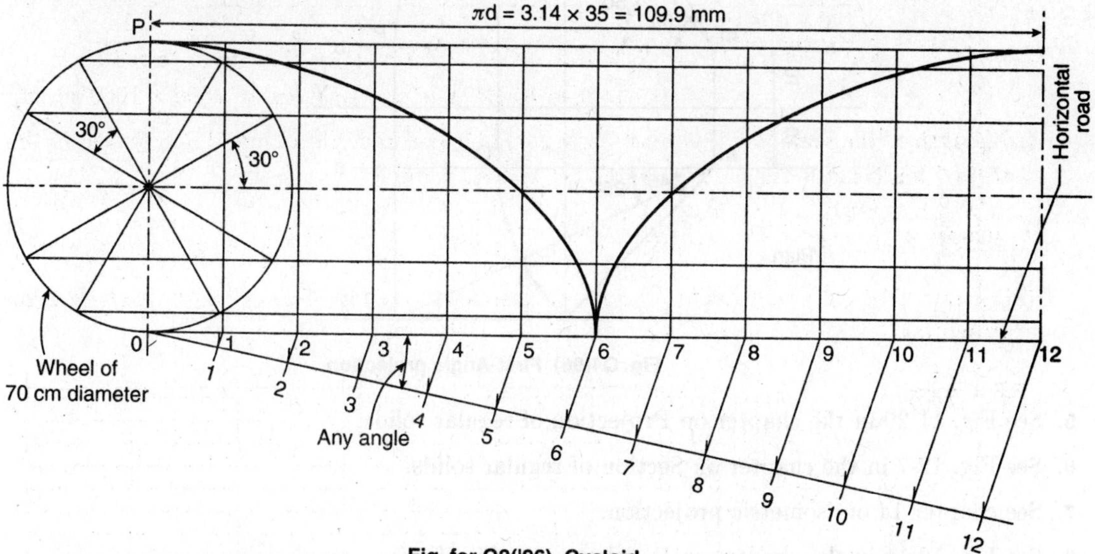

Fig. for Q2('96) Cycloid

3. Consult chapter 3 on Lettering.

Group B

4. Refer to Fig. for Q4('96).

(i) Draw the reference or ground line XY. Draw two lines *ab* and *cd* parallel to XY and 10 mm above and below XY respectively.

(ii) Take any point m' and draw the line $m'n'$ (60 mm long), which is inclined at 60° to H.P. $m'n'$ is the elevation of the straight line MN.

(iii) Project from m' to get its plan at m. Also draw a line mn, which is inclined to V.P. at 45°. Draw projection from n' to meet the line mn at the point n. mn is the plan of the line.

(iv) With centre at m' and radius $m'n'$ (60 mm), draw an arc, which cuts the line *ab* at P. From P, draw projection to meet the horizontal line through n at n_1.

(v) Join mn_1, which is the *true length* and measures 67 mm. The *true inclination* of the line with V.P. is 25°.

(vi) With centre at m and radius mn, draw an arc, which meets the line *cd* at Q. From Q, draw projection to meet the horizontal through n' at n''.

(vii) Join $m'n''$, which is the true length. The true inclination of the line with H.P. is 50°.

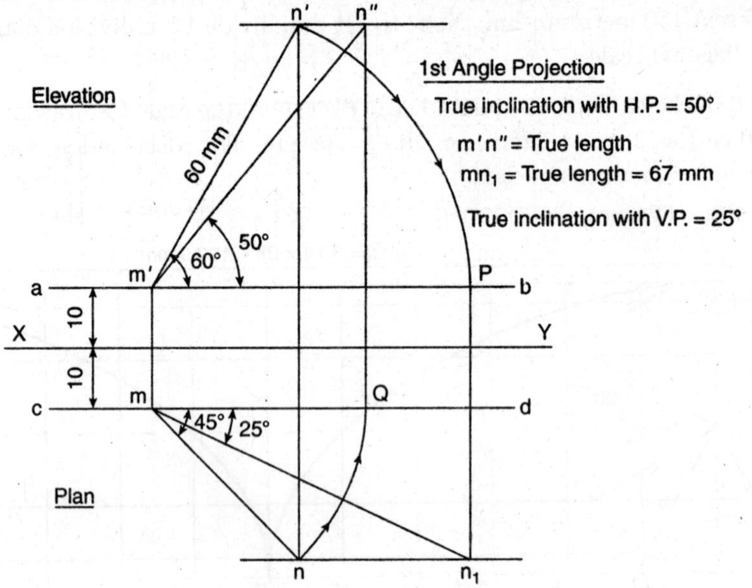

Fig. Q4(96) First-Angle projection

5. See Fig. 11-29 in the chapter on Projection of regular solids.
6. See Fig. 12-7 in the chapter on Section of regular solids.
7. See chapter 14 on isometric projection.
8. See Fig. 13-11 in the chapter on Development of Surfaces.
9. Copy Fig. 3 of the question with full-size scale. Locate the three unknown centers by trial. Write the name 'WRENCH' in single-stroke, Block-Capital, 10 mm high.

Or, (For the students of Architectural Department only)

Refer Fig. Q9, or ('96) for the sectional plan.

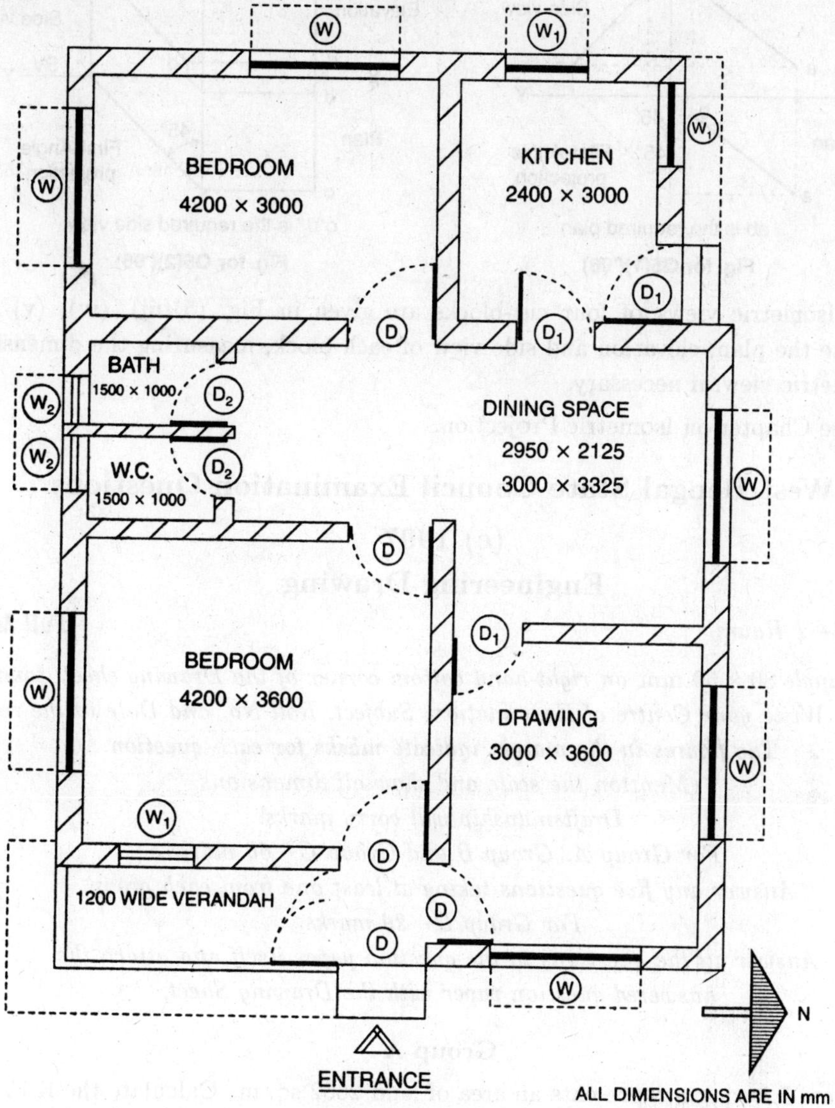

Fig. for Q9 or ('96)

10. (a) (i) cross, (ii) reduced (iii) protractor (iv) sectional plane, outer surface (v) conics, (vi) parallel (vii) behind, (viii) shortest, (ix) sphere.

(b) (i) F, (ii) T, (iii) T, (iv) F, (v) T, (vi) F, (vii) T, (viii) F, (ix) T & (x) F.

(c) (i) See Fig. for Q5(1)('96) and Fig. for Q5(2)('96).

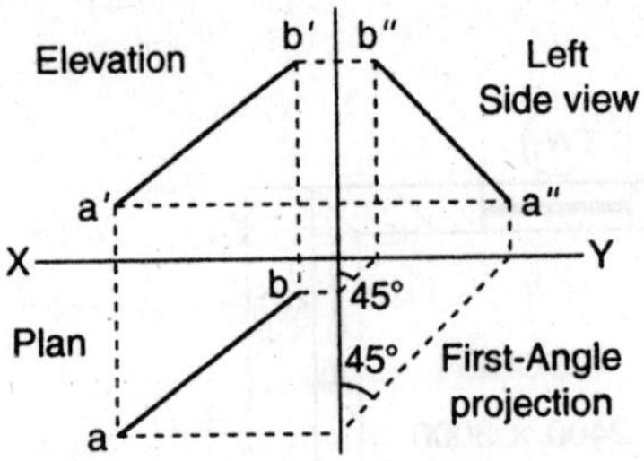

ab is the required plan

Fig. for Q5(1)('96)

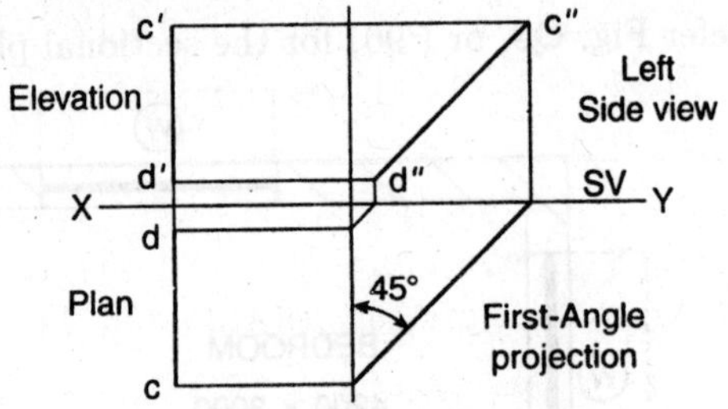

c"d" is the required side view

Fig. for Q5(2)('96)

(c) (ii) Isometric views of four cut-blocks are given in Fig. (5)(iii), (iv), (v) and (vi). Complete the plan, elevation and side view of each block, measuring the dimensions from the isometric view, if necessary.

Ans. See Chapter on Isometric Projection.

West Bengal State Council Examination Questions

(c) 1997

Engineering Drawing

Time Allowed—4 Hours *Full Marks—80*

Draw a rectangle 80 × 50 mm *on right-hand bottom corner of the Drawing sheet, touching the border line. Write your Centre of Examination, Subject, Roll No. and Date in the rectangle.*

The figures in the margin indicate marks for each question.

Mention the scale and show all dimensions.

Draftsmanship will carry marks.

For Group A, Group B and Group C—60 marks.

Answer any five questions taking at least one from each group.

For Group D—20 marks.

Answer all the questions in the question paper itself and attach the answered question paper with the Drawing Sheet.

Group A

1. On a map, 5.5 sq. cm represents an area of land 2662 sq. m. Calculate the R.F. Draw the scale to read up to single metre and long enough to show 3 hectometres at a time. Draw a rectangle 250 m × 105 m with the help of that scale. 12

2. A cricket ball, thrown from the boundary of a cricket-ground, reaches the Wicket-Keeper's gloves, following a geometrical path. Maximum height achieved by the ball above the ground is 42 metres. Assume the point of throw and point of catching position 2 metres above the ground. Radial distance of boundary from Wicket-Keeper is 75 metres. Construct the path of the ball and designate the important points and lines. Name the curve. 12

3. Print any six letters from the following letters in double-stroke block-capitals, 30 mm high, in 7 : 4 ratio: A, B, J, M, Q, R, S, W, Y, 3, 5, 9. 12

Group B

4. A line AB, 65 mm long, has its end A, 20 mm above H.P. and 25 mm in front of V.P. The end B is 40 mm above H.P. and 65 mm in front of V.P. Draw the projections of AB and show the inclinations with H.P. and the V.P. 12

5. A 30°-60° set-square of 80 mm longest side, is placed in such a way that its hypotenuse is in the V.P. and inclined at 30° to H.P. The surface of the set-square makes an angle, 45° to V.P. Draw its plan, elevation and side view in this position. 12

6. A right hexagonal pyramid, of base-side 30 mm and axis 65 mm long, is lying on V.P. on one of its longer edges. A plane containing this edge and the axis is perpendicular to V.P. Draw the projections of the pyramid, when the vertex is pointing upwards. 12

Group C

7. Draw an isometric projection of three solids, resting one upon the other, with same vertical axis, using iso-scale. The bottom one is a regular hexagonal prism of sides 25 mm each and height 20 mm. The middle one is a cube of sides 25 mm each. A hemisphere of radius 20 mm is resting centrally above the cube, as given in Fig. 1. 12

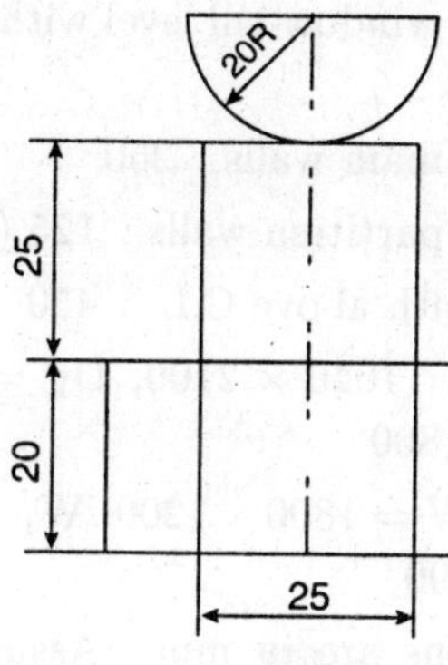

Fig. 1

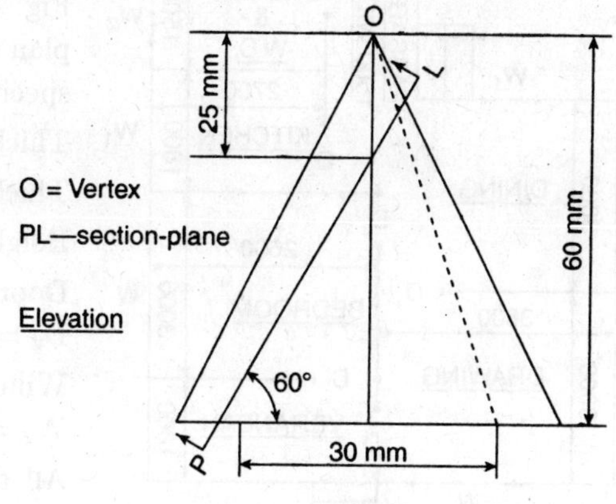

Fig. 2

8. Develop the surfaces of the lower part of the regular pentagonal pyramid that remains after section as given in Fig. 2 (Elevation only). The base-side, which is parallel to V.P., is 30 mm and height 60 mm. 12

9. Copy the 'BRACKET', as given in Fig. 3, with scale full-size. All dimensions are in mm. Print the name in single-stroke block-capitals, 10 mm high. 12

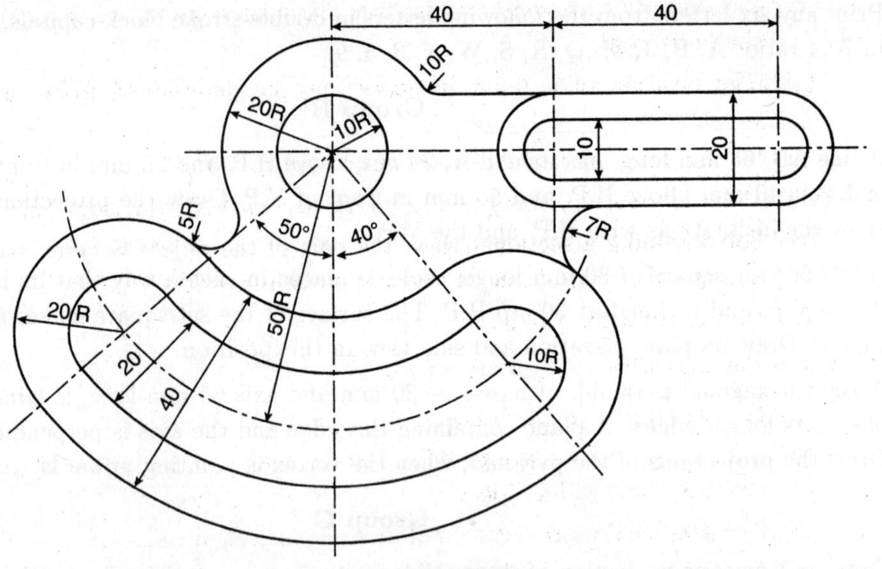

Fig. 3. BRACKET

Or, (For the students of Architectural Department only)

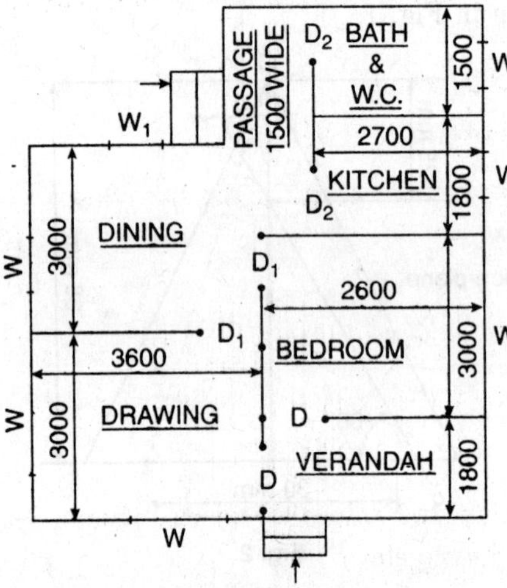

Fig. 4. Plan of a residential building

Line-diagram of a residential building is given in Fig. 4. Draw to a scale of 1 : 50 its sectional plan through window-sill level with the following specifications :

Thickness of main walls : 250

Thickness of partition walls : 125 (in bathroom)

Height of plinth above C.L. : 450

Doors : D = 1050 × 2100, D_1 = 900 × 2000, D_2 = 750 × 1800

Windows : W = 1800 × 1200, W_1 = 900 × 1200, W_2 = 600 × 90

All dimensions are in mm. Assume any other data, if necessary. 12

Group D

10. (a) Answer the following statements *(true or false)* : $\frac{1}{2} \times 10 = 5$

(i) The centre-line should end at the outline of the drawing.

(ii) For a half-sectional view, half of the object is assumed to be removed.

(iii) The distance of a point from the H.P. is marked from XY in the top view.

(iv) The base of an oblique cylinder is an ellipse.

(v) In isometric scale-construction, true lengths are cut on a line inclined at 30° to the horizontal.

(vi) The position of a point in space may be determined from any one of its projections.

(vii) Section-lines are usually drawn at an angle of 30° to G.L.

(viii) When a square prism is resting on its base on V.P., the elevation is a square.

(ix) For obtaining a sectional view, the part of the object between sectional-plane and the projection-plane is assumed to be removed.

(x) H.P. is not a fixed plane.

(b) Fill in the gaps with appropriate words: $\frac{1}{2} \times 10 = 5$

(i) For drawing of small objects, ______ scale is always used.

(ii) ______ projection is a drawing system, that uses projectors at right angles to produce views of an object.

(iii) The graphical representation of an object is called ______.

(iv) French curves are ______ of various curved shapes.

(v) Cutting plane lines are ______ lines, ______ at ends and ______ elsewhere.

(vi) In ______ system, all dimensions are placed in such a way that they may be read from the bottom of the drawing.

(vii) Dimension lines should not ______ each other.

(viii) The imaginary process of cutting the object is called ______.

(c) (i) Draw plan and elevation of the square lamina, given in the side view. $1 \times 2 = 2$

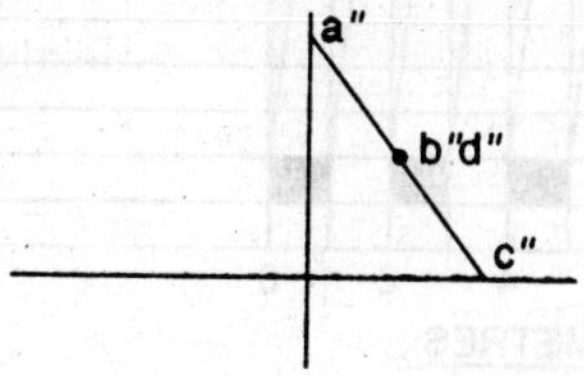

(ii) Mark the surfaces in the elevation plan and side view of the object, given in isometric. (i.e., flat surface A, shown in the elevation) $1 \times 6 = 6$

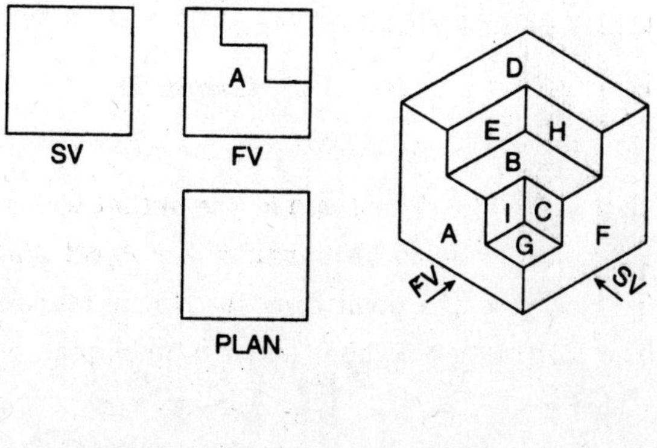

Question-Answer of West Bengal State Council Examination

(c) 1997
Engineering Drawing

Group A

All the drawings are drawn not to scale.

1. $\text{RF} = \sqrt{\dfrac{5.5\ \text{cm}^2}{2662\ \text{m}^2}} = \sqrt{\dfrac{5.5\ \text{cm}^2}{2662 \times 100 \times 100\ \text{cm}^2}} = \dfrac{1}{2200}$. [1 m = 100 cm]

Length of the scale $= \dfrac{1}{2200} \times 3$ decametre $= \dfrac{1}{2200} \times 3 \times 10 \times 100$ cm

$= 13.63$ cm (1 decametre = 10 metre)

Ref Fig. for Q1('97)

Take a line 13.63 cm long and divide it into 3 equal parts geometrically. Then, one small-division is equal to one hectometer. Now divide the left-hand division into 10 equal parts and one small division represent one decameter. To get metre, convert it into diagonal scale as shown in the Fig. for Q1('97). With the help of this scale, draw a rectangle of 205 × 105 m.

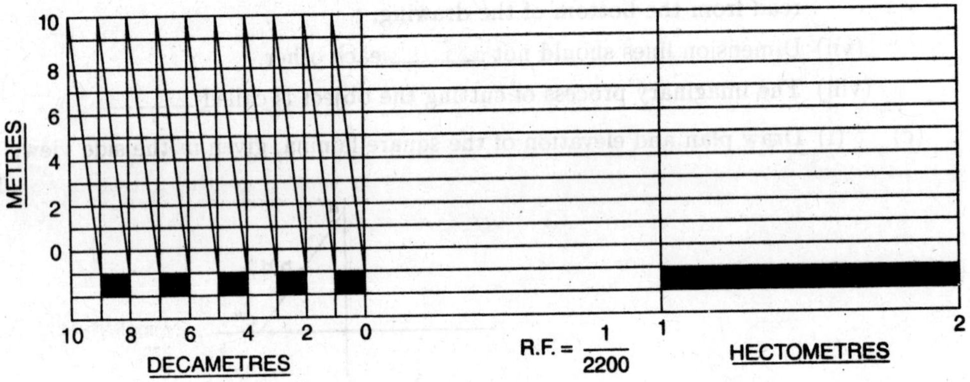

Fig. for Q1('97)

2. See Fig. Q2('97).

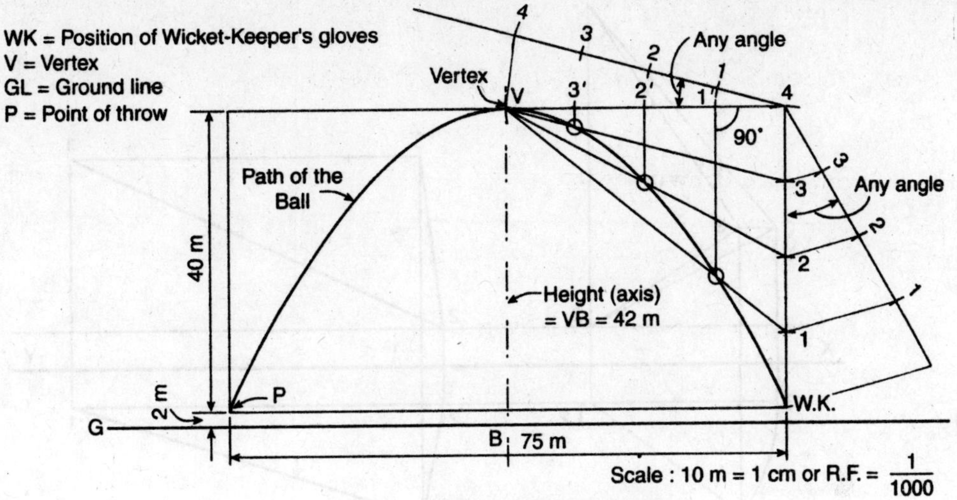

Fig. for Q2('97) Parabola

3. Consult the Chapter on Lettering.

4. See Fig. for Q4('97).

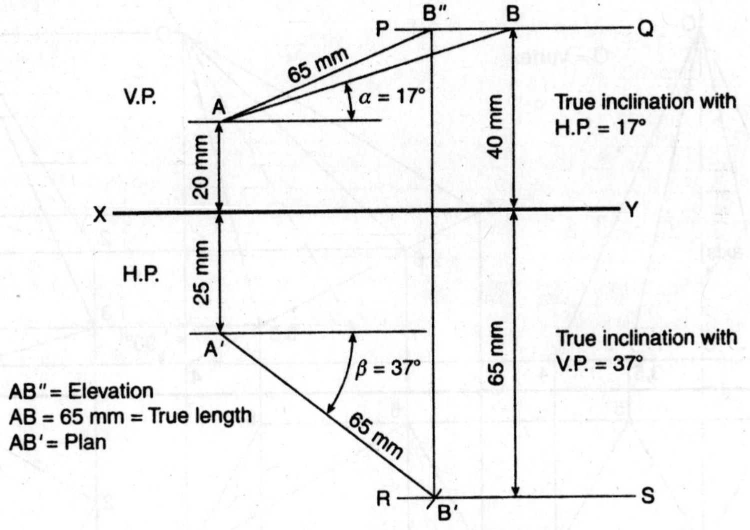

Fig. Q4('97) (First-Angle projection)

5. See Fig. for Q5('97).

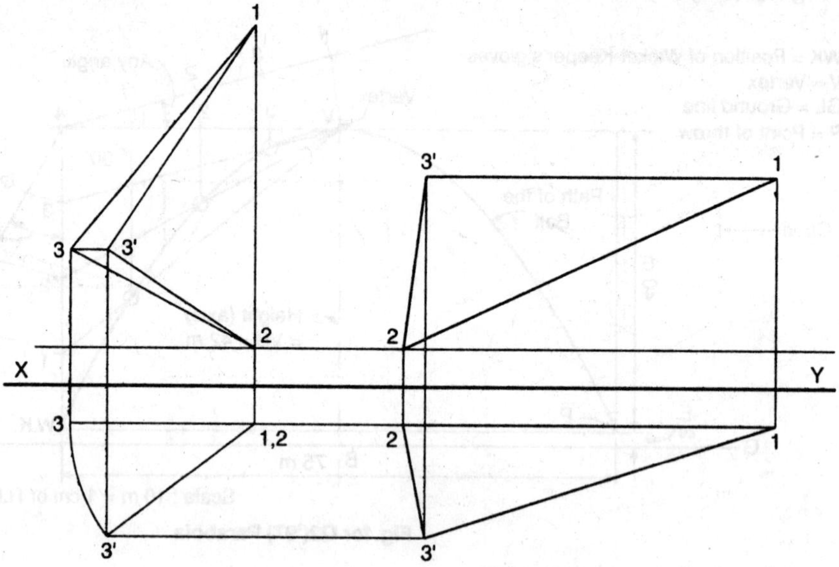

Fig. for Q5('97) First-Angle projection

6. See Fig. for Q6('97).

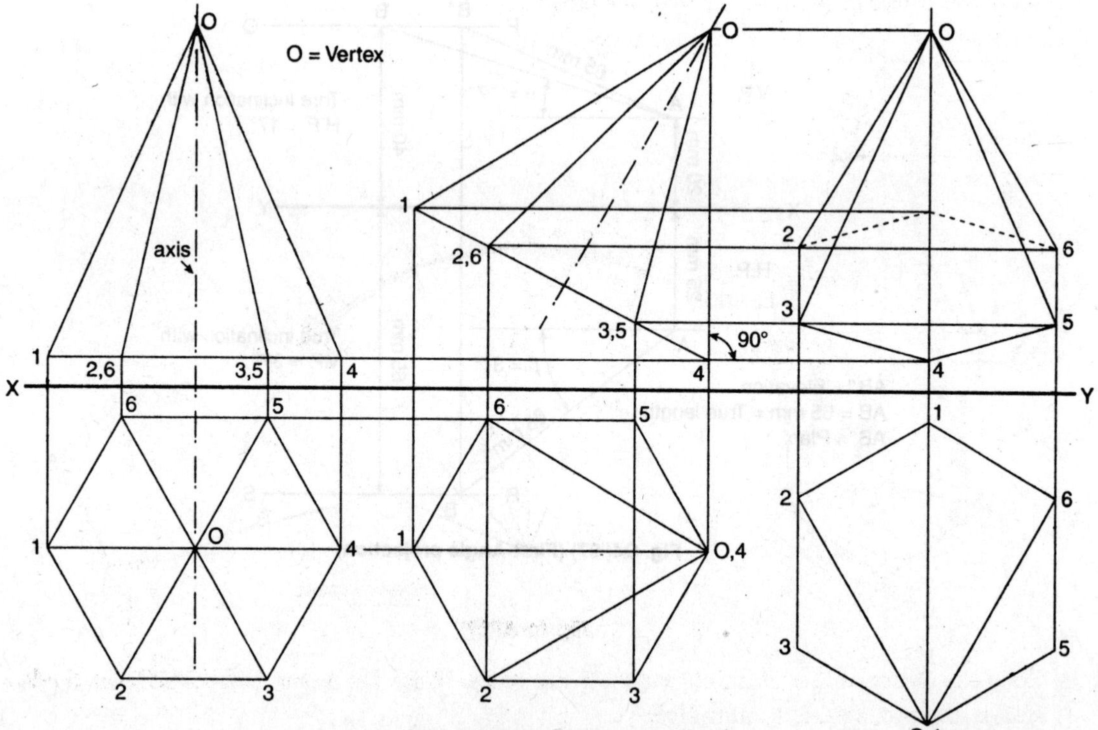

Fig. for Q6('97) First-Angle projection

7. See Fig. for Q7('97).

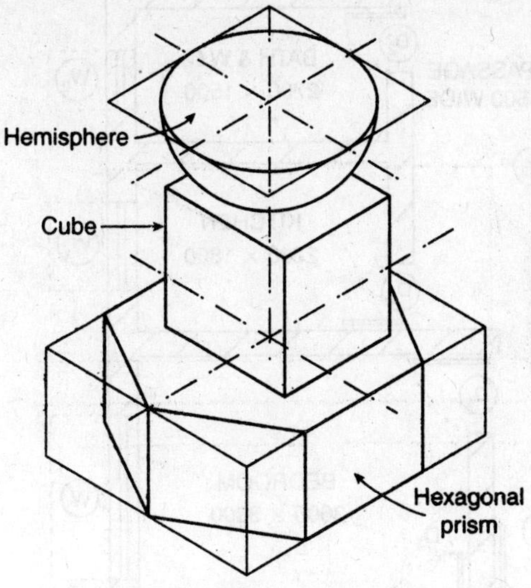

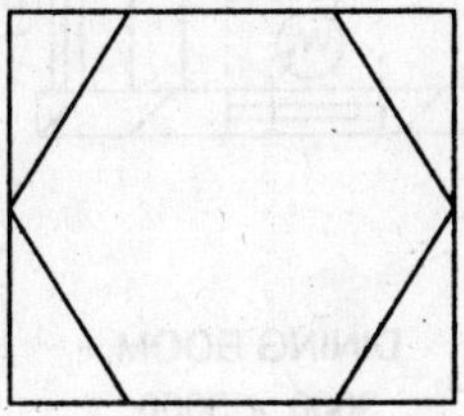

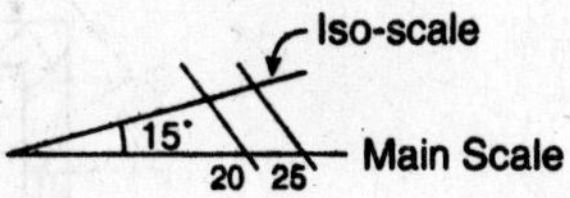

Fig. for Q7('97) Isometric View

8. Draw the top and front view to get the true edge length and number the corners and then solve the problem as shown in Fig. for Q8('97).

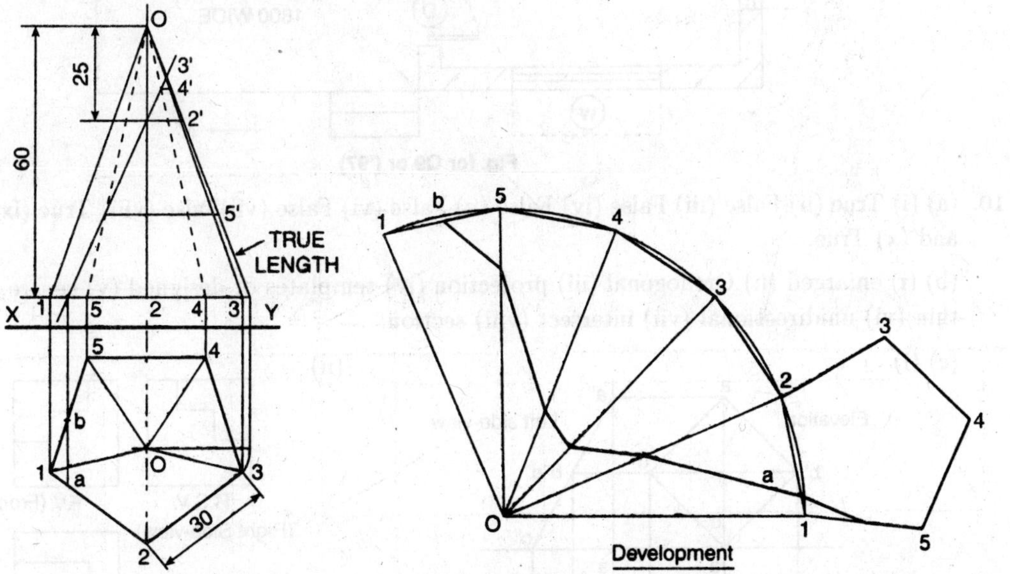

Fig. for 8('97)

9. Copy the figure of the 'Bracket' with full-size scale. Write the name 'BRACKET' in single-stroke, Block-Capital, 10 mm high.

See figure for or of Q9('97).

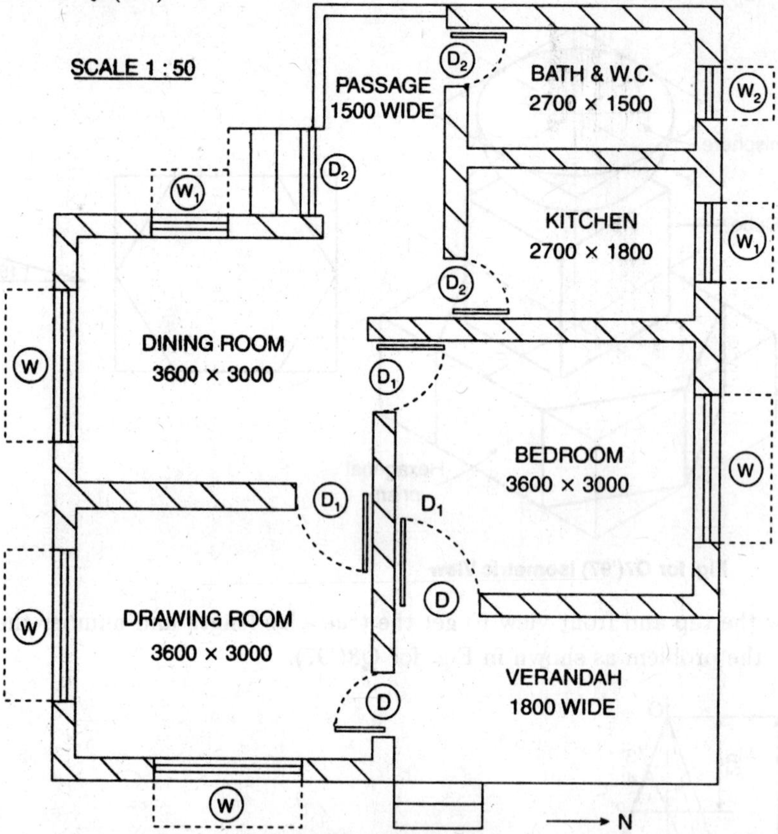

Fig. for Q9 or ('97)

10. (a) (i) True (ii) False (iii) False (iv) False (v) False (vi) False (vii)False (viii) True (ix) True and (x) True.

(b) (i) enlarged (ii) Orthogonal (iii) projection (iv) templates or designed (v) centre, thick, thin (vi) unidirectional (vii) intersect (viii) section.

(c) (i)

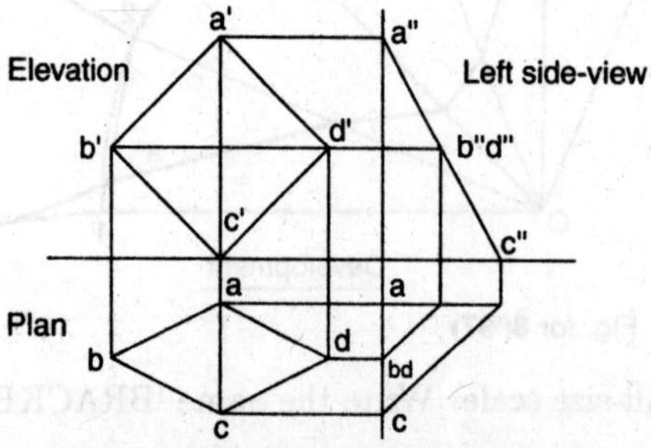

Fig. for Q10(c)(i)('97) (First-Angle projection)

(ii)

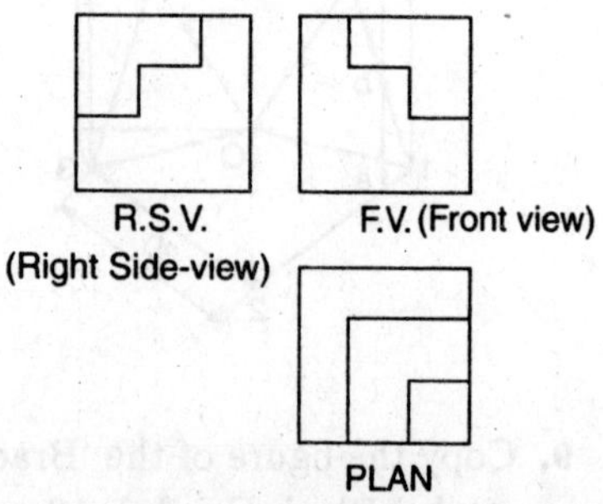

Fig. for Q10(c)(ii)('97)